PERGAMON INTERNATIONAL LIBRARY
of Science, Technology, Engineering and Social Studies
The 1000-volume original paperback library in aid of education,
industrial training and the enjoyment of leisure
Publisher: Robert Maxwell, M.C.

THE
SELF-ORGANIZING
UNIVERSE

Scientific and Human Implications of the Emerging
Paradigm of Evolution

THE PERGAMON TEXTBOOK
INSPECTION COPY SERVICE

An inspection copy of any book published in the Pergamon International Library will gladly be sent
to academic staff without obligation for their consideration for course adoption or recommendation.
Copies may be retained for a period of 60 days from receipt and returned if not suitable. When a
particular title is adopted or recommended for adoption for class use and the recommendation results
in a sale of 12 or more copies, the inspection copy may be retained with our compliments. The
Publishers will be pleased to receive suggestions for revised editions and new titles to be published in
this important International Library.

THE SELF-ORGANIZING UNIVERSE

Scientific and Human Implications
of the Emerging Paradigm of Evolution

by
Erich Jantsch

This book is an expanded version of the Gaither
Lectures in Systems Science given in May
1979 at the University of California, Berkeley.

PERGAMON PRESS

OXFORD · NEW YORK · TORONTO · SYDNEY · PARIS · FRANKFURT

U.K.	Pergamon Press Ltd., Headington Hill Hall, Oxford OX3 0BW, England
U.S.A.	Pergamon Press Inc., Maxwell House, Fairview Park, Elmsford, New York 10523, U.S.A.
CANADA	Pergamon Press Canada Ltd., Suite 104, 150 Consumers Road, Willowdale, Ontario M2J 1P9, Canada
AUSTRALIA	Pergamon Press (Aust.) Pty. Ltd., P.O. Box 544, Potts Point, N.S.W. 2011, Australia
FRANCE	Pergamon Press SARL, 24 rue des Ecoles, 75240 Paris, Cedex 05, France
FEDERAL REPUBLIC OF GERMANY	Pergamon Press GmbH, Hammerweg 6, D-6242 Kronberg-Taunus, Federal Republic of Germany

First edition 1980
Reprinted 1983, 1984

British Library Cataloguing in Publication Data

Jantsch, Erich
The self-organizing universe - (Systems science and world order library).
1. Evolution
I. Title II. Series
575   QH366.2   78-41285
ISBN 0-08-024312-6 (Hardcover)
ISBN 0-08-024311-8 (Flexicover)

Printed in Great Britain by A. Wheaton & Co. Ltd., Exeter

For ILYA PRIGOGINE

Catalyst of the self-organization paradigm

Consequently: he who wants to have right without wrong,
Order without disorder,
Does not understand the principles
Of heaven and earth.
He does not know how
Things hang together.

Chuang Tzu, *Great and Small*

Contents

Foreword

A non-technical book about evolution might, at first sight, be taken for a popularization of some of the fascinating recent insights and concepts which science has brought to this broad theme. However, the purpose of this book transcends the simple task of providing an up-to-date version of recent advances in science. My ambition with this book is far greater. It aims at a new *synthesis*, at letting appear the contours of an emergent unifying paradigm which sheds unexpected light on the all-embracing phenomenon of evolution. This new paradigm is the *paradigm of self-organization*. With it, an age-old vision is finding its scientific foundations.

In current theory, evolution is usually considered under the aspects of adaptation and survival. The dire stereotypes of the "survival of the species" and of evolution as a "game in which the only reward is to stay in the game" exert their fateful influence even on the images we hold of ourselves and of human life in general. Such a "heroic pessimism" is further enhanced by theories which view the origin of life as a mere accident, so unlikely that it perhaps occurred only once in the whole universe.

But life is more than survival and the environment to which it adapts, itself evolves and adapts. To grasp this co-evolution in a non-dualistic perspective required the development of a paradigm capable of dealing with self-transcendence, the reaching out beyond the boundaries of one's own existence, the joy of creation. The emergent paradigm of self-organization permits the elaboration of a vision based on the interconnectedness of natural dynamics at all levels of evolving micro- and macrosystems. From such an interconnectedness of the human world with overall evolution springs a new sense of *meaning*. The ultimate aim of the book is thus a profoundly humanistic one.

An "ecosystem" of new scientific concepts, most of them not older than a decade, substantiates such a unified, but non-reductionist view of self-organizing evolution. The most fascinating aspect in the development of these concepts is the dynamics with which they seem to organize themselves into the more comprehensive paradigm—which, thereby, furnishes its own proof. The ambition of the present book is to act as a catalyst in this self-organization.

My richest source of inspiration was my friendship with Ilya Prigogine of the Free University of Brussels and the University of Texas in Austin. The best part of the self-organization paradigm is his life's work. I owe him and his collaborators not only innumerable philosophical and scientific discussions, but also unpublished material. The news that Ilya had been awarded the Nobel Prize in Chemistry came while I was at work on this book. It came on a morning at which the skies above Berkeley were white, filled with self-organizing structures. Everywhere in Northern California tiny balloon spiders had hatched at the same time, climbed to the tips of grass blades and woven small silky balloons there. As if responding to a signal, they had all simultaneously let go and sailed away in the wind which blew them together until they formed lofty colonies, sometimes 500 feet long, heroically sailing toward the founding of a new home—if they did not fall into the water. It was self-transcendence, the reaching-out of evolution, made visible.

However, this book owes much to my contacts with many people. For discussions, correspondence and the exchange of publications I should like to thank, above all, the following persons, listed in alphabetical order: Ralph Abraham (University of California, Santa Cruz), Richard Adams (University of Texas, Austin), Peter Allen (Free University of Brussels), Agnes Babloyantz (Free University of Brussels), Gregory Bateson (University of California, Santa Cruz), Fritjof Capra (University of California, Berkeley), Manfred Eigen (Max Planck Institute of Biophysical Chemistry, Göttingen), Ingemar Falkehag (Westvaco, Charleston, S.C.), Paul Feyerabend (University of California, Berkeley), Roland Fischer (Esporles, Mallorca), Heinz von Foerster (Pescadero, California), Walter Freeman (University of California, Berkeley), Herbert Guenther (University of Saskatchewan), Wolf Hilbertz (University of Texas, Austin), Brian Josephson (Cambridge University), Antonio Lima-de-Faria (University of Lund, Sweden), Lars Löfgren (University of Lund, Sweden), Paul MacLean (National Institutes of Health, Bethesda, Maryland), Lynn Margulis (Boston University), Magoroh Maruyama (Wright Institute, Berkeley, California), Mael Marvin (Temple University, Philadelphia), Humberto Maturana (University of Santiago de Chile), Dennis McKenna (Honolulu), Terence McKenna (Freestone, California), Les Metcalfe (London Graduate School of Business Studies), Lloyd Motz (Columbia University, New York), Yuval Ne'eman (University of Tel-Aviv), Walter Pankow (Zurich), Karl Pribram (Stanford University, Stanford, California), Rupert Riedl (University of Vienna), Walter Schurian (University of Münster, Germany), Peter Schuster (University of Vienna), Paolo Soleri (Arcosanti, Arizona), Isabelle Stengers (Free University of Brussels), Francisco Varela (New York University), Sir Geoffrey Vickers (Goring-on-Thames, England), Conrad Waddington (Edinburgh University, died in September 1975), Christine von Weizsäcker (Kassel), Ernst von Weizsäcker (University of Kassel), Arthur Winfree (Purdue University, Lafayette, Indiana), Milan Zeleny (Copenhagen School of Economics).

The manuscript was critically read, partially or *in toto*, by Manfred Eigen, Paul Feyerabend, Walter Freeman, Lynn Margulis, Ilya Prigogine, Walter Schurian and Isabelle Stengers, as well as by the editor of this series, Ervin Laszlo (UNITAR, New York). I owe them valuable corrections and suggestions. The remaining errors are my own.

Special thanks are due to Gen Tsaconas of the University of California in Berkeley for editing the English text in the same meticulous way which she had already previously applied to three of my former books. She considers the present book the best.

When, not long ago, the Rastor Institute of Helsinki invited me to a lecture, it offered me a stipend instead of a honorarium. Work on this book benefited from it as well as from a visiting professorship at the University of Kassel in the Summer Semester of 1977.

Finally, I wish to thank the Center for Research in Management of the University of California, Berkeley—and especially its acting chairman, C. West Churchman—for the invitation to present the material of this book in the prestigious framework of the Gaither Lecture Series in Systems Science. The response of the audience seemed to bear out my hope that this new science will reach out and touch human life at many levels.

Berkeley (California), *Summer, 1979*   ERICH JANTSCH

Acknowledgements

Part I is partially based on two articles which the author published in German in the Science and Technology supplement of the *Neue Zürcher Zeitung:* "Dissipative Strukturen: Ordnung durch Fluktuation" (26 November 1975) and "Anwendungen der Theorie dissipativer Strukturen" (3 December 1975).

Chapter 16 has also been published under the same title in *Futures,* Vol. 10, No. 6 (December 1978).

The motto for the book and the motto appearing at the beginning of Part II are taken from the poems "Great and Small" and "Perfect Joy" by Chuang Tzu, translated from the Chinese by Thomas Merton and published in the book, Thomas Merton, *The Way of Chuang Tzu* (New York: New Directions, 1969), copyright 1965 by the Abbey of Gethsemane. They are reprinted here by permission of New Directions, New York, for the United States and George Allen and Unwin, London, for the rest of the world.

The motto for Chapter 3 is taken from the book: Romola Nijinsky (ed.), *The Diary of Vaslav Nijinsky* (Berkeley, California: University of California Press, 1971). It is reprinted here by permission of Simon and Schuster, New York.

The motto for Chapter 8 is taken from the book: Gregory Bateson, *Steps to an Ecology of Mind* (San Francisco: Chandler, 1972). It is reprinted here by permission of the author and Thomas Y. Crowell.

The poem "The Hidden Law", which appears as motto for Part III, is from the book: W. H. Auden, *Collected Poems,* ed. by Edward Mendelson (London: Faber & Faber, New York: Random House), copyright 1947 by W. H. Auden. It is reprinted here by permission of Faber & Faber (for the British Commonwealth) and Random House Inc. (for the United States).

The motto for Chapter 19 is taken from the poem "The Butterfly" by Joseph Brodsky, translated from the Russian by George L. Kline, which appeared in *The New Yorker* of 15 March 1976. It is reprinted here by kind permission of the author and *The New Yorker.*

The photographs used for Figs 2 and 8 were kindly provided by Arthur Winfree, Purdue University, and Hans-J. Schrader, University of Oregon, respectively.

Introduction and Summary:
The Birth of a Paradigm from a
Metafluctuation

In girum imus nocte et consumimur igni
(We circle in the night and are consumed by the fire)

Old Latin palindrome

A time of self-renewal

The relatively short period between the middle of the 1960s and the beginning of the 1970s occupies a special position in the history of our century. It was a period in which traditional social and political structures were questioned, in which protests against restrictions of human life, which first were hardly taken seriously, became powerful processes shaping an urge to find new structures. What all these processes had in common was a profound concern for self-determination and self-organization, for openness and plasticity of structures and for their freedom to evolve.

The demand for freedom of speech was only the spark which ignited the Berkeley campus in 1964/65. It was followed by a movement which spread like brushfire and which quickly embraced all essential concerns of human and social life around the world. The protest against the rigidity of the university and its alienation from reality widened to include the demand for a redesign of social reality. Ruling governments were put against the wall, especially in France and Czechoslovakia in the historic year 1968. At the same time, the Chinese cultural revolution broke up rigidifying structures; Mao Tse-Tung was the only statesman to welcome this dynamics of self-organization.

The storm blew over, the structures apparently had resisted—but the world was no longer the same. The mental and spiritual structures had changed; new values shaped new guiding images. International big power politics became increasingly despised and had to accept decisive defeats, not just in

Vietnam. The dictatorships in Greece, Portugal and Spain disappeared overnight. Even Watergate seems to have been the outcome of a moral renewal. The question of civil rights started an avalanche in America which would soon reach Africa and even the Middle East and eventually started an international discussion on human rights. The Helsinki conference unexpectedly became a boomerang for the dictatorships of Eastern Europe. But also the frozen structures of world trade, favouring one-sidedly the highly industrialized nations, were for the first time partly broken up in the oil crisis of the year 1973—and in this area there can be the least doubt that more profound changes are in the offing.

The political and economic aspects of those turbulent years remain the most visible in their consequences, but they are not the only ones weighing heavily. Of even greater importance is the intensification of human consciousness which is leading to a redesign of the individual relationships of humans with their environment—an environment of fellow humans as well as of nature. If political and economic changes represented macroscopic aspects of the systems of human life, the relationships with the environment refer to microscopic aspects; both must go together. The rising consciousness of an indivisible unity with nature—and even of human existence as an integral aspect of nature—has transformed the esotoric notion of an ecosystem into an immensely practical notion. Today, concepts of environmental protection rank nationally and internationally on an equal footing with those economic concepts with which they often do not go together very well. Together with the recognition of the limits of non-renewable resources, they are even about to force deep changes in the conventional economic processes, especially in the direction of a recycling economy instead of linear one-way (throw-away) processes. Besides the protection of nature against the consequences of technology, there is also a new spirit of consumer protection, brought about in America almost single-handedly by Ralph Nader.

Perhaps the most significant change in the consciousness of large parts of the population is the recognition that the development of technology is not an aspect of blind progress which must not be hindered, but a product of the human mind. The greatest technological triumph of the period in question was not the moonlanding, planned and carried out with incredible precision, but the withdrawal of the project for an American civil supersonic plane under the pressure of public opinion.

This new attitude toward technology was the most significant success of that explicit concern with the future which started to fascinate many people in this period. The foundations for a conscious and open design of our own future were laid by Bertrand de Jouvenel (1967) with his notion of "futuribles"—a multiplicity of possible futures—and Dennis Gabor (1963) with his concept of normative forecasting—"Inventing the Future!" With these concepts, the linearity of goalsetting was broken, although it is still powerfully present in economic thinking, and especially in econometric models. Whereas conventional economic policy is based on the permanency of economic and social structures and counts with macroscopic averages, a process-oriented attitude toward the future acknowledges the power of individual imagination, of visions which are capable to stir up resonances in many people and change the structures of reality.

Not only have the external relationships in the human world changed since the 1960s towards an increasing awareness of being connected with the environment in space and time, but so have the internal relationships of man with himself. The keen occupation with the phenomenon of human consciousness *per se*, the rising interest in a "humanistic" (that is, a non-reductionist) psychology, the techniques of a "holistic" medicine partly imported from other cultures—for example, acupuncture—the interest in non-dualistic Far-Eastern philosophies and exercises such as meditation and yoga, all this is but another important aspect of the metafluctuation which touched a large part of mankind at the beginning of the last third of this century. At least in Berkeley, where this book is written, there can be little doubt about that. Here, the many branchings of the metafluctuation may still be studied even after the big wave has run out of energy. Here, history is indivisible.

The self-renewal of science

Viewed from the outside, science seems to have weathered the turbulence of the recent past without any significant change. The trend toward interdisciplinary teaching and research, toward more relevance and closeness to reality, has subsided; special university centres and programmes which have sprung up in response to student pressure have been abolished and the power of the disciplinary departments has been strengthened. Reductionism reigns supreme.

This academic reductionism is not only an abstract thought-shrinking process, but also a phenomenon of social significance and this became clear to me when Peter Brook's stage version of an anthropological report, "The Ik", came to Berkeley. The British anthropologist Colin Turnbull (1972) had found in the mountains of Uganda a small tribe of perhaps a thousand people that was not able to cope with its enforced transplantation from the original hunting grounds to create a new basis of existence. In this situation of hunger and despair, according to Turnbull's model, human relations were reduced to the grossest type of egoism. Mothers chased their children from the fireplace, dying old people were thrown out of the house to avoid the obligation of a funeral meal, robbery and murder became practically the only strategy for survival. Everybody stood against everybody. In a discussion with faculty and students of the university, Turnbull emphasized his conviction that, in this kind of behaviour, he has discovered the "true human nature" which comes to light when the luxury of culture is shed. He even sees in the Ik the precursors to a general evolutionary trend. For this argument he let not only speak his own "conversion" to the ideology of absolute egoism and the satisfaction of physical needs, but also the fact that criminal prisoners in British jails found the play, and especially Turnbull's vision in the following discussions, of great interest. To advance from the scum to the *avant-garde* of evolution warrants some pride in one's crimes. The top of the absurd, however, was contributed to this scholarly discussion by an old professor. Deeply moved, he professed that Turnbull's vision had come as a revelation to him, the fulfilment of a life-long dream, since it now became clear that reductionist science, reducing human life to "objective" survival functions, had always been the spearhead of evolution. The parallel between science, robbery and murder hung in the air uncontradicted and the discovery of a new, deep insight made the participants shudder in awe. Instead of horror I saw shiny eyes and open mouths. . . .

And yet, in science also, a tremendous restructuring is underway. Areas which, for a long time, had been open to speculation only, especially cosmology, find empirical foundations. The discovery of the background radiation in 1965 (it had already been predicted in 1948) created for the first time an opportunity for the direct study of an effect originating in the hot, early beginning of the universe. Also in 1965 the first of four objects was discovered of which astronomers are reasonably sure that they are "black holes". They permit the direct study of the "death" phase of a star.

Equally, in 1965, micropaleontological laboratory methods were developed which permitted the discovery of microfossils in very old sedimentary rock. What has been speculation so far, namely the history of the earliest forms of life on earth, became accessible to direct observation. The oldest of the microfossils identified since then are 3500 million years old and date back to a time when our planet had reached less than a quarter of its present age.

The scope of space and time which is accessible to observation has widened immensely. The largest theoretically observable spatial dimension is limited by the so-called event horizon; it is determined by the velocity of light and is at present about 1.5×10^{26} metres.* Indeed, so-called quasars (objects with extremely intense radiation) have been observed which come close to this distance. They run away from us at 90 per cent of the velocity of light (the latter is 300,000 kilometres per second) and their light originated at a time when the universe had only reached one-eighth of its present age. The smallest observable

---

*The notation of very large or very small numbers by means of powers of 10 is of immense practical value. 10^{26} simply means a number with a one and 26 zeroes behind it. 10^{-17} means the reciprocal of 10^{17}, that is, 0,00 . . . 01, with the one appearing at the seventeenth place after the comma. 10^{26} is 10^{43} times as big as 10^{-17}, since $26 - (-17) = 26 + 17 = 43$.

length is of the order of 10^{-17} metres, corresponding to the dimensions of subatomic particles. The largest observable time span, thanks to the already mentioned background radiation, is the age of the universum which is approximately 5×10^{17} seconds. The smallest time span at present corresponds to the average life span of extremely unstable subatomic particles, about 3×10^{-24} seconds; they hardly have the character of a particle any longer and are often referred to as "resonances". The spatial span of human observation reaches over 43 powers of 10, the temporal span over 41 powers of 10. The astonishing similarity of these numbers makes one think of the hypothesis of the British Nobel laureate P. A. M. Dirac, which stipulates a correlation between macro- and microcosmos by dimensionless numbers of the order 10^{40}. In this tremendously extended space-time-continuum, interconnections and patterns emerge which are primarily of a dynamic nature and which give for the first time a scientific basis to the idea of an overall, open evolution which is interconnected at many irreducible levels.

However, it is not so much the extremes which touch our lives most directly, but the realm of direct human experience without the means of instruments. In this realm, we find the phenomena of biological, social and cultural life. The magnificent wealth of forms which we encounter here has so far been mainly the subject of empirical research. Forms were observed, classified and ordered and the particular was put into a more general context. Structures were classified according to average features distilled from a large number of single observations. This structure-oriented attitude assumed an additional time dimension with Darwin's theory of natural selection and the evolution of biological species.

Emphasis on structure, adaptation and dynamic equilibrium (steady-state flow) characterized the earlier development of cybernetics and General System Theory. These interdependent fields of study, actively developed since the 1940s, arrived at a profound understanding of how given structures may be stabilized and maintained indefinitely. This is of primary concern in technology and it was in this area that cybernetics and a specialized system theory triumphed in the control of complex machinery. In biological and social systems, however, this type of control—also called negative feedback—is only one side of the coin. No living structure can be permanently stabilized. The other side of the coin concerns positive feedback, or destabilization and the development of new forms. To arrive at a full synthesis of both aspects remained a dream for the founders of the forementioned theories, Norbert Wiener and Ludwig von Bertalanffy (1968). Their intuitively correct formulations, substantiated and carried further by Ervin Laszlo (1972) and others, are finding a firm scientific basis in our days. In the 1950s, the advent of molecular biology opened up the possibility of creating the basis for a theoretical biology. But molecular biology became hampered by a reductionist attitude and failed to connect with the phenomena of macroscopic order. The structure of DNA and the genes does not contain the life of the organism which develops by using this information.

Biological and social systems need an understanding of phenomena such as self-organization and self-regulation, coherent behaviour over time with structural change, individuality, communication with the environment and symbiosis, morphogenesis and space- and time-binding in evolution. A first step in this direction is being made by a new understanding of the dynamics of natural systems which in the 1920s, has been preceded by the process philosophy of Alfred North Whitehead (1969) and the concept of holism in evolution elaborated by the South African statesman Jan Smuts (1926).

In a concise way, this new understanding may be characterized as *process-oriented*, in contrast to the emphasis on "solid" system components and structures composed of them. These two perspectives are in their consequences asymmetrical: whereas a given spatial structure, such as a machine, determines to a large extent the processes which it can accommodate, the interplay of processes may lead to the open evolution of structures. Emphasis is then on the *becoming*—and even the being appears in dynamic systems as an aspect of becoming. The notion of system itself is no longer tied to a specific spatial or spatio-temporal structure nor to a changing configuration of particular components, nor to sets of internal or external relations. Rather, a system now appears as a set of coherent, evolving, interactive processes which temporarily manifest in globally stable structures that have nothing to do with the equilibrium and the solidity of technological structures. Caterpillar and butterfly, for example, are two temporarily stabilized structures in the coherent evolution of one and the same system. In the year 1947 Conrad Waddington had already introduced the notion of the *epigenetic process*, the selective and synchronized utilization of structurally coded genetic information by the processes of life in interdependence with the relations to the environment. It is of central importance in a process-oriented view of biology.

The decisive breakthrough occurred in 1967 with the theory and subsequent empirical confirmation of so-called *dissipative structures* in chemical reaction systems, and with the discovery of a new ordering principle underlying them. This new ordering principle, called *order through fluctuation*, appears beyond the thermodynamic branch in open systems far from equilibrium and incorporating certain autocatalytic steps. The development of this theory is the triumph of Ilya Prigogine and his collaborators in Brussels and Austin, Texas. This work has been recently presented in a comprehensive monograph (Nicolis and Prigogine, 1977).

At about the same time, work at the Biological Computer Laboratory at the University of Illinois, which functioned from 1956 to 1976 under its founder Heinz von Foerster, and paid particular attention to self-organization, led to a new formulation of the properties of living systems. A core notion, *autopoiesis*, was introduced in 1973 by the Chilean biologists Humberto Maturana and Francisco Varela and further developed together with Ricardo Uribe (Varela, Maturana and Uribe, 1974; Maturana and Varela, 1975). Autopoiesis refers to the characteristic of living systems to continuously renew themselves and to regulate this process in such a way that the integrity of their structure is maintained. Whereas a machine is geared to the output of a specific product, a biological cell is primarily concerned with renewing itself. Upgrading (anabolic) and downgrading (catabolic) processes run simultaneously. Not only the evolution of a system, but also its existence in a specific structure becomes dissolved into processes. In the domain of the living, there is little that is solid and rigid. An autopoietic structure results from the interaction of many processes. Self-reference also becomes a key notion for a new process view of brain functions (Pribram, 1971) and of human consciousness (Fischer, 1975/76).

Another important new start was made in explaining the origin of life on earth. Jacques Monod (1971) had insisted on random molecular combination which lets life appear as a highly unlikely result, perhaps unique in the whole universe; Hans Kuhn (1973) had modified this view·by proposing random reproduction by way of stereospecificity. The more exciting new view recognizes a decisive role of autocatalytic reinforcement and acceleration of processes, whose initiation may still be thought of as random. The same basic principles of self-organization which permit the formation of dissipative structures, and the same non-linear non-equilibrium thermodynamics, appear now as plausible, important factors in the formation of biopolymers from monomers (Prigogine, Nicolis and Babloyantz, 1972) and in the synthesis of complex nucleic acids and proteins in self-reproducing hypercycles (Eigen, 1971; Eigen and Schuster, 1977/78). Instead of viewing chance and necessity strictly in sequence, as Monod had done—the utterly improbable chance of a self-reproducing molecular combination being "hit" is followed by the absolute necessity of survival—chance and necessity now appear as complementary principles. Eigen and Winkler (1975) see this complementarity at work in random processes which are caught in the web of "rules of the game", or natural laws, resulting in natural selection in the sense of an undifferentiated Darwinism. The one-sided application of the Darwinian principle of natural selection frequently leads to the image of "blind" evolution, producing all kinds of nonsense and filtering out the sense by testing its products against the environment. As if this environment would not itself be subject to evolution! Evolution, at least in the domain of the living, is essentially a learning process. A more subtle view of self-organization dynamics recognizes the degrees of freedom available to the system for the self-determination of its own evolution and for the finding of its temporary optimal stability under given starting conditions (Nicolis and Prigogine, 1977; Eigen and Schuster, 1977/78). Evolution is open not only with respect to its products, but also to the rules of the game it develops. The result of this openness is the self-transcendence of evolution in a "metaevolution", the evolution of evolutionary mechanisms and principles.

Intuitive attempts to apply the same basic principles of self-organization, which are found at the levels of simple chemical and precellular systems, also to higher levels of evolution, have resulted in astonishingly realistic descriptions of the dynamics of ecological, sociobiological and sociocultural systems (Eigen and Winkler, 1975; Jantsch, 1975; Prigogine, 1976; Nicolis and Prigogine, 1977; Haken, 1977). Besides "vertical" aspects of evolution (coherence in time), "horizontal" aspects (coherence in space) also move now into the foreground of interest, including phenomena such as communication, symbiosis and co-evolution. Even the system of the biosphere plus atmosphere now appears as a self-organizing and self-regulating system (Margulis and Lovelock, 1974). The directedness of evolution now may be understood *post hoc* as the result of the interplay of chance and necessity (Riedl, 1976); necessity is introduced by the systems constraints which are themselves the result of evolution. Biological, sociobiological and sociocultural evolution now appear as linked by *homologous* principles (i.e. principles related through their common origins) and not just by analogous (formally similar) principles. This should not come as a surprise since the whole universe evolved from the same origin.

This new type of science which orients itself primarily at models of life, and not mechanical models, spurs change not only in science. It is thematically and epistemologically related to those events which I have identified as aspects of the metafluctuation which rocked the world. The basic themes are always the same. They may be summarized by notions such as self-determination, self-organization and self-renewal; by the recognition of a systemic interconnectedness over space and time of all natural dynamics; by the logical supremacy of processes over spatial structures; by the role of fluctuations which render the law of large numbers invalid and give a chance to the individual and its creative imagination; by the openness and creativity of an evolution which is neither in its emerging and decaying structures, nor in the end result, predetermined. Science is about to recognize these principles as general laws of the dynamics of nature. Applied to humans and their systems of life, they appear therefore as principles of a profoundly natural way of life. The dualistic split into nature and culture may now be overcome. In the reaching out, in the self-transcendence of natural processes, there is a joy which is the joy of life. In the connectedness with other processes within an overall evolution, there is a meaning which is the meaning of life. We are not the helpless subjects of evolution—we *are* evolution. As science, like so many other aspects of human life, is touched by the metafluctuation, it overcomes the alienation from human life and contributes to the joy and meaning of life. To convey something of this new role of science is the foremost concern of this volume.

Central to my argument is the thesis of connectedness. It cannot be grasped in a static way, but emerges from the self-organization dynamics at many levels of evolution. At each level, self-organization processes are poised on their "starting marks" to take over from random developments, if the proper conditions become established, and to accelerate or make possible in the first place the emergence of complex order. These starting conditions are perhaps relatively narrowly limited, as we suspect from our futile search for life in the solar system. But once they are given—in a particular phase of cosmic evolution, in which galaxies and stars came into being, or in the early phases of life on earth—these conditions become themselves subject to evolution. Evolution differentiates by means of a co-evolution of macroscopic and microscopic systems. That microscopic systems are just subsystems of the macroscopic ones, that the latter appear as "environment" of the former is a view which stems from a static understanding which tempts to formulate world order in dualistic terms. Life itself, in particular, creates the macroscopic conditions for its further evolution—or, viewed from the other side, the biosphere creates its own microscopic life. Micro- and macrocosmos are both aspects of the same, unified and unifying evolution. Life appears no longer as a phenomenon unfolding *in* the universe—the universe itself becomes increasingly alive.

Summary of contents

The central aspects of the emerging paradigm of self-organization are: *primo,* a specific macroscopic dynamics of process systems; *secundo,* continuous exchange and thereby co-evolution with the environment, and *tertio,* self-transcendence, the evolution of evolutionary processes. The first three parts of the book bring these three aspects consecutively into sharp focus. The last part formulates under the central aspect of creativity some of the conclusions which may be drawn for the human world.

Part I, *Self-organization: The Dynamics of Natural Systems,* deals with the typical self-organization dynamics of coherent systems which evolve through a sequence of structures and maintain their integrity as a system. Biological and social systems are of this kind. The simplest level at which this kind of dynamics may be studied is the level of dissipative structures which form in self-organizing and self-renewing chemical reaction systems.

Chapter 1, "Macroscopic Order", sketches the shift from static structure-oriented to dynamic process-oriented thinking in Western science. Classical dynamics considered the notion of isolated particles. Thermodynamics marks the transition to process thinking by introducing irreversibility, or the time directedness of processes. Time symmetry is broken, the past is separated from the future and the macroscopic world becomes historic. Finally, with non-linear non-equilibrium thermodynamics, spatial symmetry is also broken and a new level of macroscopic order is addressed. This is the level of co-operative phenomena, leading to the spontaneous formation and evolution of structures. The laws of physics are accentuated in a particular way by this macroscopic order. Where so far merely random processes have been assumed to exist, a new ordering principle comes into play, called "order through fluctuation".

Chapter 2, "Dissipative Structures: Autopoiesis", discusses the basic conditions for the dynamic existence of non-equilibrium structures. These basic conditions—partial openness toward the environment, a macroscopic system state far from equilibrium, and autocatalytic self-reinforcement of certain steps in the process chain—reappear also at other levels of self-organizing systems. Equilibrium is the equivalent of stagnation and death. A high degree of non-equilibrium which maintains the self-organizing processes is in turn maintained by continuous exchange of matter and energy with the environment, in other words by metabolism. The dynamics of such a globally stable, but never resting structure has been called *autopoiesis* (self-production or self-renewal). An autopoietic system is in the first line not concerned with the production of any output, but with its own self-renewal in the same process structure. Autopoiesis is an expression of the fundamental comple-mentarity of structure and function, that flexibility and plasticity due to dynamic relations, through which self-organization becomes possible. An autopoietic system is characterized by a certain autonomy *vis-à-vis* the environment which may be understood as a primitive form of consciousness corresponding to the level of existence of the system. For example, the size of a dissipative structure is independent of its environment, as long as the latter is big enough to permit the formation of the structure.

Chapter 3, "Order through Fluctuation: System Evolution", discusses the evolution of non-equilibrium systems through a sequence of autopoietic struc-tures. The preconditions are the same as for autopoiesis, namely, openness, high non-equilibrium and autocatalysis. The essential feature is the internal reinforcement of fluctuations (by autocatalysis) which eventually drive the system over an instability threshold into a new structure. In the transition, it is not the macroscopic averages which play their usual role, but the internal amplification and the breakthrough of fluctuations which started very small. In other words, the principle of creative individuality wins over the collective principle in this innovative phase. The collective will always try to damp the fluctuation and depending on the coupling of the subsystems, the life of the old structure may thereby be considerably prolonged. In the phase in which a new structure comes into being, the principle of maximum entropy production holds—no expenses are spared if the issue is the build-up of a new structure. However, it is not predetermined which structure will come into being. At each level of autopoietic existence, a new version of macroscopic indeterminacy comes into play. The future evolution of such a system cannot be predicted in an absolute way; it resembles a decision tree with truly free decision at each branching point. However, already at the level of chemical dissipative structures, such a system keeps the memory of its evolutionary path. If it is forced back, it retreats by the same way it has come through a sequence of autopoietic structures. The principle of order through fluctuation which underlies all coherent evolution also requires a new information theory which is based on the complementarity of novelty and confirmation in pragmatic (i.e. effective) information. The kind of information theory which has become so useful in communication technology holds only for information which consists almost totally of confirmation. In the domain of self-organizing systems, information is also capable of organizing itself; new knowledge arises.

Finally, Chapter 4, "Modelling Self-organizing Systems", gives a brief overview over the relatively successful attempts to apply the theory of dissipative structures and the principle of order through fluctuation to phenomena of self-organization in many areas. These first attempts have led to remarkable results in such fields as prebiotic evolution, the functioning of bioorganisms, neurophysiology, ecology (population dynamics) and sociobiology. Most recently, the first approaches have been made to the modelling of phenomena in the systems of human life, such as the growth and evolution of cities. For a qualitative description of the evolution of mental structures, such as scientific paradigmas, value systems, world views and religions, the same principles of autopoiesis and order through fluctuation have become valuable. This broad applicability of a theory which has first been rigorously formulated in physical chemistry does not imply a physical interpretation of biological and sociocultural phenomena, but is based on a fundamental homology (true relatedness) of the self-organizing dynamics at many levels. This homology makes it possible to view evolution as a holistic phenomenon dynamically linking many levels. Such a view is elaborated in the second and third parts of this volume.

Part II of the book, *Co-evolution: A History of Reality in Symmetry Breaks,* retells in five chapters the history of evolution, starting from the "big bang", from a particular angle of view which has rarely been employed before. This angle of view is the co-evolution of macro- and microworld, the mutual setting of the conditions for simultaneous differentiation and complexification along microscopic and macroscopic branches of evolution. In cosmic evolution, such a view is not new. Nobody imagines that the structures in the universe were built up one-sidely from the bottom up, from particles and atoms to stars, galaxies and clusters of galaxies. But in the realm of biological evolution on earth the logic usually evoked is the "build-up of higher life" in micro-evolution, neglecting the macroevolutionary branch. A systems approach emphasizing the co-evolution of both branches leads to significant new insights. It makes it also possible to distinguish sociocultural evolution which dominates the human sphere from sociobiological and ecological evolution, while simultaneously stressing their interconnectedness.

Chapter 5, "Cosmic Prelude", sketches essentially the so-called cosmological standard model, while pointing to the symmetry breaks which mark the various stages of evolution. The first of these symmetry breaks concern the four physical forces; namely, gravity, electromagnetic, strong and weak nuclear forces. With the break of their original symmetry, space and time for evolution become unfurled. Gravity acts in macroscopic dimensions, the nuclear forces in microscopic dimensions and the electromagnetic forces in an intermediary domain. In a dense and hot universe, nuclear forces come into play first. But after the production of hydrogen and helium nuclei and with the cooling of the expanding universe, the cosmic microevolution temporarily loses its momentum. Eventually, however, the configuration of microscopic parameters shifts in such a way that the gas pressure breaks down abruptly and brings gravity into play at the macroscopic branch of evolution. Gravity is primarily responsible for the production of the so-called mesogranularity of the universe which includes clusters of clusters of galaxies, clusters of galaxies, galaxies, stellar clusters and, finally, stars. In the stars, the co-evolution of macro- and microcosmos becomes visible in a particularly dramatic way. Gravity creates the conditions for a dense and hot environment which, again, brings the nuclear forces into play which continue the syntheses of heavy nuclei along the microevolutionary chain. The energy liberated in these processes of microevolution determines in turn the ontogeny of the star, its irreversible individual evolution. Another symmetry break in the starting phase of the universe concerns the excess of matter over antimatter by about 10^{-9} (one-thousandth of a millionth). This very slight excess is responsible for the formation of a matter world. The results of cosmic co-evolution, matter in various states of organization, is transferred across time and space in a kind of unordered phylogeny. Our planet earth, and we ourselves, consist to a large degree of matter which does not stem from our young sun (which is still busy with transforming hydrogen into helium), but from the outer layers and the rest from explosions of distant stars which do not exist any longer. The sun has organized this alien matter by means of gravity and its nuclear processes provide the energy for life on earth.

Chapter 6, "Biochemical and Biospherical Co-evolution", sketches the beginnings of life on earth. After the formation of organic molecules, the next step was probably the formation of dissipative, metabolizing structures which may be assumed to have played a decisive role in the formation of biopolymers and in further stages of precellular evolution. The emergence of the capability for self-reproduction may be explained by the hypercycle model which includes the principles of dissipative structures as well as of symbiosis at a molecular level. With this step, biological microevolution starts to work with information transfer instead of matter transfer, with blueprints for the organization of matter that made possible the high degree of differentiation evident in life. Single-cell life on earth started very early, probably already before the formation of a firm crust some 4000 million years ago. The co-evolution of micro- and macroworld becomes visible already in this early phase. The prokaryotes, nucleus-free single cells which represented the only life form in this phase were reponsible for a thorough transformation for 2000 million years, first of the surface of the earth by oxidation, and then of the atmosphere by enriching it with free oxygen. This transformation of the macrosystem created the prerequisites for the development of more complex life forms along the microevolutionary branch. But it also turned the bio- and atmosphere into a world-wide self-regulating, autopoietic system which has stabilized itself since 1500 million years and has ensured the maintenance of the conditions for complex life on earth. Such, at least, is the claim of the Gaia hypothesis, which has been named after the Greek earth goddess. Up to our days, the prokaryotes manage the Gaia system as tiny autocatalytic units. A part of them has joined to form the more complex eukaryotic cells, or cells with a nucleus. As organelles within these cells, the former prokaryotes still maintain a certain autonomy.

Chapter 7, "The Inventions of the Microevolution of Life", starts with a presentation of this still controversial endosymbiotic theory of the origin of the eukaryotic cell. The eukaryotes developed sexuality and with it the possibility of the systematic generation of maximum genetic variety. This invention was followed by heterotrophy, the capability to live off other bioorganisms or the material they leave after their death. This led to the emergence of complex and multilevel ecosystems which favoured the formation and explosive spreading of multicellular organisms. It seems that these multicellular organisms also have an endosymbiotic origin in the social binding of eukaryotic cells.

Chapter 8, "Sociobiology and Ecology: Organism and Environment", pursues the theme of co-evolution of the macro- and microsystems of life which has gained new aspects and process mechanisms through the inventions of microevolution. The emergence of eukaryotic cells marks the beginning of epigenetic development, the flexible and selective utilization of genetic information in line with the individual design of relations with the environment. With heterotrophy and the optimal exploitation of primary solar energy in ecosystems, the macrodynamics of life gains momentum. At the micro- and macrobranches of evolution we find now organisms and ecosystems, quite complex autopoietic systems the co-evolution of which now brings primarily new horizontal processes into play—after genetic information transfer has emphasized vertical processes. Each vertical, genetic development is being "processed" in a dense web of horizontal processes. This leads to a further enrichment of genetic evolution by epigenetic dimensions. Finally, epigenetic development overtakes genetic development in importance as well as in speed. The horizontal, cybernetic processes in societies and ecosystems become increasingly important for the evolution of groups and species. The morphological properties are not decisive, but rather the dynamic qualities, especially in young ecosystems. The advantage is with the system which advances fastest. Vertically transferred genetic information is supplemented at an equal footing by horizontally transferred metabolic information—both within complex organisms and within systems constituted by these organisms.

Finally, in Chapter 9, "Sociocultural Evolution", a ,third type of communication appears side by side with slowly acting genetic and faster acting metabolic communication. This is the very fast-acting neural communication based on a central nervous system and especially the brain. The characteristic time factor shortens from many generations through minutes to seconds and fractions of a second. In this way, symbolic expression becomes possible, first in the form of self-representation of the organism and later as symbolic reconstruction of the external reality and its active design. The concept of an evolving "triune brain" permits us to follow the stepwise emancipation of mental concepts of images from an external reality. Mental concepts, ideas and visions become autopoietic levels in their own right. Whereas genetic information transfer made the past effective in the present and epigenetic development brought the systemic nature of the present into play, mental anticipation now pulls the future into the present and reverses the direction of causality. Mind in this view is no longer the opposite of matter, but rather it is the quality of self-organization of the dynamic processes characterizing the system and its relationship with the environment. Mind co-ordinates the space-time structure of matter. Besides the neural mind, there is the more slowly acting metabolic mind which dominates in ecosystems and in single-cell organisms. Whereas the material production and distribution processes of the human world represent such a metabolic mind, the electronic age has provided the prerequisites for the emergence of a faster acting and perhaps to a higher degree self-organizing "collective brain". So far the ecology of individually conceived, ready-made ideas has dominated in the emergence of culture. But it may be expected that the fluctuations of higher individual consciousness will continue to play an important role in the future.

Part III of the book, *Self-transcendence: Toward a System Theory of Evolution*, summarizes the principles which underlie the history of evolution as told in Part II. A few possible approaches are developed which, in their thorough elaboration, may contribute to a future General Dynamic System Theory.

Chapter 10, "The Circular Processes of Life", discusses the characteristic cyclical organization of self-organizing dissipative systems. A generalized scheme sets the generation characteristics of transformatory and catalytic reaction cycles, as well as catalytic hypercycles, in relation to their degeneration and diffusion characteristics. In this way, hierarchical levels are obtained which reach from equilibrium through autopoiesis to exponential and hyperbolic growth. Hypercycles, which link autocatalytic units in cyclical organization, play an important role in many natural phenomena of self-organization, spanning a wide spectrum from chemical and biological evolution to ecological and economic systems and systems of population growth. The cyclical organization of a system may itself evolve if autocatalytic participants mutate or new processes become introduced. The co-evolution of participants in a hypercycle leads to the notion of an ultracycle which generally underlies every learning process.

Chapter 11, "Communication and Morphogenesis", attempts a synopsis of the three major phases in the co-evolution of macro- and micro-cosmos—cosmic, chemical/biological/sociobiological/ecological and socio-cultural evolution. These phases may be characterized by the ways in which different types of communication act and interact. An essential new distinction emerges in the characterization of sociobiological and sociocultural evolution. The former is based on metabolic processes in which the collective dominates, whereas in the latter these relations appear turned upside down. With the evolution of self-reflexive mind, man carries the social and cultural dimensions, the mental structures of the macroworld, within himself. He imagines, plans and turns into reality not only a new world of technological equilibrium systems, but also the autopoietic structures of his own social and cultural world. One may say, man enters into co-evolution with himself. In the self-organization of the human world, therefore, sociobiological as well as sociocultural processes are of importance. The latter dominate as long as they are able to unfold freely. With increasing speed of communication in extended and even world-wide systems of human living, they may be expected to dominate even more.

Chapter 12, "The Evolution of Evolutionary Processes", follows the interconnected branches of biological and sociocultural evolution from dissipative structures all the way to self-reflexive mind. This part of evolution is characterized by the transfer and utilization of information in the sense of stored past experience. Of particular importance here is the selective and synchronized retrieval of conservatively stored (e.g. genetic) information by dissipative processes; that is to say, by life processes corresponding to a particular semantic context, or context of meaning. Another important element is the holistic system memory which already appears in chemical dissipative structures. It makes it possible for the system to link backward to its own origin and thereby experience its own total evolution which provides guidance for the partial self-determination of the future evolutionary path. If, at the same time the output of an autopoietic structure serves as input for another level of autopoietic existence, self-transcendence, the reaching out beyond the dynamics of the system proper, becomes possible. In this way, the evolution of complex life forms and mental capacities may be described as the evolution of evolutionary processes, or metaevolution, which links a continuous chain of autopoietic levels.

Chapter 13, "Time- and Space-binding", develops the idea that one important result of evolution may be seen in the increasing intensification of autopoietic life in the present by means of including the experience of the past and anticipations of the future. Biological evolution makes the experience of an entire phylum, starting from the formation of the first biomolecules, effective in the present. The emancipation of a mental reality, our inner world, from external reality makes visions of the future and plans effective in the present. In a certain sense, the whole universe concentrates to an increasing extent in the individual. The individual, in turn, increasingly assumes an integral responsibility for the universe in which it lives.

Chapter 14, "Dynamics of a Multilevel Reality", presents the results of evolution, and in particular man, as a multilevel reality in which the evolutionary chain of autopoietic levels of existence appear in hierarchic order. However, it is essential that this hierarchy is not a control hierarchy in which information streams upward and orders are handed from the top down. Each level maintains a certain autonomy and lives its proper existence in horizontal relations with its specific environment. The organelles within our cells, the descendants of the ancient prokaryotes, go about their business of energy exchange in a highly autonomous way and maintain their horizontal relationships within the framework of the world-wide Gaia system. There are many levels of self-organizing systems of cell populations, such as neuronal systems generating the rhythm of motor activities or managing perception and apperception of an environmental situation. Cancer may be understood as the dynamic régime of a cell population. Each of these self-organizing cell systems is co-ordinated from a higher level from which it is inhibited or activated, or both alternately. The mind of an individual presents those levels of co-ordination which refer to the organism as a whole. But man is no "higher" than other organisms in the sense that he stands at a higher level. Rather, he lives simultaneously at more levels than life forms that appeared earlier in evolution. We contain the entire evolution within us, but it is orchestrated to a fuller and richer extent than in less complex life forms.

Part IV, *Creativity: Self-organization and the Human World*, limits its ambition to elaborating in five short chapters some of the essential perspectives which emerge for the human world in the light of modern process thinking. The chances for true creativity are seen in overcoming a dualism which separates the created from the creator.

Chapter 15, "Evolution—Revolution", points to a profound dilemma in which the human world finds itself thanks to order through fluctuation. With increasingly flexible coupling of the subsystems, due to communication and transport technology, the metastability of political, social and economic structures is enhanced. But this also means an increasing danger that the fluctuations get bigger and bigger and may become disastrous. We have ourselves prepared some fluctuations of potentially destructive force, such as the arsenal of nuclear weapons. The simple scheme of occasional, massive restructuration in clearly defined quantum jumps of social and cultural organization seems to become modified at our level of complexity which includes the capabilities for self-reflexion and anticipation. A monolithic idea of culture dissolves into a cultural pluralism which may permit smoother or "gliding", transitions. A prerequisite for such non-destructive transitions would be the dismantling of social control hierarchies and strengthened autonomy of the subsystems.

Chapter 16, "Ethics, Morality and System Management", discusses the possibilities for providing these prerequisites. Ethics is but a behavioural code in tune with evolution and morality is the live experience of such a tuning-in. In a multilevel reality, ethics, too, is multilevel. Such a multilevel ethics is very complex in the human world because here the individual shares integral responsibility for society and culture which ultimately are his own creations. The task is to combine individual ethics with an ethics of whole systems and a general ethics of overall evolution. A multilevel systems approach seems to point to the possibility of matching flexible long-range planning and evolutionary dynamics to a good extent.

Chapter 17, "Energy, Economy and Technology", starts with a discussion of present energy technology viewed as the utilization of energy storages stemming from removed phases of evolution that are ever farther removed in the past. This kind of energy utilization is thus an incident of time-binding. Since in the sociocultural phase of evolution, man re-creates the world not only in mental, but also in physical constructs, time- and space-binding are extended to the physical world. In contrast to this, there is the possibility of an autopoiesis which is based on circular processes, in particular on the tapping of the solar energy flux and on a recycling economy. Perhaps there will be an interpenetration of autopoiesis and evolution in the near future of mankind. Evolution, or the opening up of new "niches", seems possible in an internal as well as an external sense—with the latter perhaps referring to space colonization. But an internal evolution of human consciousness appears indispensable in any case.

Chapter 18, "The Creative Process", deals in more detail with the self-organizing systems which emerge in the inner reality of man and act in the outer reality. Artists who are at the same time theoreticians of their own art are beginning to discover the self-organizing dynamics of their works of art, a dynamics which follows the same rules of openness, non-equilibrium and autocatalysis as hold for physical self-organization. The same may be said of the evolving structures of science. The creative process may perhaps be best described by using a model which its author calls the "revolving stage of consciousness". This revolving stage offers two equally valid ways to higher, visionary levels of consciousness, namely, ecstasis and meditation. But the creative process does not only consist of the perception of a vision, but also in giving it a valid form. Therefore, it depends on a multilevel, richly orchestrated consciousness, a dynamic régime in which many levels vibrate.

Chapter 19, "Dimensions of Openness", summarizes the implications of time- and space-binding in the present phase of human evolution. Time-binding results in overcoming historical time. In self-reflexion, we may experience evolution directly in terms of a genealogical tree as well as a root system. But only in the image of a rhizome which finds its most perfect expression in a dissipative structure, the totality of evolution may be experienced in the present. However, such an experience is no longer a sequence, but forms associative patterns. In this way, meaningful connections, dispersed over space and time, become visible.

In an *Epilogue: Meaning*, finally, the central theme of the dynamic connectedness of man with an unfolding universe is re-evoked. In a world which is creating itself, the idea of a divinity does not remain outside, but is embedded in the totality of self-organization dynamics at all levels and in all dimensions. This self-organization dynamics has been identified in an earlier chapter with mind. God, then, is not the creator, but the mind of the universe.

PART I

Self-organization:
The Dynamics of Natural Systems

> Time is a river which sweeps me along,
> but I am the river; it is a tiger which
> destroys me, but I am the tiger; it is a fire
> which consumes me, but I am the fire.
>
> Jorge Luis Borges

Self-organization is the dynamic principle underlying the emergence of a rich world of forms manifest in biological, ecological, social and cultural structures. But self-organization does not only start with what we usually call life. It characterizes one of the two basic classes of structures which may be distinguished in physical reality, namely, the dissipative structures which are fundamentally different from the equilibrium structures. Thus, self-organization dynamics becomes the link between the realms of the animate and the inanimate. Life no longer appears as a thin superstructure over a lifeless physical reality, but as an inherent principle of the dynamics of the universe. In dissipative structures of chemical reaction systems we are given the opportunity to study self-organization in its simplest, "purest" form. The same conditions which also reappear at more complex levels—openness, high non-equilibrium and internal reinforcement of fluctuations—may be recognized here in great clarity and simplicity. This is the reason why this book starts with a concise discussion of the theory of dissipative structures, as it has been developed primarily by Ilya Prigogine and his school of thought.

1. Macroscopic Order

> Were there not this unborn, unoriginated,
> uncreated, unformed, there would be no
> escape from the world of the born,
> originated, created, formed.
>
> Gautama Buddha

Overcoming reductionism

From every-day experience we know what happens if we open a water tap. At first, the water jet is smooth, perfectly round and transparent; the physicist calls this *laminar* flow. But if we open the tap further and thereby increase the water pressure, this image changes abruptly at a certain point. The water jet forms strands and presents itself in a dynamic structure which somehow conveys the impression of being "muscular". This is the typical appearance of *turbulent* flow which remains unchanged for a while and, if the tap is opened even wider, changes over abruptly to other, similar structures. The beautiful regularity of the laminar jet, which almost seemed to stand still, is destroyed and disorder seems to rule.

But appearances betray truth. It is precisely in turbulent flow that a higher degree of order rules. Whereas in laminar flow the movement of the individual water molecules follow a random statistical law, turbulent flow groups them together in powerful streams which, in their overall effect, permit an increase of throughflow. The frequently cited "limits to growth" are overcome by the evolution of the dynamic structure; they reappear as the widened limits of a new structure. In nature, processes often have more than one structure available for their unfolding. In our example, the increase in water pressure and throughflow leads to the instability of the laminar flow structure and the appearance of a turbulent structure, which in turn evolves through a sequence of turbulent structures. What we are calling here structure is nothing solid, composed of the same components, but a dynamic régime which puts ever new water molecules through the same strands. It is a *process structure*.

Another example from the field of hydrodynamics demonstrates the spontaneous appearance of macroscopic order perhaps even more impressively. If we heat a big, shallow pan—bigger than usually found in kitchens—uniformly from below, regular hexagonal cells, the so-called *Bénard cells,* appear. At first, the temperature in the layer of liquid is practically uniform, or in other words, the system is in its thermodynamic equilibrium. In this state, the heat from the bottom of the pan spreads through *heat conduction,* a mechanism in which the molecules get into thermal vibrations and transfer a part of this thermal energy in collisions with their neighbour molecules without moving from their place. But if the bottom of the pan becomes hotter and the temperature gradient in the liquid layer steeper, thermal non-equilibrium increases. At a certain gradient, *convection* starts, or in other words, heat transfer by movement of molecules. At first, smaller convection streams are suppressed by the environment. Beyond a critical temperature gradient, however, the fluctuations become reinforced rather than suppressed and the dynamic régime abruptly switches from conduction to convection. Macroscopic molecular streams form which include more than 10^{20} molecules—a degree of order which cannot be explained by the familiar thermodynamic principles. A new macroscopic order emerges which may also be understood as a macroscopic fluctuation, stabilized by energy exchange with the environment. This order becomes manifest in the Bénard cells (Fig. 1). From the perspective of the molecules, this phenomenon of structuration corresponds to a higher level of co-operation. As will be discussed later, the high non-equilibrium is of decisive importance here.

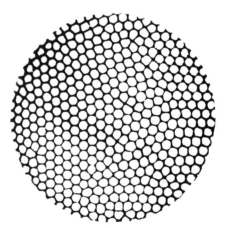

Fig. 1. An example of a hydrodynamic dissipative structure. The Bénard instability occurs beyond a critical temperature gradient, if a layer of liquid is heated from below. Co-operative behaviour in the form of macroscopic convection streams becomes dominant. Viewed from above, a pattern of regular hexagonal cells develops.

Macroscopic order plays an important and even dominant role in our everyday experience. This even holds for the realm of the inanimate. From water vortices and sand dunes all the way to stars and galaxies we recognize structures expressing macroscopic order principles. The rich world of forms has always exerted a certain aesthetic lure which sometimes has led to deeper thought and systematic exploration. In particular, the photographic studies of the Swiss physician Hans Jenny, who died in 1973, and known under the term "cymatics" (Jenny, 1967 and 1972) have become the precursor of empirical research interested in macroscopic order. More recently, Ralph Abraham (1976) is continuing these studies with his "macroscope" built at the University of California at Santa Cruz. In the macroscope, a varying energy throughput is produced in a viscous layer of a liquid by means of acoustical vibrations. A process structure appears which evolves with higher frequency through a sequence of instabilities and new structures.

However, physics has so far restricted itself to the study of a few specific phenomena only, which may be surprising in view of the wide open area of interesting research. General theoretical approaches developed in the 1940s and triggered by interest in biological phenomena, resulted in the broad frameworks of General System Theory (Bertalanffy, 1968) and cybernetics (by Norbert Wiener and others). Both approaches succeeded in acquiring a good understanding of the stabilization of given structures which becomes very fruitful in the domain of technology. But the original vision of achieving an understanding of macroscopic order across the boundaries between the animate and the inanimate world remained vague and inconclusive.

For biology and the social sciences, the macroscopic dynamics of coherent systems—systems whose structure does not remain rigid, but evolves in a coherent way—became ever more important and urgent. Organisms of all kinds and ecosystems are coherent systems, as are also cities, communities and the institutions of societies. But the consideration of macroscopic dynamics ran into the resistance of the ruling reductionism of Western science whose aim it is to reduce all phenomena to *one* level of explanation—a level which physics hopes to find in the microscopic, in the basic structure of matter. It is not without irony that the Heisenberg indeterminacy relation—the recognition that it is impossible to observe simultaneously with high precision position as well as velocity of a particle—has first been formulated at the subatomic, microscopic level. In physics, it is of importance there because the impact of the observer on the observed phenomenon can no longer be neglected. For such an observation in the microscopic domain, the comparison has been made with the watch-maker who tries to repair a delicate wrist watch with a huge, heavy hammer.

But in the domain of life and to an even higher degree in the domains of social and psychological relations, the inclusion of the observer is even much more evident. With every action, every thought—and also with every observation and theory—we interfere with the object of our study. It appears as strange, therefore, that many physicists today search for the interface between the physical and the psychic, and even for the free will, exclusively in the microscopic realm of quantum mechanics. The complementarity between the concepts of the independent movement of individual particles and the holistic behaviour of the entire system of the atomic nucleus, as first formulated by Niels Bohr, has always been evident in the macroscopic realms of biological and social relationships. But to really see it, a new basic attitude was needed toward the phenomena which matter to us in our life.

The traditional reductionism of Western physics is not only based on a belief in the "simplicity of the microscopic" (as Prigogine puts it), but also on a static view which is primarily interested in spatial structure. A rigid structure may be easily disassembled into its pieces and reassembled again. In many cases, such a structure may be understood in terms of combinations of a few basic elements. Macroscopic properties, such as weight, stability, or strength, may be explained by the properties of components and their configuration.

In a true system, however, not all macroscopic properties follow from the properties of components and their combinations. Macroscopic properties often do not result from static structures, but from dynamic interactions playing both within the system and between the system and its environment. An organism is not defined by the sum of the properties of its cells. In chemical reaction systems, certain molecules which do not participate in the reaction may act as catalysts and thereby influence the overall dynamic system in a decisive way. A human being falling in love—perhaps only once in a lifetime—changes the life of the community of which he or she is part. Such considerations already hint at the fact that a systemic view of necessity leads to a dynamic perspective. Quite generally, a system becomes observable and definable as a system through its interactions. The hydrodynamic structures used as an example at the beginning of this chapter do not even consist of durable components, but are essentially process structures.

Three levels of inquiry in physics

Such a systematic and dynamic view helps to overcome the ruling reductionism in physics (not to speak of the domain of the living). Following a suggestion by Ilya Prigogine, we have to distinguish in physics at least between three levels of inquiry and description which are irreducible to each other.

Classical or *Newtonian dynamics,* the development of which is linked to names such as Laplace and Hamilton, makes statements in terms of mechanics, such as position or velocity of particles. It reduces the world to trajectories, or space-time lines, of single material points. The movement of a particle from point A to point B is perfectly reversible. The impulse for the movement has to be provided from outside; there is no self-organization. The particles crossing the world on their lonely tracks, do not interact with each other. Thus, classical dynamics becomes the idealized case of "pure" motion of a particle or a wave packet, a mere thought model which nevertheless is useful for many considerations. But the "dirty" reality includes encounters, collisions, exchanges, mutual stimulation, challenges and coercions of many kinds. The collective with all its complexity can hardly be denied anywhere.

With the development of *thermodynamics* in the nineteenth century, a macroscopic view was introduced which considers whole populations of particles. Thermodynamics makes statements in terms such as temperature and pressure which represent the average effects of movements of a large number of molecules. This level of description addresses processes, or in other words, the order of change in the macroscopic values. The order of the processes themselves, or the evolution of the systems characterized by them, found its first valid formulation around 1850 in the well-known *second law of thermodynamics* (Clausius, basing on Carnot): The so-called entropy of an isolated system can only increase until the system has reached its thermo-dynamic equilibrium. It may suffice here to introduce the complex term entropy as a measure for that part of the total energy which is not freely available and cannot be used in the form of directed energy flow or work.* In other words, entropy is a measure for the quality of the energy in the system. In contrast to the mechanical description, *irreversibility* or the directedness of processes is introduced as characteristic of this new level of description. Any future macroscopic state of the isolated system can only have equal or higher entropy, any past state only equal or lower entropy. The reverse is impossible. All irreversible processes produce entropy. More than a century ago, Ludwig Boltzmann interpreted the increase in entropy as increasing disorganization, as evolution toward the "most probable" state of maximum disorder. If the world had appeared as a stationary machine in the mechanical view, it now seemed to be doomed to the "heat death", a notion which influenced pro-

---

*This is usually written in the following way: $E = F + TS$, where E is the total energy, F the freely available energy, T the absolute temperature in degrees Kelvin (corresponding to degrees Celsius plus 273.15) and S the entropy. Entropy shows its effect, for example, in the limitations of converting heat in a power plant or in a thermal machine, such as a car engine. According to Carnot, the ideal conversion efficiency of heat into work, without friction or other losses, is given by $(T_2 - T_1)/T_2$, with T_1 as the lower and T_2 as the upper temperature in the total process. If the thermal medium is heated to 600°K and the reject heat cooled away at normal environment temperature (300°K), the maximum efficiency for the conversion of heat into work is $(600 - 300)/600 = 0.50$. Modern power plants reach a conversion of 42 to 44 per cent; the rest goes as "thermal pollution" to water or air in the environment.

foundly the pessimistic philosophy and art of the turn of the century up to our days.

Classical thermodynamics deals with *equilibrium systems* or systems near equilibrium. An isolated system without environment shows a particular type of self-organization (or, to be precise, self-disorganization). It will evolve in the direction of its equilibrium state and remain there. The more general and realistic case is a partially open system which exhibits similar behaviour. This is generally the case for systems not far from equilibrium. For all such systems, the Boltzmann ordering principle holds which stipulates irreversible evolution toward equilibrium.

With irreversibility, the notions of process and history enter. Time is given a direction from the past into the future. The system evolves through a sequence of thermodynamic states which may be ordered along the scale of a single macroscopic parameter, entropy. In a microscopic view, the system gains the experience of innumerable encounters and exchanges between system components (molecules), but in a macroscopic view all that changes is the relation between free energy and entropy in the total system. Although the time scale is not given in an absolute way, but depends on the internal system processes, the path and the goal of the macroscopic system evolution are unambiguously predetermined. Within the experience of a system of this kind, the equilibrium state determines the origin of the system as well as its death. It constitutes the only point of self-reference of the system.

In thermodynamics, irreversibility often leads to the destruction of structures. But this does not hold in an absolute sense. At low temperatures, and if binding forces act at the same time, the approach to the equilibrium state may also give rise to structures. Crystals, snow flakes and biological membranes are such equilibrium structures with higher entropy than their corresponding liquid state from which they emerged. The emergence of form at increasing entropy also plays a role in condensation models of the universe (as will be discussed in Chapter 5). But there is also another way in which new structures may form, a way which is of central interest in this book. This is the spontaneous formation of structures in open systems which exchange energy and matter with their environment. Such systems constitute the other basic class of physical systems, namely, non-equilibrium systems of a particular kind.

With such considerations, a third level of inquiry and description is opened up, which may be called the level of coherent, evolving systems or (to anticipate what will be explained in a moment) the level of *dissipative structures*. Open systems have the possibility of continuously importing free energy from the environment and to export entropy. This means that entropy, in contrast to isolated systems, does not have to accumulate in the system and increase there. Entropy can also remain at the same level or even decrease in the system. The "accounting" includes the environment. The second law of thermodynamics becomes extended for open systems and the entropy change dS in a specific time interval is split into an internal component d_iS (entropy production due to irreversible processes within the system) and an external component d_eS (entropy flow due to exchange with the environment): $dS = d_eS + d_iS$, where the internal component d_iS can only be positive or zero (as in the case of the isolated system), but never negative ($d_iS \geq 0$). The external entropy flow d_eS, in contrast, can be both positive and negative. Therefore, the total entropy may also decrease, or be stationary ($dS = 0$). In the latter case, we have $d_iS = -d_eS \geq 0$. The internal production of entropy and the entropy export to the environment are in balance. Since, for the equilibrium state, both components would approach zero, it already becomes clear that an open state of order may be maintained only in a state of *non-equilibrium*. There has to be exchange with the environment and the system renews itself continuously. Being and becoming fall together at this level.

Dissipative structures constitute the simplest case of spontaneous self-organization in open evolution. They will be discussed in much detail below. However, it is important to recognize that even without the phenomena of life, the description of a dynamic reality requires at least three levels of inquiry. The reduction to one level, the ancient dream of physics, is no longer possible.

Symmetry-breaking as a source of order

The three levels of inquiry are irreducible to each other because the transitions between them are characterized by symmetry breaks (Prigogine, 1973). For the transition from the mechanical to the thermodynamic level this is evident. Irreversibility implies a *break of the time symmetry* between past and future, a symmetry which is still conserved in the equations for the evolution of a classical mechanical system. At the thermodynamic level, the Fourier equation for the macroscopic phenomenon of thermal conduction expresses the irreversibility by stating that a non-uniform temperature distribution will change toward a more uniform distribution in the future. The direction of time cannot be reversed. Never will a non-uniform distribution result from a more uniform one if the system is left alone. If we pour hot water from one side and cold water from the other into a bowl, the result will be lukewarm water; but lukewarm water can never separate by itself into hot and cold water.

The symmetry break between past and future, or between the "before" and the "after", results in temporal order, or *causality* in a strict sense. Since a process which is described at the thermodynamic level can only run in the direction of uniformity and equilibrium, any non-uniform initial condition (such as the non-uniform temperature distribution in our example) is introduced as random fluctuation. There is no ordering principle at this level which would be able to account for it.

But such a new ordering principle is obviously needed to account for the self-organizing of evolving systems at the third level of inquiry. It requires instability of the thermodynamic order which leads to a *break in time and space symmetry.* Physicists used to believe that non-equilibrium states do not contain any interesting physical information. Thermodynamic non-equilibrium was treated as a temporary disturbance of equilibrium, rather than as a source of something new. But it was precisely Ilya Prigogine's conviction, pursued over three decades, that non-equilibrium may be a source of order, of organization, that became the foundation for a non-linear thermodynamics of irreversible processes now permitting the description of phenomena of spontaneous structuration. The new ordering principle, which has been clearly recognized only since 1967, has been called *order through fluctuation.* It describes the evolution of a system to a totally new dynamic régime. This dynamic régime represents a spatial and temporal order which would contradict the second law of thermodynamics if it were near the equilibrium.

This development permits us to discuss the *relations between the three levels of inquiry.* Instead of a sterile reductionism and a vague antireductionism, the consideration of physical reality may now proceed at all three levels simultaneously, with each level being ascribed a clearly defined domain of phenomena.

In the following two chapters I shall describe in more detail the dynamics of self-organizing dissipative structures. This dynamics underlies both the globally stabilized, temporary "existence" in specific structures and their evolution to new structures.

2. Dissipative Structures: Autopoiesis

Spontaneous structuration

The examples used in the preceding chapter to demonstrate the spontaneous formation of hydrodynamic structures (turbulent flow and Bénard cells) were characterized by outside imposition of the energy penetration by means of water pressure and heating of the liquid, respectively. The emergent structures represent the manner in which the system copes with increased energy and mass penetration. Of much greater interest, however, are those physical-chemical reaction systems which themselves maintain energy and matter penetration by way of exchange with the environment and which give rise to the self-organization of globally stable structures over extended periods of time. These structures are the *dissipative structures* in the narrow sense of the word. They have been called that way because they maintain continuous entropy production and dissipate the accruing entropy. They emerge from dissipative self-organization, in contrast to conservative self-organization which uses the static (attracting or repelling) forces in the system itself.

The most frequently cited example for a dissipative structure is the so-called Belousov-Zhabotinsky reaction, discovered 1958 and named for its Russian investigators (Zhabotinsky, 1974). It involves the oxidation of malonic acid by bromate in a sulphuric acid solution and in the presence of cerium (or also iron or manganese) ions. If certain conditions are met, concentric or rotating spiral waves may be observed which lead to interference patterns (see Fig. 2). In this and similar reaction systems pulsations of great regularity may be observed which may last for many hours, so that one also speaks of "chemical clocks". Or there are periodical bursts of

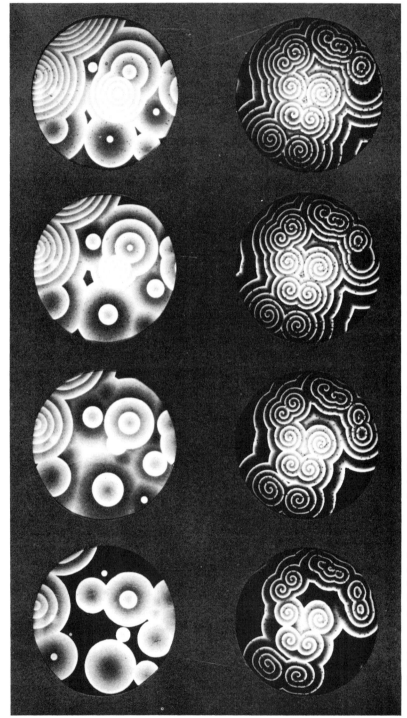

Fig. 2. An example of a dissipative structure. In the Belousov-Zhabotinsky reaction malonic acid is oxidized by bromate in the presence of cerium ions. If the reaction is carried out in a shallow dish, spiral waves develop. Photographs by A. Winfree (1978).

sudden chemical activity which, as "chemical vectors", may prefer certain spatial directions, or the welling-up of concentric chemical waves and other dynamic phenomena which sometimes become even more spectacular by the play of bright colours by which they are accompanied.

For the spontaneous formation of such structures in chemical reaction systems, a "generalized" thermodynamics by Glansdorff and Prigogine (1971) stipulates precise conditions. They include *openness* with respect to the exchange of energy and matter with the environment, *far from equilibrium* conditions and auto- or crosscatalytic steps in the reaction chain. The last point means that certain molecules participate in reactions in which they are necessary for the formation of molecules of their own kind (autocatalysis), or first for the formation of molecules of an intermediate kind and subsequently of their own kind (crosscatalysis). The result is a type of behaviour which is called non-linear and which is characterized by runaway processes. In technological cybernetics, such a behaviour is called positive feedback—a deviation from a given reference value is not eliminated, but increases. The global population explosion and other growth factors in the human contemporary world are examples for autocatalytic non-linearities. In our case, however, they are not scary, but constitute an essential factor in the creative act of gestalt formation.

Dissipative structures exhibit two different types of behaviour: near their equilibrium, order is destroyed (as it is in isolated systems), but far from equilibrium, order is maintained or emerges beyond instability thresholds. The latter type of behaviour is called coherent behaviour. As long as dissipative structures exist, they produce entropy. However, this entropy does not accumulate in the system but is part of a continuous energy exchange with the environment. It is not the statistical measure of the entropy share in the total energy of the system at a given moment that characterizes a dissipative structure, but the dynamic measure of the *rate of entropy production* and of the exchange with the environment—in other words, the *intensity* of energy penetration and conversion.

Whereas free energy and new reaction participants are imported, entropy and reaction end products are exported—we find here the *metabolism* of a system in its simplest manifestation. With the help of this energy and matter exchange with the environment, the system maintains its inner non-equilibrium, and the non-equilibrium, in turn, maintains the exchange processes. One may think of the image of a person who stumbles, loses his equilibrium and can only avoid falling on his nose by continuing to stumble forward. A dissipative structure continuously renews itself and maintains a particular dynamic régime, a globally stable space-time structure. It seems to be interested solely in its own integrity and self-renewal.

There is a remarkable parallel to a new theory of subatomic particles which has been founded by Geoffrey Chew (1968) in Berkeley and which is also called the "bootstrap model". It is based on pure process thinking and considers the so-called "hadrons" (which include, in particular, the protons and neutrons in the atomic nucleus) as temporarily stable configurations which result from the interaction of processes. Hadrons may transform themselves into each other and help other hadrons in their transformations. They may appear as composite particles, constituents of other particles, or binding forces. The actually unfolding process chains and the resulting process webs are unpredictable, but they obey certain rules. These rules are based on a single fundamental principle, *self-consistency*. Whatever comes into being has to be consistent with itself and with everything else. A reduction of physical reality to basic building blocks or even to basic laws is not possible according to this concept which is in full development. As we shall see later, the open evolution of the macrocosmos, too, may be understood as being based on nothing else but self-consistency.

A hierarchy of characteristic system aspects

By its very nature, a system cannot be described by the sum of single properties. However, it is possible to make significant distinctions by looking at certain aspects of the whole system. The most essential system aspects include the following ones which, in the order in which they are enumerated, also form a hierarchy in ascending order.

With respect to its *relations with the environment*, a system is called open that maintains exchange with its environment—especially exchange of matter, energy and information—and that is open toward the new and unexpected (toward novelty, as we shall call it later). Systems without exchange with their environment are called isolated. An exchange with the environment can be maintained by the system itself only when its *internal state* is in nonequilibrium; otherwise, the processes would die down.

The (logical) *organization* of a system refers to the characteristic pattern in which processes are linked in the system. It may be represented by a flow scheme. Of particular importance is cyclical (closed circular) process organization which will be discussed in greater detail in Chapter 10. The dissipative systems which interest us here in the first line are organized in *hypercycles*, as Manfred Eigen called this particular organization. A hypercycle is a closed circle of transformatory or catalytic processes in which one or more participants act as autocatalysts. The above-mentioned Belousov-Zhabotinsky reaction, for example, may be represented as a hypercycle formed by the intermediary products *X*, *Y*, *Z* (see Fig. 3). In order to maintain a specific sense of rotation—in Fig. 3 in a clockwise sense—there has to be non-equilibrium. The "inner" process circle renews itself continuously and, as a whole, acts like a catalyst which transforms starting products into end products.

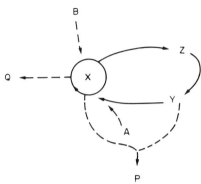

Fig. 3. The cyclical organization of the Belousov-Zhabotinsky reaction with an autocatalytic step which is marked by the arrow in a closed circle. Starting products $A = B = [BrO_3^-]$; intermediary products $X = [HBrO_2]$, $Y = [Br^-]$, $Z = [2\ Ce^{4+}]$; end products P, Q. The intermediary product X reproduces autocatalytically and thereby holds the cycle in motion; it decays continuously into Q, so that the system is globally stabilized.

Another important aspect of system organization concerns the arrangement of processes at one or more levels. Among multilevel systems, *hierarchic* systems are of particular importance. In such systems, each level includes all lower levels—there are systems within systems within systems . . . within the total system in question. As will be shown in later chapters, evolution leads to differentiation in multilevel, hierarchic systems.

The *function* of a system embraces the total characteristics of its processes, including the relations with the environment and the system organization, but beyond that the kinetics of the individual processes also and their interaction. The logical scheme of relations appears here in the framework of time.

The function of *autopoiesis* (from the Greek for "self-production") occupies a special place. It has been introduced in the early 1970s by the Chilean biologist Humberto Maturana and has been further developed by him in collaboration with Francisco Varela and Ricardo Uribe. A system is autopoietic when its function is primarily geared to self-renewal. A biological cell, for example, is autopoietic in its balanced self-renewal through the interplay of anabolic and catabolic reaction chains; over longer periods, it does not consist of the same molecules. An autopoietic system refers in the first line to itself and is therefore also called *self-referential*. In contrast, an allopoietic system, such as a machine, refers to a function given from outside, such as the production of a specific output.

The *structure* of a system has long been understood primarily in terms of its spatial structure. In connection with dynamic systems, however, the notion of a *space-time structure* is of importance, or in other words a structure given in a particular moment of time which represents not only the spatial arrangements, but also the kinetics effective in this moment at each spatial point. Such a space-time structure includes the function of the system, and thus also its organization and its relations with the environment. The co-operative principle of dissipative self-organization becomes manifest in the spatio-temporal order of interactive processes—in the space-time structure. It is the dissipative structure which is responsible for ordering the processes in such a way that there is balance between generation and degeneration, that the autocatalytic self-reproduction in the system does not blow it up into pieces and keeps it imprisoned in its own tread-mill.

A time sequence of space-time structures yields the *total system dynamics*. It may be organized from outside the system, such as in the case of a machine operated from without, or it may be self-organizing. In self-organization, as has already been mentioned, conservative and dissipative modes may be distinguished. Conservative self-organization may lead to static or dynamic (steady-state) equilibrium systems; an example for the latter class is the solar system with its rotating planets. In this book, we are primarily interested in dissipative self-organization.

These hierarchically ordered, six system aspects may be summarized so as

Table 1.

An overview of the hierarchy of characteristic system aspects makes the distinctions visible between two fundamentally different classes of systems. Structure-preserving systems are at their equilibrium or approach it irreversibly. Evolving systems are far from equilibrium and evolve through an open sequence of structures

Characteristic system aspect	Structure-preserving systems		Evolving systems
Total system dynamics	Static (no dynamics)	Conservative self-organization	Dissipative self-organization (evolution)
Structure	Equilibrium structure, permanent	Devolution toward equilibrium structure	Dissipative structure (far from equilibrium), evolving
Function	No function or allopoiesis	Reference to equilibrium state	Autopoiesis (self-reference)
Logical organization	Statistical oscillations in reversible processes	Irreversible processes in direction of equilibrium state	Cyclical (hypercycle), irreversible sense of cycle rotation
Internal state	Equilibrium	Near equilibrium	Non-equilibrium
Relationship with environment	Isolated or open (growth possible)		Open (continuous, balanced exchange)

to define with their help two fundamentally different classes of systems, structure-preserving and evolving ones (see Table 1). Structure-preserving systems may be subdivided into systems which have already reached their equilibrium, and systems which are only on their way there and whose dynamics is already geared to the equilibrium they are aiming for. This dynamics may be called *devolution* since it runs in the opposite direction of evolution.

Characteristics of dissipative structures

It may be interesting to ask whether a dissipative structure is to be understood as a material structure organizing the flow of energy, or as an energetic structure organizing the flow of matter. At this level of self-organization, both descriptions are equally valid. They constitute two sides of a complementarity. At higher levels of self-organization, however, to an increasing extent a description will suggest itself which views energy systems manifesting themselves in the organization of material processes and structures.

A deeper study of these new and fascinating phenomena became possible by means of model calculations which, due to the decisive role of non-linearities, usually require the use of big computers. Many specific cases have to be calculated before the overall characteristics of non-linear behaviour become discernible. A large part of such theoretical-mathematical work was carried out by the Brussels school of Ilya Prigogine by using a model of a cross-catalytic chemical reaction which in the literature was accordingly given the name "Brusselator". There are also other, similar models, for example the "Oregonator" (developed at the University of Oregon), which represents a simplified version of the Belousov-Zhabotinsky reaction. The Brusselator, however, as may be demonstrated, represents the simplest case of a reaction which leads to the formation of dissipative structures. It may be schematically written in the following way:

$$A \rightleftharpoons X$$
$$B + X \rightleftharpoons Y + D$$
$$2Y + Y \rightleftharpoons 3X$$
$$X \rightleftharpoons E$$

where A, B, D, E are starting and end products and X, Y are intermediate products whose development over time and spatial distribution are to be studied. In the numerical calculations, the reversed reactions (in the scheme from right to left) are usually neglected. It is, above all, the autocatalytic third step in the reaction system which introduces the non-linearity which is primarily responsible for the particular behaviour of the system. Since in this third step, three molecules of the species X are formed, one also speaks of the "trimolecular model".

The system can assume a single homogeneous stationary state, for which the corresponding concentrations are given by $X = A$ and $Y = B/A$. This stationary state, however, becomes instable if the concentrations A, B and the diffusion coefficients (according to the so-called Fick law) D_X, D_Y and perhaps also D_A satisfy certain conditions. Then, the behaviour of the system may vary in the following ways:

1. If D_X and D_Y are very large so that the system may be viewed as being nearly homogeneous, there may be stable periodic oscillations around a stationary state, so-called limit-cycle behaviour (Fig. 4). For somewhat

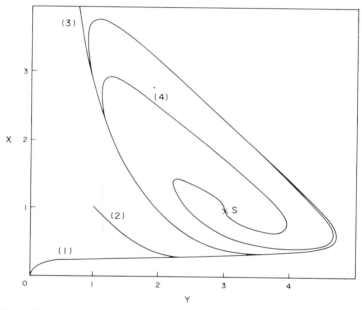

Fig. 4. Formation of limit-cycle behaviour in the chemical reaction system "Brusselator" according to numerical integration for various initial conditions: (1) $X = Y = 0$; (2) $X = Y = 1$; (3) $X = 10$, $Y = 0$; (4) $X = 1$, $Y = 3$; where always $A = 1$, $B = 3$. The point S corresponds to the unstable stationary state (X_s, Y_s). Whatever initial conditions are chosen, the system will tend toward a single well-defined solution of periodic oscillations. After R. Lefever (1968).

smaller values of D_X and D_Y, a spatio-temporal régime may emerge which shows the spreading of concentration waves or of stationary chemical waves. The evolution of such a régime is shown in Fig. 5.

2. Far from equilibrium (for example very small concentrations of D and E), the system may evolve in the direction of a new stable, stationary state with an inhomogeneous distribution of X and Y. It is possible to define a wavelength of the dissipative structure which is directly proportional to the spatial extension of the system; thereby, the macroscopic character of the emergent order is emphasized. Figure 6 shows such a stable

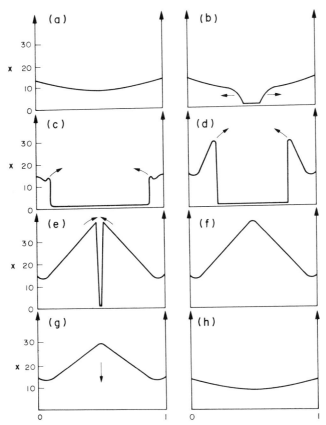

Fig. 5. Evolution of a dissipative structure of the type "Brusselator". Characteristic steps are shown in the development of the spatial distribution of the intermediary product X. They were calculated under the following assumptions: $X(0) = X(1) = 14$; $B = 77$; $D_X = 0.00105$, $D_Y = 0.00066$, $D_A = 0.195$. After M. Herschkowitz-Kaufman and G. Nicolis (1972).

dissipative structure which also exhibits the characteristic spontaneous occurrence of polarity (higher concentration of X on one side) which a system exhibits under the influence of a disturbance. This becomes even clearer in Fig. 7 which is drawn in polar co-ordinates. Such a spontaneous formation of polarity is particularly important for developmental biology (the development of the embryo from an originally homogeneous cell mass which results from the division of the zygote). A comparison of Fig. 7 with the microscopic skeleton of a freshwater alga (Fig. 8) shows the same basic structure.

3. If also the diffusion of the starting product A within the system can no longer be neglected, dissipative structures form beyond a critical instability (Fig. 9). In this case, the spatial organization is restricted to a

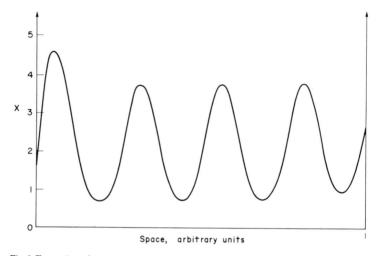

Fig. 6. Formation of a stable dissipative structure far from equilibrium in the chemical reaction system "Brusselator". The concentration of X was calculated under the following assumptions: $A = 2$; $B = 4.6$; $D_X = 0.0016$, $D_Y = 0.0080$. The higher concentration of X on the left side indicates spontaneous polarization. After M. Herschkowitz-Kaufman (1973).

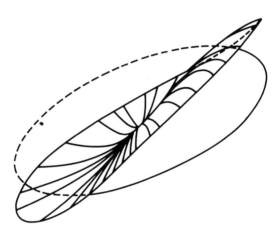

Fig. 7. A two-dimensional, spatial dissipative structure of the chemical model reaction "Brusselator", shown in polar co-ordinates. The calculations were based on the following assumptions: diameter 0.2; boundary penetration zero; $B = 4.6$; $D_X = 0.00325$, $D_Y = 0.0162$. After T. Erneux and M. Herschkowitz-Kaufman (1975).

certain volume outside of which the thermodynamic order continues to rule.

In these modes of behaviour, rhythm may be understood as the manifestation of a break in time symmetry, whereas the formation of a field points to a break in spatial symmetry. The emergence of these phenomena is only

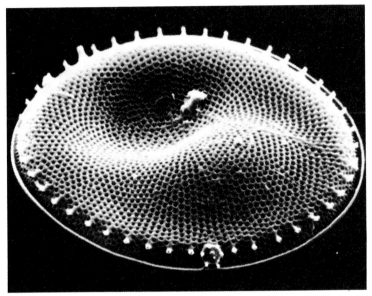

Fig. 8. Photograph of the unicellular alga *Coscinodiscus lacustris Grun.,* made with the electron microscope of the Geological-Paleontological Institute of the University of Kiel (Germany). Original diameter 0.08 millimetre. This alga lives suspended near the surface of water with various salinity and is frequently found in the Baltic Sea. Photograph: Hans-J. Schrader.

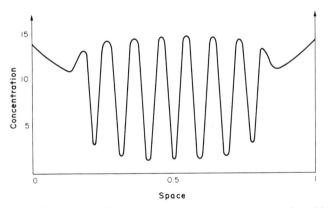

Fig. 9. Formation of a localized dissipative structure in the chemical model reaction "Brusselator". The concentration of X was calculated under the following assumptions: $B = 26$; $D_A = 0.197$; $D_X = 0.00105$, $D_Y = 0.00526$. Outside the dissipative structure, the classical thermodynamic order reigns. After M. Herschkowitz-Kaufman and G. Nicolis (1972).

possible in macroscopic, coherent media because they require a large number of interactive processes.

The behavioural modes deduced from theory are recently finding their empirical confirmation in numerous non-linear oscillation phenomena of physical, chemical, biochemical, electrochemical and biological nature (Faraday Symposium, 1974; Nicolis and Prigogine, 1977). The best studied dissipative structures belong to the already mentioned Belousov-Zhabotinsky reaction and the glycolytic cycle which is responsible for the energy conversion in the biological cell.

The case described under point 3 above already points to the importance of the size of a system for the formation of dissipative structures. A system which is too small will always be dominated by the boundary effects. Only beyond certain critical dimensions do the non-linearities find an opportunity to unfold their characteristics and may bring a selection of new structures into play resulting in a certain autonomy of the system with respect to its environment. A dissipative structure comes only into being when a specific critical size can be realized. But then, there is no difference in the structure, whether the environment is scarcely sufficient or is vast (except for the duration of the life of the system which, of course, depends on its environment for "food" in the form of free energy and new reaction participants). If there is the additional factor of spatial concentration of catalytically active molecules, such as found in membranes of biochemical systems, there may be dissipative structures of very small size. Biological cells contain such microscopic dissipative structures.

Self-reference and environment

An autopoietic régime includes the expression of a particular individuality, a particular autonomy from the environment. Other than a crystal (an equilibrium system) which grows indefinitely when placed into a suitable solution, a dissipative structure finds and maintains its proper form and size independent from the "nourishing" environment. Structure and function are realized the more characteristically the more degrees of freedom there are available to the system. The natural dynamics of simple dissipative structures teaches the optimistic principle of which we tend to despair in the human world: The more freedom in self-organization, the more order!

If conciousness is defined as the degree of autonomy a system gains in the dynamic relations with its environment, even the simplest autopoietic systems such as chemical dissipative structures have a primitive form of *consciousness.* Maturana's (1970) description of the feedback relations of an autopoietic system with its environment as constituting a cognitive domain fall not very far from this insight. And a dissipative structure "knows" indeed what it has to import and to export in order to maintain and renew itself. It needs nothing else but the reference to itself.

From another angle of view, this autonomy appears as an expression of the fundamental *interdependence of structure and function* which is one of the most profound laws of dissipative self-organization. The spontaneously emerging structure corresponds to the systemic function (the totality of the processes), and vice versa. This plasticity underlies the possibility of attaining a true autopoietic balance on the one side, but also co-evolution (the joint evolution of a system together with its environment). Even complex biological, social, psychological and cultural systems, which partly make use of the transfer of information brought into a rigid form, are highly malleable. In contrast, equilibrium systems and machines are not malleable in this sense.

The complementarity of structure and function may generally be regarded as an expression of process thinking: well-defined spatial structures result from the interaction of processes in a specific dynamic régime. The circularity of many of these process chains calls for a dynamic formulation in terms of macroscopic notions referring to the system as a whole (for example, self-renewal in autopoietic régimes). A cell contains several thousand biochemical processes in a very small volume, and many of them are interlinked by complex, intermeshing feedback loops. As Varela (1975) has shown, the microscopic equivalent of following all these individual process interactions would require infinite time. The resulting diagram resembles a decision tree which branches indefinitely. However, if one searches for "rules of the game" for all these processes and for criteria of holistic system behaviour—in other words, if the semantic level gets introduced—one may hope for a simpler representation. However, this simplicity is then no longer the above-cited "simplicity of the microscopic", but a new "simplicity of the macroscopic" which is yet to be discovered. Pure self-reference in an autopoietic system provides a striking example. The prize offered for the development of a truly systemic view is tremendous.

The dynamic existence of non-equilibrium structures is not only characterized by continuous oscillation and self-renewal, but also by the impossibility of ever achieving absolute stability. There is always a possibility of forcing a certain dynamic régime—a specific autopoietic dissipative structure—into a new régime. In the same way in which the first dissipative structure forms spontaneously beyond an instability of the thermodynamic (equilibrium) branch, the structure may again become instable and switch to a new structure. The discussion of system evolution in the following chapter will provide an opportunity to point out the decisive role of fluctuations in the emergence of new order.

3. Order through Fluctuation: System Evolution

> When I was a boy and my father wanted
> to teach me to swim, he threw me into
> the water, I fell and sank to the bottom. I
> could not swim, and felt that I could not
> breathe. . . . I do not know how I walked
> under the water, and suddenly saw the
> light. Understanding that I was walking
> towards shallow water, I hastened my
> steps and came to a straight wall. I saw
> no sky above me, only water. Suddenly I
> felt a physical strength in me and jumped,
> saw a cord, grasped it, and was saved.
>
> Vaslav Nijinsky, *Diary: Life*

Evolutionary feedback

The dissipative structure, which forms spontaneously beyond the thermodynamic order, is not the end of dynamic development. It is in principle stable as long as the energy exchange with the environment is maintained and as long as the continuously occurring fluctuations are absorbed within the framework of the given dynamic régime. But, quite generally, no structure of a non-equilibrium system is stable by itself. Any structure may be driven beyond a threshold into a new régime when the fluctuations exceed a critical size. This corresponds to a *qualitative* change in the dynamic existence of the system. The transition to a new dynamic régime renews the capability for entropy production, a process which may be viewed as life in a broad sense. Life always carries on.

The fluctuations referred to here are not fluctuations in concentration or other macroscopic parameters, but fluctuations in the mechanisms which result in modifications of the kinetic behaviour (e.g. reaction or diffusion rates). Such fluctuations may hit the system more or less randomly from without, as through the addition of a new reaction participant or changes in the quantitative ratios of the old reaction system. But they may also build up within the system through positive feedback which, in this case, is called evolutionary feedback:

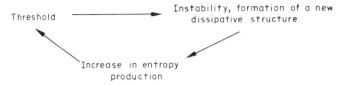

This cycle may repeat itself in many steps.

Such an evolutionary feedback occurs in chemical reaction systems, for example, when the kinetics is modified and a new non-linearity is introduced thereby, as may happen particularly with the formation of new substances. The specific energy dissipation per mass unit increases and the system is forced through an instability to a new régime which corresponds to a higher level of interaction between system and environment. These concepts have become of great importance for recent theories of prebiotic evolution which will be discussed in Chapter 6.

There are certain overlaps between the theory of dissipative structures and an alternate description of evolving systems, *catastrophe theory* developed by René Thom (1972), and taking a topological approach. The essential difference lies in the limitation of catastrophe theory which, in its present state of development can only represent switches from a postulated equilibrium state (represented by an "attractor") to another equilibrium state which represents a new dynamic régime. The switches are called "catastrophes". There is no true self-organization and self-reinforcement of fluctuations. Both the motion of the system and the "morphogenetic field" in which it occurs are given from the outside. One may think of the image of a golf ball which is driven up a steep slope. If it falls to the earth before it reaches the ridge, it will run down the same slope and end up close to the player. If it falls down even barely beyond the ridge, however, it will roll into another valley—or get stuck in a hole higher up. The theory of dissipative structures, in contrast, would describe the player as one who walks up the slope on his own and selects his resting places himself.

The older concept of *ultrastability* developed by Ross Ashby (1960) describes the step-wise adaptation of a system to its environment until equilibrium is reached. In contrast to evolutionary feedback, there is no intensification of the interactions between system and environment, but their eventual termination instead. The self-organization of systems, the inner dynamics which drive them to reconstitute themselves in new structures, cannot be adequately described by such models. But this is precisely what the theory of dissipative structures can do by virtue of considering the self-reinforcement of fluctuations within non-equilibrium systems. It also presents morphogenesis, the emergence of new form, in the light of a partial "self-determination of the system".

Table 2 presents a systematic overview of the applicability of approaches to dynamic system modelling. All these approaches are useful—if their inherent limitations are recognized and made explicit. This is rarely done, however.

The role of fluctuations: the micro-aspect

Autopoietic global stability represents only a special case of an evolving dynamic system—that case, to be precise, in which fluctuations are absorbed by the system or, in other words, damped by the environment in which they occur. The same conditions which lead to autopoeisis—openness, non-equilibrium and especially autocatalysis—also underlie the possibility of internal self-amplification of fluctuations and their ultimate breakthrough. Without such internal self-amplification there is no true self-organization. The possible consequence is the evolution of the system through an indefinite sequence of instabilities each of which leads to the spontaneous formation of a new autopoietic structure. It becomes clear now why the new ordering principle beyond the Boltzmann principle is called order through fluctuation.

Autopoiesis and evolution, global stability and coherent change, appear as complementary manifestations of dissipative self-organization. Whereas autopoiesis, as has been discused in the preceding chapter, may be described in terms of the complementary pair structure ⟷ function, a three-fold correspondence holds for self-organization which includes evolution:

In other words, the dynamic system as a whole may also be understood as one gigantic fluctuation.

The discussion of the relations between these three levels of description may take a microscopic or a macroscopic angle of view. The microscopic description, also called stochastic (time-dependent) description, follows the formation of fluctuations and their destiny close to the transition threshold between an old and a new structure. It takes chance into account by generally considering the occurrence and the kind and size of fluctuations as random. The macroscopic description, in contrast, emphasizes a deterministic element in describing how the system as a whole is forced into a new structural-functional order. This new order is not predetermined in an absolute way, however; the system may choose among at least two new structures, as will be dis-

Table 2

Domains of application for various dynamic system approaches. Recent developments only permit the modelling and the study of self-organization dynamics as it characterizes biological, social and cultural life

Type of system structure	Type of system dynamics	Microscopic processes ("causes")	Macroscopic system dynamics ("effect")	
			Continuous (same dynamic régime)	Discontinuous (evolution through a sequence of dynamic régimes)
Equilibrium	Equilibration	Continuous (Law of Large Numbers holds)	Differential equations (a) Linear: econometrics (b) Non-linear: simulation of feedback systems (Forrester's "System Dynamics", Club of Rome studies)	(a) Ultrastability (Ashby) (b) Catastrophe theory (Thom)
Non-equilibrium	Self-organization	Discontinuous (Law of Large Numbers suspended)	(a) Theory of metastable dissipative structures (Prigogine and Nicolis) (b) Tesselation games—autopoiesis (Maturana and Varela)	(a) Order through fluctuation (Prigogine) (b) Tesselation games—evolution (Eigen and Winkler)

cussed below. The fluctuations themselves may be of random origin; but their result is no longer purely random. Only both descriptions together give a realistic picture. *Chance and necessity* appear here as *complementary* principles, that is to say, as integral aspects of one and the same process. Quite generally spoken, a complementary view is akin to process thinking.

The microscopic (stochastic) description, developed by Prigogine and his collaborator Nicolis, allows us to understand the formation of new dissipative structures by a *nucleation process*. Such processes are familiar from the formation of rain drops which form around a nucleus, such as a grain of dust, and grow until they are sufficiently heavy to drop to earth. In the nucleation of new dissipative structures from fluctuations, a characteristic "nucleation length" plays an important role. It is determined exclusively by the inner dynamics of the system and is independent of the reaction volume. Only fluctuations of sufficient spatial extension (beyond a "critical" size) are capable of driving the system into instability and into a new régime. In this case, the "law of large numbers"—stating that an adequate description of a heterogeneous system is possible by means of average values—is rendered invalid. Here, fluctuations which may appear small in comparison with the system as a whole may change the average values in a decisive way. In the instability phase, any modelling approach breaks down, as is also the case in phase transitions (such as from water to steam), which show a surprising similarity in the basic equations. But the kind of activity which dominates in the instability phase introduces a directedness, a vector which already indicates in which direction the new structure may be expected.

For the chemical reaction system "Brusselator", which has served us as a model so far, the result of the competition between fluctuation-amplifying chemical reactions and fluctuation-damping diffusion (which introduces a tendency toward homogenization) is depicted in Fig. 10. If the characteristic chemical parameter k lies below the "macroscopic" threshold value k_o all fluctuations are damped. Beyond this threshold, fluctuations become amplified and lead to instability of the overall system only if they exceed the critical nucleation length. One may put it this way, that the fluctuations continuously "test" the stability of the structure. If they are too small, the system remains even beyond the "macroscopic" threshold in a state of *metastability*.

This result may also be interpreted in such a way that the environment of an "innovative", individualistically deviating subsystem will always try to damp the fluctuations and keep the total system stable even if, from a point of view of macroscopic theory, it should already have become unstable. This is an example for the complementarity of "chance", incorporated in the fluctuations, and "necessity", arising from the coupling of the subsystems.

Besides the dimensions of the total volume available for the development of

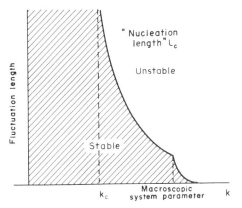

Fig. 10. Metastability of the chemical model reaction "Brusselator" in case the fluctuations beyond the "macroscopic" critical parameter k_c do not attain the spatial dimensions of a critical "nucleation length" l_c. In such a case, the environment of a fluctuating subsystems enforces stability even when the system ought to be unstable from the point of view of the macroscopic theory. After G. Nicolis and I. Prigogine (1971).

structures, packing density and the degree of coupling of the subsystems play an important role in the formation of dissipative structures. A number of isolated revolutionaries will not bring about a revolution, nor will a well-organized group of revolutionaries which are incapable of keeping their plans secret from the environment (high diffusion rate). The penetration of fluctuations and the formation of new dissipative structures depend on sufficiently dense packing on the one hand and on flexible, not too strong and rigid coupling on the other.

Macroscopic indeterminacy

The macroscopic description attempts to identify instability and transition thresholds between two structures but assumes infinitely big fluctuations so that in reality the old structure will still be maintained beyond that instability threshold. On the other hand, the macroscopic description attempts to discern which new dynamic régimes are available beyond the instability. For the model reaction of the "Brusselator", Nicolis and Herschkowitz-Kaufman in Brussels and Auchmuty at Indiana University were able to show with the help of *bifurcation analysis* that in every symmetry-breaking transition at least two new possible régimes, or structures, are offered for choice. The dissipative structure depicted in Fig. 6 (p.38), for example, could also have formed as its own mirror image (with the higher concentration of X on the other side). It is interesting that, depending on the symmetry properties of the critical state, different versions of transitions are possible. For even critical wave numbers, there is symmetry break and "smooth" transition to two alternative stable dissipative structures (Fig 11a); for odd critical wave numbers, there is symmetry break as well as bistable behaviour, coupled with hysteresis (Fig. 11b).

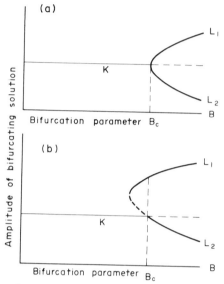

Fig. 11. Bifurcation diagram for the transition from the thermodynamic branch K to two possible dissipative structures L_1 and L_2 when instability is reached, as indicated by the critical bifurcation parameter B_c (concentration of B). (a) For even critical wave numbers, the transition to one of the two new solutions L_1 and L_2 is "smooth"; (b) for odd critical wave numbers, the transition occurs by way of a bistable domain in which the new solution L_1 is separated by a jump from the thermodynamic branch K. After G. Nicolis (1974).

This bifurcation, in principle, repeats itself at each new critical value of the bifurcation parameter (in Fig. 11 the concentration of B) of which there is an infinite number. At each transition, two new structures become spontaneously available from which the system selects one. Each transition is marked by a new break of spatial symmetry. The path which the evolution of the system will take with increasing distance from thermodynamic equilibrium and which choices will be made in the branchings *cannot be predicted.* The further the system moves away from its thermodynamic equilibrium, the more numerous become the possible structures. The possible paths of evolution resemble a decision tree with branchings at each instability threshold (Fig. 12). The obvious analogy to the microscopic process model of an autopoietic structure—which, as mentioned in the last chapter, also resembles an indefinitely branching decision tree—poses the question whether a holistic description will also become available here, a description not only of a single structure, but of the total system evolution through many structures.

A new element of indeterminacy enters the realm of physics at the level of coherent system behaviour. Side by side with the quantum-mechanical in-

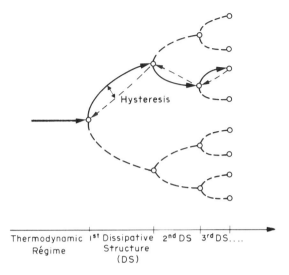

Fig. 12. Macroscopic indeterminacy in the evolution of a dissipative structure. At each instability threshold there is choice among several (at least two) possibilities. If non-equilibrium, however, is diminished again, the structure retreats along the same path which it has come, except for the so-called hysteresis effect which is due to the work invested in restructuration. The structure "remembers" the initial conditions.

determinacy which is expressed in the Heisenberg indeterminacy relation, there is now the *macroscopic indeterminacy* in the formation of structures. The far-reaching consequences of this discovery will be discussed in later chapters. Some scientists start to talk of dissipative structures in terms of "macroscopic quantization" because the properties of the emerging, qualitatively different structures may be expressed by only a few "quantum numbers", which represent the influence of the kinetic constants and the diffusion coefficients as well as of spatial symmetry and boundary conditions.

If a dissipative structure is forced to retreat in its evolution (for example, by a change in the non-equilibrium), as long as there are no strong perturbations it does so along the same path which it has come, except for hysteresis loops which reflect the work invested in changing a structure (see Fig. 12). This implies a primitive, holistic *system memory* which appears already at the level of chemical reaction systems. The system "remembers" the initial conditions which made a particular development possible, the beginnings of each new structure in its evolution. We may say, the system is capable of *re-ligio*, the linking backward to its own origin. In linking backward, the system "relives" its own experience—not in separable details, but in a sequence of holistic autopoietic régimes. In a specific autopoietic régime, the system is self-referential with respect to a specific space-time structure. In a broader perspective, we may now characterize an evolving system as being *self-*

referential with respect to its own evolution—that is to say, with respect to itself as a dynamic system with the potential of manifesting itself in a variety of structures, not in random order, but in coherent, evolutionary sequences.

The levels of global stability or autopoietic existence reached along such an evolutionary path are not predetermined, but result partly from the interaction between system and environment. In this respect, they represent true *experience*. We may also say that knowledge is expressed by the system's finding of its own stability with respect to fluctuations and, further, that this knowledge is nothing else but the experience of the interaction between system and environment, cast into a specific reference frame. In this sense, all knowledge is experience; objective and subjective knowledge become complementary.

For chemical dissipative structures, the criterion for optimal stability seems to be the possibility for high energy exchange, or high energy penetration of the system. In Chapter 12, criteria will be named for other levels of evolution.

A particular aspect of this self-determination is the *principle of maximum entropy production* which holds near the instability phase, in which a new structure forms. During the transition, entropy production increases significantly, whereas close to an autopoietic stable state it tends toward a minimum. In other words, the system does not spare any expense for the creative build-up of a new structure—and justifiably so as long as an inexhaustible reservoir of free energy is available in the environment. Only an established system, going for security, has to economize. This apparently does not only hold for dissipative structures, but for all evolving systems. The specific heat development in fertilized chicken eggs was measured at 0.32 watt per gram on the fourth day, but at only a sixth of this value on the sixteenth day. The frequently cited principle of minimum entropy production does not generally hold for natural processes, but refers only to fully established structures. But even then it refers to structures which are organized so as to assure a relatively high energy penetration.

Novelty and confirmation

The discussion of order is often couched in terms of information. This is of particular value for the discussion of self-organization because the general paradigma embraces not only material structures, but also mental structures, such as ideas, concepts or visions. For biology, P. Fong (1973) has defined information as any non-random spatial or temporal structure or relation and Carl Friederich von Weizsäcker (quoted in: Ernst von Weizsäcker, 1974) calls information that which generates new information. In this definition, the self-organization motive is already evoked.

But the conventional, mathematically elaborated information theory founded by Claude Shannon and Warren Weaver (1949) is primarily geared to equilibrium and the stabilization of structures. Just as in Boltzmann's thermodynamic ordering principle there is only one direction possible, the direction toward equilibrium structures, in the theory by Shannon and Weaver new information is also primarily considered to reconfirm and strengthen existing information structures. The amount of information is given; it can only decrease due to the inevitable noise effect, as in equilibrium thermodynamics order can only decrease. This type of information theory considers only the syntactic level, the arrangement of signs, which is already of considerable value in the development of machine codes.

In the domain of dissipative self-organization, and especially life, information is not transferred in one-way processes, but is exchanged in circular processes and is born new. This exchange occurs in a semantic context, that is, in the context of a particular meaning. If I say "I want to eat", a variety of actions follow depending on whether I am in my bachelor flat, in a restaurant, a plane or a fruit garden. But information exchanged among autopoietic systems is more than semantic; it is pragmatic in the sense of being geared to make a certain effect. That information is exchanged in a particular semantic context becomes meaningful only by its effectiveness. "The semantics of semantics is pragmatics", as Ernst von Weizsäcker (1974) puts it. Pragmatic information, however, changes the receiver. A machine may receive some news and afterwards still have the same expectation for receiving identical or similar news. But a human being will change his expectations. If in the radio we hear of a developing tropical hurricane, half a day later the news of considerable destruction in the area in question will not come as a surprise, but the news of destruction by an unexpected earthquake will.

Exchange with the environment, as it characterizes autopoietic structures, means also that each structure is at the same time sender and receiver of information. Since pragmatic information changes the receiver, it also changes the potential sender in the same structure. Therefore, we may now modify with Ernst von Weizsäcker (1974) the above-quoted definition of information proposed by his father: "Information is what generates information potential."

Ernst and Christine von Weizsäcker (1974) have given a model of pragmatic information which seems to be tailor-made for the discussion of order through fluctuation. Pragmatic information (and perhaps other types of information also) are composed of two complementary aspects, novelty and confirmation. The relationship is sketched in Fig. 13. Pure novelty, that is to say, uniqueness, does not contain any information; it stands for chaos. Pure confirmation does not bring anything new; it stands for stagnation or death. In between, however, there must be a finite maximum, depending on the complexity of exchanged information including the complexity of sender and

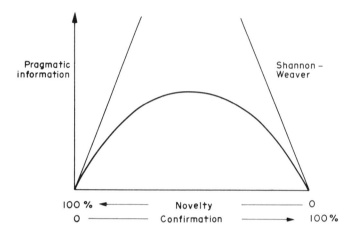

Fig. 13. Pragmatic (effective) information is composed of the two components novelty and confirmation and reaches a maximum when both components are balanced. After E. von Weizsäcker (1974).

receiver. This model also makes it clear that the Shannon-Weaver theory is applicable only to information characterized by a high degree of confirmation and very little novelty.

We may now easily establish the connection between this model of pragmatic information and the ordering principles at work in equilibrium and non-equilibrium structures (Fig. 14). A hundred per cent confirmation corres-

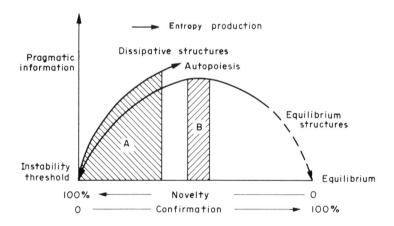

Fig. 14. Dissipative structures transform novelty into confirmation, whereas equilibrating structures tend toward maximum confirmation. Dissipative structures may evolve through states characterized by maximum novelty (instability thresholds) to new states characterized by a balance between novelty and confirmation (autopoiesis). In this transition, the entropy production reaches a maximum (area A), whereas in autopoiesis it tends toward a minimum (area B).

ponds to a system in thermodynamic equilibrium. That pragmatic information becomes zero at this point is the correlate of the impossibility of bringing about any directed effect in equilibrium. A hundred per cent novelty, in contrast, may be interpreted as the instability phase in which stochastic processes cease to confirm the old structure and have not yet established the new structure. Everything happening in this phase is novel. In between, in the balance between novelty and confirmation, we find the domain of autopoiesis.

The scheme according to Fig. 14 also allows the representation of the change in entropy production occurring when a new dissipative structure is born. Entropy production, in this context, is nothing else but the production of structure, implying at the same time more information and more confirmation. Immediately beyond the "chaos" of the instability threshold maximum entropy production is needed to attain a certain degree of confirmation. Area A in Fig. 14 has to be "won" very quickly by hard work. After the formation of an autopoietic structure, however, the system oscillates in a balance between novelty and confirmation and has to do work only to the extent that novelty must be coped with continuously, as exemplified by area B in the time unit. This work, or entropy production, never becomes zero because the structure is "kept busy" by novelty entering through the exchange with the environment. In the scheme, it is pushed toward the left so that maintaining the balance requires ever new work (movement toward the right in the scheme). In this way, novelty is continuously transformed into confirmation. Cognition is not a linear process, but a circular process between the system and its environment.

Autopoiesis, in this scheme, implies an existence near the maximum of exchangeable pragmatic information—a conclusion which intuitively appears correct. In the following, we shall frequently employ the representation of self-organization in the graphical terms of novelty and confirmation.

System dynamics and history

Generally speaking, the theory of dissipative structures describes the particular spatio-temporal self-organization of energy conversion in systems in exchange with their environment. It may also be considered as an elementary description of the evolution of historical systems—systems with *history*—whose development depends on the past history of each of its subsystems. The question may be asked to what extent such a description, geared to the simplest phenomenological level, grasps the fundamental characteristics of a dynamics which is also basic to the understanding of biological and social systems. In other words, do we have here a special or a general dynamic system theory?

If we have a general theory, it becomes evident that considerations concerning the nucleation of new structures, the role of fluctuations and the maintenance of metastable states is of great importance for biological and social systems with their flexible coupling of subsystems (especially through the exchange of information). The same may be said with respect to the functions of the brain and the nervous system. The old theme of the connection between complexity and stability and the question whether there is a natural limit to complexity—so important in our days—appear in a new light. An increase in complexity certainly does not always imply a loss of stability as mathematical models suggest which have been elaborated under equilibrium assumptions (May, 1973). Ecological and social systems often seem to bear out the contrary.

In the following chapter, applications of this young theory to a broad spectrum of fields will be summarized. The results seem to confirm that the theory of dissipative structures is capable of forming the core of a future general dynamic theory of natural systems.

4. Modelling Self-organizing Systems

As above, so below; as below, so above.

Law of correspondence of Hermetic philosophy

Homologous dynamics of natural systems

Let us recapitulate the kind of dynamics which is described by the theory of dissipative structures. The three basic conditions for the formation of dissipative structures in chemical reaction systems were openness toward the environment and exchange of energy and matter, great distance from equilibrium and the inclusion of autocatalytic steps. Any system which satisfies these conditions and which has relatively stable "reaction participants", may be described as a "reaction system" whose dynamics is described by the same basic equations even when the processes are other than chemical in nature.

The biophysicist Aharon Katchalsky, who lost his life in the spring of 1972 in a terrorist attack at the airport of Tel-Aviv, had recognized this some time ago. He postulated (Katchalsky, 1971) that any system which includes a large number of non-linear elements which are coupled diffusely and therefore interact almost like in a continuum, may be driven into non-equilibrium by increased energy penetration and will then exhibit the typical behaviour of dissipative structures, namely, autopoiesis and system evolution. Katchalsky acted as an important catalyst for early applications of the generalized theory to biological and neurophysiological systems.

Autopoietic existence and evolutionary self-organization by way of self-amplification of fluctuations indeed characterize many biological, sociobiological and sociocultural systems. The relatively successful attempts to describe their dynamics with the help of the same formalism which has been developed for dissipative structures must not be misunderstood as physicalism, as reduction to the level of purely physical processes. The nature of the processes and the spontaneously forming structures is independent of the dynamics which underlies them. The abstract relationships in the equation system which has been called "Brusselator", generate a wealth of modes of dynamic behaviour which often find their empirical confirmation only at a later stage. In the same way, these mathematical relationships may be applied to the modelling of dynamics at other than the chemical levels. A certain imagination is required, of course, to find the corresponding abstract and real occurrences. In an ecosystem, for example, the autocatalytic step consists of the self-reproduction of a species in the presence of a sufficient supply of food in the environment. But it may also depend to a large extent on the autocatalysis of another species, especially if we are dealing with carnivores which depend on specific prey species; such a coupling will determine the total system behaviour in a significant way.

The models which may be composed in such a way are characterized by the same complementarity of stochastic and deterministic elements, chance and necessity, which has been discussed for chemical dissipative structures. Close to the instability threshold marking the transition from one structure to another, the model breaks down. In this phase it is not the general which is of decisive importance, but the particular—in short, the creative. The precise moment of the occurrence of an essential genetic mutation, or of a new species in an ecosystem, cannot be predicted; nor can relationships with the environment. But once these properties are known, it is possible to predict whether they will ultimately dominate or not.

The interrelationship of the self-organization dynamics of material and energetic processes from chemistry through biology to sociobiology and beyond seems to point to the existence of a general dynamic system theory which is valid in a very wide domain of natural systems. The kind of general system theory which has been developed over the past few decades has searched for connecting features primarily with respect to the preservation and stabilization of structures (by means of negative feedback control). With a generalized theory of dissipative structures, the dynamic aspect of a general system theory moves into the foreground and the macroscopic quantization of structures becomes of importance as well as the creative role of fluctuations.

But if this basic type of autocatalytic self-organization dynamics underlies observable phenomena in such a wide range, we have not only analogy or formal similiarity, but true homology or relationship in kind. Although these phenomena belong to very different levels of reality which are irreducible to each other, they are *connected by way of homologous dynamics*. This recognition marks a triumph for process thinking. Whereas it follows the unfolding of reality into multilevel, irreducible patterns of existence and co-ordination, it unifies at the same time wide domains of this reality through dynamic concepts. Whether it will ever become possible to include in this picture the processes of differentiation apparent in cosmic evolution and the synthesis of particles and atoms, remains to be answered.

In this chapter, some of the first applications of the theory of dissipative structures to more complex systems will be briefly sketched. These studies have, of course, been undertaken with greatly simplified assumptions. However, they let us feel the *quality* of a world which gives birth to ever new variety and ever new manifestations of order against a background of constant change. Prediction is not the aim of these studies, but a deeper understanding of the behaviour of natural systems. Before enumerating some of these studies, an alternative approach will be briefly discussed which is based on the application of catastrophe theory. It has little to say about the stochastic processes in the system itself, but it may provide valuable qualitative conclusions concerning the type of new structures beyond instability thresholds.

Catastrophe theory as alternative

Applications of catastrophe theory in a very wide spectrum are primarily due to the imagination and enthusiasm of the British mathematician Christopher Zeeman (1977). They range from applications to physics, such as the twinkling of stars and the stability of ships, to biological, psychological and social-psychological phenomena of high complexity, including mental disturbances and prison riots. Of special importance is the application to problems of developmental biology (the development of the embryo). Here, catastrophe theory meets with biological notions introduced by Conrad Waddington, such as the epigenetic landscape (the topology of the developmental process), chreods (development lines fixed by evolution and guiding ontogeny) and canalization (the development along chreods).

In the application of catastrophe theory it is essential to recognize that it is always discontinuous effects of *continuous* causes which are to be modelled. For example, the decision situation in a country facing external threat may be characterized by the group of "hawks" on the one hand, and the group of "doves" on the other (Fig. 15). With increasing threat, the doves will eventually give up their régime of thought and adopt the régime of the hawks. The reverse will happen if the threat decreases. But the transition from one régime to the other is not continuous and does not happen in both cases at the same size of threat. Each group will try to stick to its thought régime as long as possible (in a kind of metastability) until there is an abrupt switch-over to the other régime. This simplest of all catastrophes is called the fold.

Of great importance are topological stability considerations for models of ecological and epigenetic evolution. The basic issue may be illuminated by the comparison with a ball which rolls along a valley or remains at its lowest point, but may also come to rest at a hole half-way up a slope—if it can get there somehow (for example, by external fluctuations, such as gusty winds). If

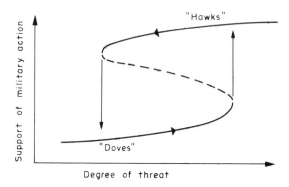

Fig. 15. The simplest example of a catastrophe (discontinuity): the fold. If the threat continues to increase, a point is reached at which the "doves" switch abruptly to the behavioural pattern of the "hawks", and vice versa. After C.A. Isnard and E.C. Zeeman (Zeeman, 1977).

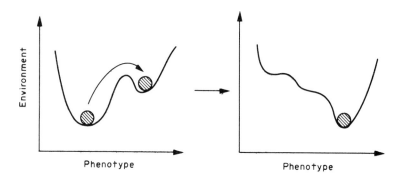

Fig. 16. Epigenetic evolution in a topological view. An accidentally reached stable position is built into the new ecological niche and determines the further evolution of the species.

a biological phenotype (an individual) happens to get from its well-defined ecological niche into such a secondary stable position, this position will eventually become a major niche (Fig. 16).

It is essential to recognize that this type of modelling does not explain the motion itself. It is imposed from the outside, in the first example, perhaps by the actions of a threatening country or by the media's interpretation of the situation. True self-organization by means of the self-amplification of fluctuations generated in the system itself cannot be modelled with such an approach. Internal reinforcement by autocatalytic steps is not included in the formalism. But this is precisely what the theory of dissipative structures can do. It is therefore always important to distinguish between the modelling of true self-organization and morphogenesis with given dynamics. In this book, we are primarily interested in self-organization.

Physical-chemical systems

In dissipative structures self-organizing matter builds up and maintains order without any additional anti-entropic force (such as a life force, or "*élan vital*", which has been assumed in earlier theories). It seems tempting to build an all-embracing *cosmology* on this principle. However, attempts in this direction have not gone very far. The thermodynamic way of looking at things evidently is insufficient for the development of a holistic view of cosmic evolution. In particular, it remains an open question how entropy may be defined with the inclusion of interactions at large distances, such as is effected by gravity. Perhaps it will prove altogether impossible to order the states of the universe along the scale of a single parameter, such as entropy. The entropy of the universe has been linked with the still mysterious phenomena of "black holes". But this raises new difficulties since in the singularities of such "black holes" the laws of physics are disorderly.

Nevertheless it may be surmised that thermodynamic non-equilibrium and dissipative structures are of local importance for keeping evolution going. We may tentatively view dissipative self-organization as an "intermediary" phenomenon which reaches to the far ends of neither space nor time. Things are certainly much more complicated than suggested by the image of the general decay of structures in a "heat death" as has been concluded from equilibrium thermodynamics. But they can also certainly not be explained by a simple condensation model of structuration. This problematics will be touched upon in the following chapter.

In the area of *geology* the continental drift is suspected to be a phenomenon caused by a dissipative structure. A somewhat surprising idea concerns the facilitation of exploiting oil in porous rock by reducing the surface tension through a forced evolution of the system. *Meteorological* dissipative structures range from local convection thunderstorms and hurricanes to vast weather systems with instability fronts. A particularly interesting case in which an originally small fluctuation breaks through and causes dramatic change in the macroscopic system behaviour is provided by the "seeding" of clouds with dry ice or silver iodide—be it with the intention of causing rain or making hurricanes deviate from their path.

Finally, a number of *electrical* phenomena may be discussed in terms of dissipative structures. They range from the spreading of the so-called Alfvén waves in magnetogasdynamics and in northern lights to certain effects in plasma physics. However, there have not been any essential attempts to model such electrical phenomena so far.

Biological systems

Biological systems are characterized by properties favouring the formation of dissipative structures in a particular way:

—They are linked with their environment by energy exchange which permits the maintenance of the structure far from equilibrium.

—They include a large number of chemical reactions and transport phenomena, the regulation of which depends to a high extent on non-linear factors of molecular origin (such as activation, inhibition, direct autocatalysis, etc.).

—They are in high non-equilibrium not only from the point of view of energy but of matter exchange since the reaction end products are either eliminated from the system or are transported to other locations in order to fulfil their functions there.

Biological systems find themselves indeed in the focus of empirical and theoretical research concerned with dissipative structures. It is important to realize that the aim is not to reduce the phenomena of organic life to the level of physical-chemical analogies, but the formerly assumed separation between "antientropic" life and "entropic" inanimate world also has vanished. The task is now to understand the structuration of biological systems as *particular accentuation*, as specific co-ordination, of the laws of physics and to stipulate precise conditions for their occurrence.

The theory of dissipative structures is in a position to contribute significantly to the understanding of *prebiotic evolution*, the emergence of life from organic molecules (Prigogine *et al.*, 1972). Chapter 6 will deal with this stage in more detail. But it may already be said here that in the light of dissipative self-organization life does no longer appear as the highly unlikely accident it is in a reductionist view. Microscopic chance and macroscopic necessity have to be understood as *complementary* aspects in the emergence of life— and not sequential as Jacques Monod (1971) presented it.

Another research complex is building up around the role of dissipative structures in the *functioning of bioorganisms*, or the maintenance of life. One of the most significant results in this area was obtained by Boiteux and Hess (1974) who proved experimentally that dissipative structures occur in the glycolytic cycle, a biochemical reaction system which is also of importance for intercellular communication. In close agreement with the theory of Goldbeter and Lefever (1972), limit-cycle behaviour and chemical waves were observed in several stability regions as well as the transitions between these regions. Quite generally, biological rhythms of various types seem to be connected with dissipative structures

A German group of researchers has developed a model which is capable of simulating morphogenesis in the development or regeneration of simple multicellular organisms, in particular the formation of extensions and tentacles in freshwater polyps (Gierer, 1974). The model contains as its most important assumptions an experimentally found activator substance which is the carrier of morphogenetic information and which, in freshwater polyps, appears to consist of small protein molecules and is exchanged in the tissue between cells; an inhibitor substance which diffuses faster in the tissue than the activator substance, is degraded more slowly and is capable of suppressing the diffusion of the activator; and finally an autocatalytic mechanism with whose help the activator may accelerate its own formation and distribution. Starting from an originally homogeneous mixture of activator and inhibitor substances in high non-equilibrium, a pattern of discrete, stable regions with high activator concentration appears which already anticipates the bodily forms to emerge. This process refers to the formation of body parts in which precision in the form and number of these parts is of no importance. In more highly evolved animals the canalization of development lines, or chreods, plays a much greater role.

From plant physiology it is known that growth and morphogenesis is essentially determined by the interaction between growth-inducing hormones (whose concentration decreases from the tip of the sprout downward) and growth-inhibiting hormones (whose highest concentration occurs in the upper part of the roots). From this interaction, standing waves emerge which anticipate the morphogenesis of the plant.

A number of research projects study dissipative structures in the *central nervous system,* experimentally as well as theoretically or in a combination of both. At a microscopic level it has been found that the excitable *membrane* of a nerve cell, which is much older in evolution than the cell itself and may be found in single-cell protozoa, functions as a dissipative structure. The non-equilibrium in a polarized state is maintained by the opposite ion charges on both sides of the membrane. This may lead to an instability which evolves toward a cyclical depolarization. Recently, the experimental synthesis of bimolecular lipid membranes has been achieved as well as the stimulation of articifial nerve impulses by canalizing molecules in these membranes (Baumann, 1975).

At a macroscopic level, the behaviour of neuron populations is studied up to a number of about 10 million neurons, especially by Walter Freeman (1975) in Berkeley. The prerequisites for a modelling approach on the basis of the theory of dissipative structures are given in more than one way. At the synapses, the interfaces between the dendrites of a neuron with the axon tip of another neuron—the structure of the neuron will be explained in more detail in Chapter 9—non-linear transformations may occur. Furthermore, the numerous feedback connections between densely packed neurons almost create a continuum. And finally the flow of activity within the neurons and across the synapses acts like diffusion in a chemical reaction system. At sufficently high non-equilibrium the result is the occurrence of "active" states between big interactive groups of neurons. With further positive feedback between these groups, the active states become unstable and may give rise to dissipative structures. In particular, one may observe a kind of limit-cycle behaviour which is linked to a decisive step in the coding of sensory input.

Perhaps it will also become possible one day to understand the intuitively developed approaches to a so-called *holistic medicine* as the stimulation of specific co-operative modes of behaviour and transitions in psychosomatic dissipative structures. Bioenergetic techniques such as acupuncture, Esalen massage, psychic healing, yoga and the chanting of special mantras as well as the mental techniques of hypnosis and meditation in various versions generate effects which may best be described in terms of transitions between different dynamic régimes—but the nature of the processes involved is not well understood at present. Nevertheless some of these techniques, especially meditation, have come to be used in American cliniques, especially cancer cliniques, in connection with conventional therapy—with striking success. Western medicine seems primarily focused on a basic régime of the body/mind system whereas healing in other régimes may act much faster and bring about stunning results.

Within an organism, there are many self-organizing systems maintaining some semi-autonomous dynamics but generally geared to the autopoietic régime of the organism as a whole. An example is provided by the rhythmic *motor activity* of vertebrates, for example in walking or running (Pearson, 1976). It is generated by an oscillating cell system and temporarily "coupled into" the organism; in between, the rhythm idles. Self-organizing systems beyond a critical size may also become decoupled and threaten the organism. This is the case in the transition from the normal state of "microcancer" to the pathological state of *"macrocancer"*. Our organism always contains a number of cancer cells, whether they have been inherited or have been generated by environmental influences. They may be considered as fluctuation which, below a critical size, is damped by the healthy cell environment, but beyond this size, becomes amplified. Prigogine's collaborators Lefever and Garay (1977) have developed a model which reflects new qualitative insights.

In parenthesis it may be mentioned that the theory of dissipative structures also promises important contributions to the development of *industrial biochemistry*. In particular, the immobilization of catalytic enzymes on membranes—a trick also used by nature—may greatly enhance non-linear processes and thereby increase the yield significantly. In this area, the work of Daniel Thomas in Compiègne (France) is of great importance.

Sociobiological systems

The mathematical modelling of the emergence of macroscopic order in animal population is not very difficult where it is based on simple chemotaxis,

the attraction due to certain chemical substances (Prigogine, 1976). Chemotaxis may be observed in striking examples with single-cell organisms, such as
the amoebae which, at certain times, join to form the slime mold
Dictyostelium discoideum. In times of food scarcity, the amoebae stop to divide
and form spontaneous aggregations, and every 3 to 5 minutes send out
chemical pulses of cyclical AMP (adenosinemonophosphate), a signal leading
to rhythmical chemotaxis (Fig. 17). As theory and observation well agree, the

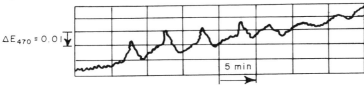

Fig. 17. Rhythmic chemotaxis in the aggregation of amoebae forming the slime mold. The
rhythm in the secretion of cyclical AMP is based on limit-cycle behaviour of the enzyme system
responsible for the production. The increasing aggregation was measured by means of light
scattering. After A. Boiteux and B. Hess (1974).

autocatalytic enzyme system responsible for the production of cyclical AMP
becomes instable and enters into limit-cycle behaviour. The chemotactic
aggregation system is itself autocatalytic. In the further development, the
pseudoplasmodium forms, a cell mass of between 10 and 500,000 cells which
moves along the earth resembling a worm of 0.1 to 2 millimetres length. This
mass undergoes further structuration and shapes into a flat basis, a "foot",
consisting of cells with high cellulose contents, and a big, round "head"
consisting of cells with high polysaccharide contents. This whole development may take 20 to 50 hours after which the multicellular body dissolves
again into individual single-cell amoebae which reproduce by cell division—
until the cycle starts all over again (Bonner, 1959).

Besides the slime mold, Prigogine and his Brussels group have also
modelled other chemotactically triggered phenomena, such as the marching
order of ants (in which instabilities become visible as branchings) and the
construction of termite nests (Prigogine, 1976). The latter start obviously as
unordered deposits of matter until random fluctuations in the distribution of
matter turn chemotactical and mechanical stimulation into an autocatalytic
phase of co-ordinated activity. A complex example is also provided by the
formation and emigration of new swarms of bees in overpopulated beehives.
Here, the instability occurs when an enzyme diffusing through the hive and
inhibiting the production of a new queen is no longer capable of suppressing
the latent fluctuations when the dimensions of the hive exceed a certain size.
Of particular interest is the simultaneous effect of mechanical factors (certain
wind streams) and chemotactic factors in a highly autocatalytic system which
results in the formation of gigantic swarms of locusts over Africa extending
over hundreds of cubic kilometres volume.

Ecological systems

The kinetic equations for the reproduction of living organisms are the same as for the autocatalysis of non-living systems. Therefore, the application of the mathematical formalism of self-organization dynamics to ecological systems combining the interactions of several animal and plant species yields impressive results (Allen, 1976). The model of an ecosystem may be made more realistic in several steps. An autocatalytic system in an environment of unlimited resources—for example, herbivores whose food requirements are by far exceeded by plant growth—will in principle exhibit exponential growth (Fig. 18a). The growth rate always corresponds to the existing amount (e.g. 10 per cent per year of the existing amount). If the environmental resources are limited, a logistic growth curve results (Fig. 18b) which approaches asymptotically a saturation value. As is known from laboratory experiments, this characteristic development is also found with unlimited food supply due to sociobiological limitations of other kinds (reproduction, "crowding").

Things get more interesting if the additional assumption is made that mutants of the same species, or new species (x_1, x_2), may enter the system with a capability of exploiting the given environment differing from that of the original population x_0. These newly added "competitors" may be considered as ecological fluctuations. It may be shown that these fluctuations get through and replace the old population when the new mutants or species have a better capability of exploiting the same resources, or the ecological niche (Allen, 1976) (Fig. 18c). These combined factors—reproduction, variety and selection—correspond to the simplest case of Darwinism, or the principle of the "survival of the fittest". It is incomplete to the extent that it does not take into account any interaction between the competing species or mutants or any change in the environment. For the simplest case of co-evolution, additional assumptions are required.

The extremely important notion of *co-evolution* was coined in the same "magical" year 1965 by the American biologists Paul Ehrlich and Peter Raven—this, at least, is the version which the journal *CoEvolution Quarterly* gives in its impressum. Ehrlich and Raven noted that certain plants contain large amounts of alcaloids, the production of which requires a great deal of energy. The purpose is evidently protection against caterpillars for which these alcaloids are poisonous. A few species of caterpillar, however, such as the caterpillar of the Monarch butterfly, were able to adapt and now can digest the poison, as well as scare away with them birds for which they themselves represent the prey. Monarchs are not eaten by anyone which is also the reason for their being imitated by other, non-poisonous butterflies, such as the Hypolimnus. This imitation occurs as well with respect to the slow see-saw type of flight as to the bright orange colour and the wing patterns (although

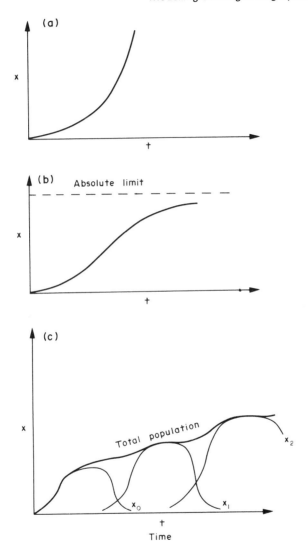

Fig. 18. Different growth characteristics in ecosystems. (a) Exponential growth in the presence of unlimited resources; (b) logistic growth with natural limitation of resources; (c) increasingly better use of the resources by the introduction of mutants or new species. After P.M. Allen (1976).

this can hardly be explained by simple mutation and Darwinian selection). The birds, in turn, learn to distinguish between the original and the imitation and the plants diversify their alcaloid combinations so that caterpillars have to specialize in certain plants only. This, in turn, leads to a diversification in the caterpillars, and so forth.

There are many predator-prey stories of a similar kind; for example, bird and insect species which grow increasingly smarter. Such a "battle" is not won by any side—for the predator, this would even be the worst of all possible outcomes. But both sides are spurred to ever newer developments—they co-evolve. Besides the predator-prey relationship, the following chapters will bring to light further manifestations of co-evolution, such as symbiosis on the one hand and the co-evolution of a system with its environment on the other, in particular under the aspect of the co-evolution of macro- and microcosmos.

But back to the predator-prey relationship. Already in the 1920s, Alfred Lotka (1956) and Vito Volterra (1926) have described such a dynamic system in terms of two autocatalytic steps which may be written in the simplest form as follows:

$$A + X \rightarrow 2X$$
$$X + Y \rightarrow 2Y$$
$$Y \rightarrow E$$

where X, for example, signifies a herbivorous prey population, Y a carnivorous predator population preying on X, A plant growth providing primary energy and matter, and E the exit of members of population Y which die of old age (and whose elements ultimately benefit A, in turn). In phase space (in which each point corresponds to a specific combination of the numbers of X and Y), this system of equations yields an infinite variety of closed concentric orbits (Fig. 19a). The system "jumps" between these orbits and cannot be stabilized. With the slightest fluctuation, it becomes unstable. But this only reflects the simplified, unrealistic assumptions. If time delays, non-random (or contagious) attacks of predators and a minimum density of the prey population for reproduction are introduced as further assumptions, the general result is that of a "domain of attraction" (Holling and Ewing, 1971). Within this domain, all orbits tend toward an equilibrium state: without the domain they spiral away and lead sooner or later to the extinction of one or both populations.

A comparison of theory and empirical observation has led to the insight that a "healthy" and resilient ecosystem is generally far from its equilibrium which may be represented by a single point within the domain of attraction. It is generally characterized by large spatial and temporal fluctuations, is always on the move in phase space and prefers an existence near the boundaries of the domain of attraction. It is precisely this continuous local instability which furthers the global stability of the autopoietic régime. Typically, limit-cycle behaviour (Fig. 19c) or more complex types of behaviour are observed. One may even speak of a new non-equilibrium ecology which is developed primarily by Holling (1976) in Vancouver. The closer the system gets to equilibrium, the less resilient it becomes. Any random fluctuation, such as climatic fluctuations or the appearance of a new species, may destroy the system completely.

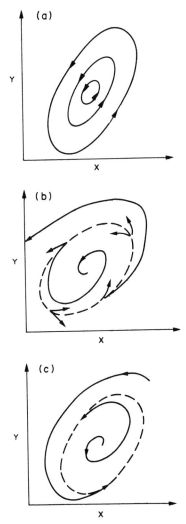

Fig. 19. Representation of dynamic system behaviour by means of phase diagrams. Animal populations X and Y, which may represent a predator-prey relationship, have been chosen here as characteristic parameters of an ecosystem. The three cases shown are: (a) neutrally stable orbits; a (generally unrealistic) Lotka-Volterra system jumps between such orbits without ever becoming stable; (b) domain of attraction, indicated by the dotted, elliptical boundary; within such a domain, all developments tend toward an equilibrium, outside all developments lead to instability; (c) stable limit-cycle; all orbits tend to a globally stable régime of periodic oscillations, indicated by the dotted ellipse (see also Fig. 4, showing the limit-cycle behaviour of a dissipative structure).

Such extreme cases have indeed been observed with systems which have approached their equilibrium due to human resource management. Examples are fish populations in the North American Great Lakes whose catastrophical decline and partial extinction had seemed inexplicable. In many places of California, the reproduction of the giant Sequoia trees, which may reach an age of 2000 to 3000 years, has practically stopped. Young trees only grow on a "clean" forest floor without underbrush, a prerequisite normally fulfilled by the natural rhythm of forest fires (every 8 years on the average). The thick bark protects the trees. But human interference—and who would have doubted that fighting forest fires was a good thing from any possible angle of view!—has interrupted this rhythm. Even if one considers now the possibility of controlled burning, the natural rhythm cannot simply be re-established. The underbrush has by now grown so high that the flames will partly damage the sensitive leaf crowns of the trees. These and more examples point to important conclusions to be drawn for management strategies, not only in the area of resource management, but also in the sociocultural domain (Holling, 1976).

The evolution of such systems may now be followed under the additional assumption that mutants are introduced. Let us again consider a simple predator-prey relationship. Mutations in the prey species mean better exploitation of the available resources and an improved capability of escaping the predator. Mutations in the predator population, in turn, lead to an enhanced capability of exploiting the prey species and of reducing the predator's death rate. The result is a slow increase in the predator/prey ratio. But the birth rate of the prey species increases at the same time as the death rate of the predator population decreases (Fig. 20). Co-evolution ultimately benefits both species (Allen, 1976).

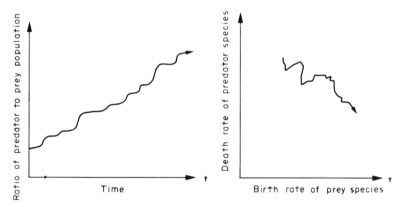

Fig. 20. Co-evolution, in principle, benefits predator as well as prey species. After P.M. Allen (1976).

Two further aspects shall be briefly mentioned which result from the same criteria for evolution. One aspect refers to the development of "specialists", geared to one or few types of resources, and "generalists", capable of exploiting a large variety of resources. A rich milieu in which all resources occur in large amounts, favours the development of specialists whereas a scarce milieu favours generalists (Allen, 1976). This has been observed with the Galapagos finches. The larger the environmental fluctuations (e.g. climatic variation), the more clearly the niches of the generalists have to be separated from each other and the less numerous the number of co-existing species. The tropics are much richer in specialized species than the regions near the poles with their relatively few generalist species.

In the "ecosystem" of the world economy, the reverse is true. The wealthy countries are generalists and the poor ones specialists living from the export of one or two products only. The result is a system which is becoming increasingly less viable and which can only be prevented by force from becoming instable. But for how long?

Another aspect concerns the possibility of securing chances for survival not by general species properties in every individual, but above all by *group* properties—for example, the properties of an insect colony. This possibility is created where the complexity of the group permits the division of functions, in particular division of labour, as well as hierarchical relationships and mechanisms of population control (as in the above-mentioned example of the splitting beehive). This is an expression of the co-evolution of the macroscopic and the microscopic which will be discussed at length in the following chapters.

Sociocultural systems

The theoretical treatment of systems in which man participates in an important way, meets with more difficulties. Here we no longer have highly specialized competing systems, as is the rule in the subhuman world, but the complex interactions of individuals with multifaceted capabilities. Also, there is much greater variety and complexity of communication mechanisms. But above all there is now, in addition to the exchange of physical energy, the exchange of social and spiritual energy equivalents which, in this combination, characterize the human world. The capability of self-reflexion, as Chapter 9 will show, turns many things upside down. But on the other side, the empirical description of many non-linear phenomena in the human realm shows astonishing similarities with the evolution of physical non-linear non-equilibrium systems. It may therefore be permitted to hypothesize that the theory of dissipative structures provides a general description of the dynamics of self-organizing systems where the parameters characterizing the space-time structures may be of a physical nature as well as of a social and mental nature.

The autocatalytic principle appears in the human world in many ways, ranging from population growth to the economic principle of the production of money by money. In many cases the transition from exponential growth curves shown in Fig. 18a (see p. 65) to logistic curves of the type shown in Fig. 18b comes as a severe shock in our days. In these transitions natural system boundaries become expressed, especially with respect to the limitation of resources which are utilised in a one-way economy. Generally, these limitations are of a material type and with the help of an ever-improving technology one may expect enhanced exploitation corresponding to Fig. 18c.

Of particular interest is the similarity in the evolutionary behaviour of mental processes. The best example is again provided by technology, both by its development and its diffusion (innovation) (Jantsch, 1967). The evolution of technological performance parameters, such as speed, temperature resistance or conversion efficiency, in the exploitation of a specific "resource", i.e. a specific scientific-technological principle, generally follows a logistic curve corresponding to Fig. 18b. But in many cases, the "system" of a particular technological performance evolves through many "mutants" or different "species", i.e. through the development and exploitation of a variety of different scientific-technological principles. The result is an envelope curve— a "big S" riding on the evolving small S-curves (Fig. 21). The top velocity of

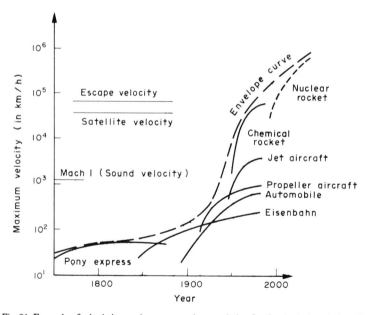

Fig. 21. Example of a logistic envelope curve, characteristic of technological evolution. By the introduction of ever-newer technological principles, the inherent limitations in the same basic parameter (in the figure the maximum velocity of person transport) are pushed closer to absolute limits which determine the envelope curve. In our case here, the velocity of light presents such an absolute limit.

person transport, for example, has been increased tremendously in its evolution through many technologies, from the pony express through railroad, car, piston and jet aircraft to rockets. Ten years ago, the corresponding envelope curve still resembled exponential growth, promising the crossing of the speed of light in the year 2010! But in technology, earlier than in other areas of human concern, the lesson has been learned that growth processes run into natural limitations—a simple truth still not fully recognized by conventional economics. The development of technology is already considerably influenced by rethinking processes which bring higher levels of social and psychological limitations into play. In the realm of technology we recognize the first effects of a type of normative, long-range planning which tests material and technological growth against the consequences for society at large. In Chapter 16, such an approach to planning conducted in an evolutionary spirit will be discussed.

The application of technology and its wide diffusion—in short, technological innovation—follows to a large extent the same curves of system evolution as has been obtained for the successively improved exploitation of an ecological niche by mutants or new species (see Fig. 18c on p. 65). Instead of mutants, we have here technological products or principles. But just as in ecology, it is essential here also that every technology which replaces an old one is not only capable of doing the same, but generally also generates new opportunities. The transistor, for example, has not simply replaced the vacuum tube but has initiated a tremendous development of microelectronics. The portable radio was only the beginning. It exemplified the extension of functions as well as the generation of new problems.

Another, particularly interesting example for the application of the theory of dissipative structures is the study of the evolution of cities and regional agglomerations with approaches developed by Prigogine's Brussels group (Allen *et al.*, 1977). It is assumed that a district which attracts the most inhabitants, has the more developed economic functions. The economic function here plays an autocatalytic role, but at the same time depends on local demand—and the demand elsewhere which, however, is limited by the costs of transportation which increase with distance—as well as on the competition by production facilities in neighbouring districts. The economic function may also be regarded here as a fluctuation which forces the originally homogeneous population distribution into a markedly heterogeneous structure (Fig. 22). The structure itself is not predictable, because it depends on which specific fluctuations (economic functions) break through. It is also possible to draw general conclusions such as the favouring of activity centres of limited size by improved transportation conditions. The idea that economic activity acts in an autocatalytic way is not new. The Swedish Nobel laureate Gunnar Myrdal (quoted in Allen *et al.*, 1977) already in the 1950s proposed models of

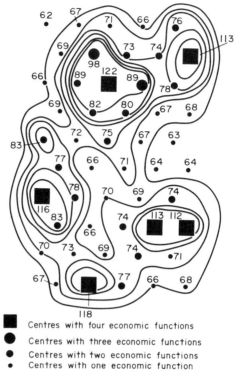

■ Centres with four economic functions
● Centres with three economic functions
● Centres with two economic functions
• Centres with one economic function

Fig. 22. Evolution of a city. After some time, a heterogeneous pattern develops with respect to the distribution of the economic functions as well as to population density. The population density is indicated above each point. After a computer simulation by P.M. Allen (personal communication, 1978).

inter-regional development based on this quality. But only today are more detailed computer studies starting. In a similar approach, one may study the evolution of cities with respect to heterogeneous habitation patterns, separating higher-income from lower-income classes. The result is the typical pattern of American cities on the one hand, in which the poor population is found in the centre of the city and in a second ring around the city with a wealthy suburban ring. On the other hand, there is the typical pattern of the modern French city with the well-to-do population in the centre of the city (except for a few poor "ghettoes") and the poorer population in the suburbs on one side of the city and a less segregated suburb on the other side.

Other examples of areas in which the theory of dissipative structures may lead to interesting qualitative conclusions range from the settling of new land to the organization of space in geopolitical evolution, from local social change to revolution, from individual perception and apperception to the overall system of science in the sense of Thomas Kuhn's (1970) theory of scientific revolutions, from individual creativity to the great currents in art, from individual personality development to the evolution of all-embracing cultural guiding images and religions (Jantsch and Waddington, eds., 1976). The common denominator is always an open system far from equilibrium which is driven by fluctuations across one or more instability thresholds and enters a new co-ordinated phase of its evolution.

These approaches mark a profound break with the dominant tradition of describing human systems. They are of immense importance for an improved understanding of our current approach to instability and for the design of the future of mankind. The conventional behaviouristic approaches used for recent world models start from global or regional homogenized equilibria for which any fluctuation and any positive feedback appears as a threat to the structure. They postulate mechanistic systems which cannot evolve and cannot change their structure. The norm of a forced stabilization of an equilibrium structure which is deduced from such models is nothing but a circular conclusion which generates fatal misunderstandings.

Globally viewed, mankind gets further and further away from equilibrium and seems to urge a new structure which may only be reached after a major instability. There is no lack of fluctuations (oil crisis, recession) or autocatalytic reactions (escalation of tensions). But interestingly, it is precisely the potentially strongest autocatalytic factors, such as the preparation of a huge arsenal of nuclear weapons and strategies of "mutual strikes", which may also act in a strongly inhibiting way. Generally, fluctuations which threaten to touch on systems boundaries and limitations may also be damped by these boundaries and limitations in an *anticipatory* way. This is certainly the case with the exploitation of non-renewable resources. Herein lies an important distinction between self-reflexive and merely physical autopoietic structures. Many of these questions will be taken up again in later chapters.

The decisive question now concerns the available and newly acquirable degrees of freedom. Can a global autopoietic system live at all three levels, physical, social and cultural levels? What is the environment with which it is then in exchange in order to maintain its non-equilibrium? To what extent is it permissible to supplement the natural solar energy flux by liberating energy stored within the system? Quite generally, a "cybernetic" recycling technology seems to move into focus, replacing a one-way technology which is increasingly recognized as "unnatural". Perhaps we ought to interest ourselves at the social and cultural level more in the symbiosis of subglobal autopoietic systems, in pluralism and non-equilibrium, than in a Utopian world government and world culture. And we must not be afraid of evolution because, as the philosopher Alfred North Whitehead (1933) wrote long ago: "It is the business of the future to be dangerous. . . . The major advances in civilisation are processes that all but wreck the societies in which they occur. . . ."

But the ultimate answer may perhaps be found again in a complementarity: as we elevate pluralism to our creative principle, we embed the totality of human history meaningfully into the dynamics of an overall evolution which acts as an unfolding unity. As we realize ourselves as wholes, we become an integral aspect of a universal whole. As we live to the fullest extent, we overcome cosmic cold and loneliness. Whether we shall soon establish contact with extraterrestrial intelligence or not—we are never alone.

PART II

Co-evolution of Macro- and Microcosmos:
A History of Reality in Symmetry Breaks

Heaven does nothing: its non-doing is its serenity.
Earth does nothing: its non-doing is its rest.
From the union of these two non-doings
All actions proceed,
All things are made.

Chuang Tzu, *Perfect Joy*

The evolution of the universe is the history of an unfolding of differentiated order or complexity. Unfolding is not the same as building-up. The latter emphasizes structure and describes the emergence of hierarchical levels by the joining of systems "from the bottom up". Unfolding, in contrast, implies the interweaving of processes which lead simultaneously to phenomena of structuration at different hierarchical levels. Evolution acts in the sense of simultaneous and interdependent structuration of the macro- and the micro-world. Complexity emerges from the interpenetration of processes of differentiation and integration, processes running "from the top down" and "from the bottom up" at the same time and which shape the hierarchical levels from both sides. Microevolution (such as the emergent forms of biological life) itself generates the macroscopic conditions for its continuity and macroevolution itself generates the microscopic autocatalytic elements which keep its processes running. This complementarity marks an open evolution which reveals ever new dimensions of novelty and exchange with the environment. It is not adaptation to a given environment that signals a unified overall evolution, but the coevolution of system and environment at all levels, the co-evolution of micro- and macrocosmos. Such an overall evolution is indeterminate, imperfect and prefers dynamic criteria in the choice of its strategies before morphological ones. It is self-consistent and creative.

75

5. Cosmic Prelude

Ultimately what we can do here on the earth will be limited by the same laws that govern the economy of astronomical energy sources. The converse of this statement may also be true. It would not be surprising if it should turn out that the origin and destiny of the energy in the universe cannot be completely understood in isolation from the phenomena of life and consciousness.

Freeman J. Dyson, *Energy in the Universe*

Evolution as a symmetry-breaking process

For more than 2000 years the main interest of Western physics has been devoted to the recognition of structure. From Democritus to our days the search was on for the ultimate building stones of matter, whether they were called atoms, subatomic particles or—according to a contemporary concept—quarks. Physics hoped to trace the properties of matter back to these ultimate building stones and to reduce the whole to its parts. But in the recent past doubts arose that the fundamental principles of matter may be completely deduced from its components. One idea which received much attention stipulated an ultimate level of basic symmetries as suggested by Werner Heisenberg in the last years before his death. This idea no longer corresponds to an atomistic view but represents a systemic view which focuses on the relations *between* components, not on their individual properties. The most interesting aspect of this idea is the compelling way in which it leads from a static view—the dissection of matter—to a dynamic one. To the spatial dimensions the time dimension is added. Instead of a timeless structure of matter, the processes of evolving matter move into the foreground—or, to be even more precise, the evolving organization of matter.

It appears feasible to represent the physical universe in terms of basic symmetries—not in the present, but by tracing its evolution over time back to its origin, or at least to the beginning of the expanding phase in which it finds itself today. Usually, according to the so-called "standard model", an open universe is assumed which started from a "big bang" some 15,000 to 20,000 million years ago and which expands into the infinite. Very recently, however, increasing evidence is found for the existence of vast masses of dark, gaseous matter between the luminous structures of stars and nebulae. These dark masses may account for such a large share of the total mass of the universe that the latter may be sufficient (i.e. at least 10 to 20 times as heavy as the luminous masses) to justify including the possibility of a pulsating universe. The American cosmologist Lloyd Motz (1975) assumes a pulsation period of 80,000 million years. A pulsating universe would make it possible to explain the high number of photons (about 1000 millions per matter particle) as the cumulative effect of a kind of inner friction in the universe over many cycles of expansion and contraction. The number of photons is indeed interpreted as a measure for entropy in the universe. In line with these assumptions, it would increase with each cycle. Other recent models assume a gradual increase in the cycle length which may ultimately lead to an open universe (Hönl, 1978).

In the past few years, direct measurements of the escape velocity of very distant galaxies whose light is several thousand millions of years old when it reaches us, seem to support the model of a closed universe with slowing expansion velocity. It also appears that the motion within the universe may be more complicated than hitherto assumed. The present situation is somewhat confusing because at the same time, the arguments for an open universe gain support (Huber and Tammann, 1977). However, as long as we are not after an ultimate, all-embracing paradigm in this book, but after an "intermediary" one, all this is not of crucial importance. What we usually call evolution refers to a development which is enclosed within the present phase of expansion of the universe.

Whether an open or a closed universe is assumed, the beginning of the expansion phase is characterized by an extremely dense and hot universe which is ruled by different principles than our environment. There, in the mould of unimaginable temperatures, the simplicity and the unit of nature manifests itself directly. There, symmetries hold between particles and forces which have been lost in the expanding, cooling universe. The original simplicity and symmetry has become "frozen out" and distorted until it is no longer recognizable. This lost simplicity and unity of nature can only be reconstructed with the help of abstract mathematics, the so-called "gauge field" theories which are based on certain invariances in transformations. But the difficult details of these theories do not have to concern us here. What is essential, as Steven Weinberg (1977) puts it, is that there is a parallel between the history of the universe and its logical structure. Such a parallel, it may be added, implies the reality of evolution or the emergence of structures from historical processes.

The basic symmetries may be found in the linking backward to the historic origin. This means, that in the reverse direction evolution, the unfolding of history, is characterized by a *sequence of symmetry breaks*. Such fundamental symmetry breaks may indeed be found not only in the physical-cosmological history of the universe, but also in the history of biological and mental life in our local world. This will become clear from the detailed description of evolution in this and the next few chapters. Symmetry breaks introduce new dynamic possibilities for morphogenesis and signal an act of self-transcendence. Complexity becomes possible only through symmetry breaks. The world which emerges from them becomes increasingly irreducible to a single level of basic principles which can only be grasped in the common origin and in an abstract way. The reality which emerges is co-ordinated at many levels.

The asymmetrical origin of matter

One of the symmetry breaks near the origin of the expanding universe constitutes the prerequisite for the development of a matter world. This is the *symmetry break between matter and antimatter,* or more precisely, between the number of corresponding matter and antimatter particles. Whereas this symmetry break is the most evident one in its consequences, it is still the most mysterious in its logical origin and in its place in time. We do not even know whether it is to be understood in such a way that in a more or less homogeneous universe—in the first moments of its birth, or as cumulative effect over many expansion-contraction cycles—an excess of matter emerged somehow, or in such a way that besides our local matter world there is another antimatter world which is spatially separated from ours, or perhaps not and only prevented by unknown reasons from interacting with our world. The meeting of matter and antimatter leads to mutual annihilation; matter becomes pure radiation. Or a model might be more realistic which assumes continuous formation and disappearance of matter and antimatter in the universe, with a slight excess of matter. The continuous formation and disappearance of matter is the subject of speculations around so-called black and white holes.

In white holes the state of the beginning of the universe would be reconstituted. In black holes matter would disappear but at the same time new matter would be generated. Black holes might conceivably be also responsible for the asymmetry between matter and antimatter. In the universe, there is always some spontaneous formation of pairs of matter and antimatter particles. According to quantum mechanical considerations (Hawking, 1977) such an event near a black hole may lead to one particle becoming imprisoned in the black hole whereas the other escapes and cannot join its twin brother again. But such an effect would not in itself explain the observed asymmetry. A closed universe would end its contraction phase in a black hole; the old symmetry would in such a case become reconstituted.

Several theories which deviate from the standard model (see Benz, 1975) assume a beginning of the universe at relatively low pressure. This, for example, is assumed for the model of the Swedish Nobel laureate Hannes Alfvén (1966). Sharply separated volumes of the size of super-clusters (clusters of clusters of galaxies) or even bigger, containing either matter or antimatter, are then assumed to expand independent of each other. There is no empirical support for such a separation.

The American astronomer Edward Tryon (1974) has made the startling proposal that the universe be regarded as a gigantic vacuum fluctuation which is a phenomenon to be expected by quantum field theory and actually observed in small size. Such a vacuum fluctuation represents a spontaneous plus/minus polarization of the vacuum in such a way that the sum of *all* physical values over the whole fluctuation always remains zero. The symmetrical emergence of matter and antimatter may be explained in such a way, but also the universe would always have zero net energy content because the energy represented in mass and motion would be balanced by the negatively accounted gravitation energy so that the sum total of the energy would always be zero— an assumption which is quantitatively not impossible as calculations have shown. Such a universe would, in contrast to the standard model, emerge from the spontaneous *creation* of symmetries out of nothing. However, immediate local symmetry breaks would be required in order to prevent annihilation and make further evolution possible. Matter and antimatter would have to separate very quickly and go about their own evolution—until they meet again and would reconstitute the nothingness out of which they emerged. Such a theory appears tempting in so far as it renders any further assumption about the beginning and the end of the universe superfluous. Before and after there is literally nothing, a nothingness which unfolds symmetrically and closes again without leaving a trace. The standard model, in contrast, has to deal with the messy problems not only of an excess of matter but also of an initial energy and an impulse—and yet cannot discuss the before and the after in a meaningful way. Except for the difficulties with the initial conditions, however, the standard model is receiving increasing support from recent results of observations. Perhaps one day a synthesis will appear possible between these directions of thought which, at present, seem to exclude each other.

The basis for the discussion of cosmic evolution in this chapter will be the standard model. Steven Weinberg (1977), in his book *The First Three Minutes,* has given it an extraordinarily clear and readable presentation which forms the backbone for much of the rest of this chapter.

In the framework of the standard model we believe to know with some certainty that in a very early and hot phase of the universe—beyond a threshold temperature of about 6000 million degrees Kelvin, corresponding to the first 8 seconds in the history of the expanding universe—there was a mixture of electrons and positrons (the antiparticles of the electrons) as well as the massless particles photons, neutrinos and antineutrinos. As long as the temperature was far from the threshold, this mixture was in thermal equilibrium. Each species of particles counted about the same number and they continuously collided. The collisions between photons produced new electron-positron pairs whereas the collision between electrons and positrons led to annihilation and thus back to photons. Radiation and matter/antimatter became continuously transformed into each other.

When the temperature was still at least 1000 or 10,000 times higher, during less than the first hundredth of a second of the expanding univese, the heavier protons and neutrons formed and disappeared in the same way together with their corresponding antiprotons and antineutrons. The events in this very early phase are so complex that we know little for certain. In particular we know little about the symmetry break in the number of these nuclear particles and their corresponding antiparticles. Had the symmetry in the pair-wise formation and annihilation of matter and antimatter in the form of these particles been maintained perfectly, there would not have been any rest left over after the crossing of the temperature threshold of their spontaneous formation from radiation. There would never have been atoms in later phases. Had also the symmetry in the formation and annihilation of electrons and positrons been maintained perfectly, the universe would consist only of radiation.

In reality, at least in our corner of the universe, there was a tiny excess of matter particles which may be estimated at one proton or neutron per thousand million photons. Only a thousandth of a millionth of the original mass of matter has survived. Out of this almost immeasurable remainder which since then has undergone little change, except for transformations between neutrons and protons (with a resulting higher share of the protons), came our matter world with all its wealth of forms, came nuclei, atoms, molecules, stars, galaxies, and finally living and mental structures in the universe. Matter is more or less durable and may be regarded as some form of "frozen" energy. In nuclear fission and fusion, part of this energy stored in matter is set free again, but the largest part remains bound in the new atomic nucleus. The energy of radiation, in contrast, decreases with decreasing temperature and therefore also with the cooling of an expanding universe. Out of a small "pollution" in a universe whose energy appeared almost totally in the form of radiation, a matter-dominated universe emerged eventually whose energy is invested primarily in matter.

The transition from a radiation- to a matter-dominated universe occurred at a temperature of about 4000°K (degrees Kelvin), by accident or some un-known logical coincidence very close to that temperature of 3000°K at which the free electrons—equal in number to the left-over protons in order to equalize the charge—were captured by the atomic nuclei with complete atoms resulting. With this effect, the universe became transparent for radiation and the radiation pressure, until then enormous, became ineffective. Since each particle or photon had contributed about equally to the total pressure, the loss of pressure due to the photons, which were more abundant by a thousand millions, implied a dramatic lowering of the total pressure by the same factor. This brought gravity into play for the first time as a morphogenetic factor; only now it was capable of overcoming the internal pressure. The so-called Jeans mass, named after the astrophysicist Sir James Jeans who was active in the first part of our century, indicates the mass at which, with given pressure and density, gravitational clumping starts. Within a very short time, the Jeans mass decreased from a million times the mass of a big galaxy to a ten-millionth of the mass of such a galaxy. (Since the Jeans mass is proportional to the power 3/2 of the pressure, and the latter decreased by a factor of 10^9, the Jeans mass decreased by a factor of more than 10^{13}.)

The universe had an age of about 700,000 years when its macrostructures started to form in this way. If the symmetry break between matter and anti-matter was the prerequisite for the occurrence of matter at all—in other words, for the start of cosmic microevolution—the continuation of this microevolution (the synthesis of heavier elements) as well as the formation of macrostructures (galaxies, stars, etc.) is due to another, much earlier symmetry break of equally fundamental importance. This is the symmetry break between the different physical forces.

Symmetry break between physical forces: The unfurling of the space-time continuum for the unfolding of evolution

According to our present knowledge there are four physical forces. In our everyday life we notice only two of them, the electromagnetic and the gravita-tional forces which both decrease with the square of the distance. Whereas the electromagnetic forces are proportional to the sum of the electric charges, however, gravity is proportional to the product of the masses attracting each other. In the microscopic domain, gravity is much, much weaker than the electromagnetic forces—in the case of an electron-proton pair, for example, by the unimaginable factor of 2×10^{39}, which comes very close to the already mentioned correlation factor of 10^{40} between macro- and microcosmos, hypothesized by Dirac. In the macroscopic domain of big masses, however, gravity becomes dominant and acts at very large distances. In the domain of our everyday life and except for the uniform attraction of the earth, electromagnetic forces are responsible for phenomena of structuration from atoms with their positively charged nuclei and negatively charged electron shells to molecular binding and thus for all chemistry and biology, and further to the formation of crystals. In short, electromagnetic forces are responsible for structuration in the intermediary region between the subatomic and the cosmic extremes.

But there are two more types of physical forces which act only at extremely short distances. These are on the one side the so-called strong nuclear forces which act between the particles bound together in the atomic nucleus, the protons and the neutrons. They also act on other particles belonging to the group of "hadrons", but these particles are of no interest here. The range of these strong nuclear forces is essentially limited to the dimensions of an atomic nucleus, i.e. to 10^{-13} centimetres. On the other side, we also find in this microscopic, subatomic domain the so-called weak nuclear forces which are responsible for certain radioactive decay processes but also play a role in the formation of atomic nuclei.

Whereas the strong forces at very short distances exceed the electrical repelling force between two positively charged protons about a hundred-fold (a fact responsible for the formation of atomic nuclei from a multiplicity of positively charged protons and electrically neutral neutrons), the weak forces are in typical reactions a million-fold weaker than the electromagnetic forces. However, the weak forces are responsible for a stable sun steadily transforming hydrogen into helium. Heavy hydrogen, called deuterium, has a nucleus consisting of one proton and one neutron. In the hydrogen bomb or in future fusion reactors it may be transformed explosively into helium—an effect due to the strong nuclear forces. The proton-proton reaction of ordinary hydrogen, in contrast, takes a detour dominated by the weaker forces and thus is approximately 10^{18} times slower than a reaction based on the strong forces at the same density and temperature. (The importance of this delay may be imagined if one recognizes that the ratio between one second and the entire age of the universe is of the order 10^{18}.) The strong forces, as always are active between the two protons, but they are insufficient by a few per cent only for bringing about a bond between them. This fact, which appears accidental, is of decisive importance for the full unfolding of evolutionary mechanisms in the universe. The differentiated interplay of highly heterogeneous, asymmetrical physical forces results in the regulation of the unfolding of evolutionary processes in time. We may also say, such an interplay *generates* cosmic time. As we shall see immediately, it also determines—or generates—the space for the unfolding of cosmic evolution.

The already mentioned gauge field theories, which have been elaborated over the past few years, deal with the basic symmetries between these four physical forces. However, the symmetries become directly effective only at extremely high temperatures, and therefore only in extremely early phases of the expansion of a hot universe. It seems, according to these theories, that the electromagnetic and weak nuclear forces were practically equal beyond a temperature of $3 \times 10^{15\circ}$ K. Both decreased then with the square of the distance and both were approximately equally strong. At even higher temperatures, the energy of particles in thermal equilibrium becomes so important that the gravitational forces between them—which are caused not only by mass but by all forms of energy—become equally strong with the strong nuclear forces. According to contemporary approaches to a quantum theory of gravity it may be estimated that this is the case beyond a temperature of $10^{32\circ}$ K, corresponding to a time of 10^{-43} seconds after the start of the expansion of an infinitely dense universe. It is not clear whether such a moment ever occurred in the real history of the universe. According to the theory by Dirac and Canuto (Maeder, 1978), the changes occurred gradually with expansion and continue today. For our considerations, this is not so important. It is not even of decisive importance whether the far-reaching symmetry between the physical forces has ever existed in reality or whether it is only a logical extrapolation against the background of that tremendous singularity, the beginning of space and time with the expansion of an originally infinitely dense and hot universe. The consequences of these symmetry breaks in the history of the universe are what is important here.

The decisive and earliest symmetry break between the strong nuclear and the gravitational forces implies that structuring forces were made available for a simultaneous evolution at microscopic and macroscopic extremes. The symmetry break between the weak nuclear and the electromagnetic forces adds structuring factors for intermediate domains. The electromagnetic forces were to play an especially central role in the later evolution of the most complex systems which we know, the biological and mental systems. But it is above all the *interplay* between these forces which determines the evolution of the universe. We may perhaps say that only the symmetry break which led to their "fanning out" was responsible for the unfurling of space and time for evolution. It is significant that the symmetry break between the forces which act in extremely microscopic and extremely macroscopic domains occurred at a time at which space, in our understanding of the term, did not yet exist at all; a signal proceeding with the speed of light would not yet have left the volume of a single subatomic particle and the total observable universe would have been limited to one particle.

Augustinus already spoke of the emergence of time *with* the creation. In Western thinking, space and time were often understood as metaphysical categories in Kant's sense, that is to say as *a priori* existing empty space which eventually fills with the forms brought about by evolution, and as absolute time scale into which evolution pours like water into a dried-out river bed. This understanding of space and time can no longer be maintained in a world view of self-organization. Self-organization implies the generation of the space-time continuum for system evolution *by the system itself*. There is a connection between energy density and time. But Kant may be partially vindicated by the fact that the unfurling of the space-time continuum by a heterogeneous, interactive system of physical forces precedes evolution. It creates conditions which come into play one after the other. But this is precisely the characteristic of a space-time continuum in which space and time are inseparable.

Gravity, for example, does not play a morphogenetic role until after the break-down of the gas pressure after 700,000 years; until then, it only acts as a brake for the expansion. But then, it dominates the macroscopic development until it enters anew into interaction with the strong nuclear forces in the development of stars and the hot centres of galaxies. The electromagnetic forces act at high temperatures merely in the direction of delaying morphogenetic processes by opposing the strong nuclear forces and by forcing the more numerous photons to exist as gas in thermal equilibrium with free charge carriers. But much later, they become the dominant factor in the synthesis of molecules and complex biological and mental structures.

The immediate consequence of the symmetry break in the physical forces is the simultaneity of macro- and microevolution in the universe. Macroscopic structures become the environment for microscopic structures and influence their evolution in decisive ways, or make it possible at all. Vice versa, the evolution of microscopic structures (nuclear, atomic and molecular synthesis) becomes a decisive factor in the formation and evolution of macroscopic structures. This interdependence constitutes nothing but an aspect of *co-evolution*, of the same principle which plays such an important role in the domain of the living. This principle implies that every system is linked with its environment by circular processes which establish a feedback link between the evolution of both sides. This holds not only for systems at the same hierarchical level, such as the predator-prey relationship discussed in the preceding chapter, or in symbiosis; the entire complex system plus environment evolves as a whole. But the reverse hold equally, that the environment cannot be one-sidedly adapted to a powerful system, a lesson we are learning today in the relationships of technological man to his environment.

Interlude: Structuration by condensation

In the development of the universe, the co-evolution of macro- and microcosmos gets into gear relatively slowly. The macroscopic branch of cosmic evolution shows with the expansion of the universe a gradual change in the macroscopic physical conditions, such as pressure, temperature and thermal equilibrium, whereas microevolution stagnates for the time being. There are no pluralistic ecosystems of particles yet, and matter in the expanding, homogeneous universe consists essentially of hydrogen nuclei (free protons) and helium nuclei in a mass ratio of approximately 22 to 28 per cent helium, and of free electrons whose number corresponds to the total number of protons. The free protons or hydrogen nuclei have been left over as excess matter in the matter/antimatter annihilation and the helium nuclei have formed below a temperature of 900 million degrees Kelvin (corresponding to an age of the universe of 3 minutes and 46 seconds) by fusion of protons and neutrons until practically no free neutrons were left. Although with further cooling the conditions would be met for the synthesis of heavier atomic nuclei, nothing happens since almost all neutrons are bound in helium nuclei, a fact which also has been favoured by a zone of instable isotopes following after helium. The microscopic evolution of matter would have arrived at its end—had not the already mentioned breakdown of gas pressure after 700,000 years brought gravity into play as a decisive, if not sole, structuring factor in macroevolution. With this turn, the conditions of a hot and dense universe are re-created and even stabilized over longer periods of time and microevolution starts rolling again.

If in the first few minutes in the life of the universe, the microscopic branch of evolution was most active, it is the macroscopic branch which initiates new activity after the long pause of 700,000 years. Hierarchic levels of structure form "from the top", the top here signifying the direction of "higher" levels which include the "lower" ones in their scope. The largest structures of the universe which we are able to recognize today are galaxies, clusters of galaxies, super clusters (clusters of clusters of galaxies) and perhaps even some super super clusters. Galaxies include in the average 100,000 million stars (our Milky Way system has apparently up to about 400,000 million) and have diameters between 5000 and 500,000 light years (LY)* (our Milky Way system about 100,000 LY). Half of the 10,000 million galaxies in the observable universe belong to clusters of galaxies which have diameters between 1 and 25 million LY. The regular type of these clusters contains thousands of galaxies, in particular of the elliptic, non-spiral kind. The irregular type contains about 20 to 2500 galaxies of all kinds. Our Milky Way system forms part of an irregular cluster with only twenty members and 3 million LY diameter. Super clusters, which represent still another hierarch-

---

*One light year is the distance which light travels in one year. It is practically 10^{13} (ten million million) kilometres.

ical level of structures in the universe, contain tens of thousands of galaxies and measure from 150 to 300 million LY in their longest linear dimension. According to recent results (Longair and Einasto, eds., 1978), super clusters seem to form interlocking, lace-like cells of a superstructure with voids between them in which the number density is down by a factor of 1000.

Had these structures formed immediately at the beginning of the expansion of the universe, the tidal effect would have ripped them apart instantly. But after 700,000 years the temperature has dropped to 3000 degrees Kelvin which does not only imply the possibility of stable atoms (complete with electron shells), but also brings the macroscopic binding force of gravity into play. With gravity, the further development of the universe may be described macroscopically by the *condensation model* proposed by Carl Friedrich von Weizsäcker (1974). This model holds that in the presence of a binding force and at sufficiently low temperature, structures form in a system at thermodynamic equilibrium. Ebert (1974) has been able to show this for cosmic conditions with the effects of gravity taken into account. The essential point in this conclusion is the formation of structures even near the thermal equilibrium and when entropy increases. High entropy is not to be equated with structureless uniformity. At sufficiently low temperatures the heat death would not resemble a soup but an assembly of complicated skeletons, to use the graphical comparison of Weizsäcker.

Thus, we may imagine that in this early phase of cosmic evolution macroscopic structures are "frozen out" just as snow crystals are frozen out of water vapour. The preceding short microevolution up to the formation of hydrogen and helium nuclei, too, would resemble such a "freezing out" of matter from the original mixture of radiation, matter and antimatter.

Self-organization of cosmic structures

But condensation does not remain the only effective mechanism. The macroscopic differentiation into regions from which eventually super clusters, clusters of galaxies and galaxies are to emerge, does not result in equilibrium systems. This becomes partly evident already at the level of galaxies in which very dense cores may form, perhaps due not only to the contraction effect of gravity. In so-called quasars and in Seyfert galaxies (galaxies with very bright and turbulent core), such dense cores are the stage for unimaginably huge explosions the origins of which are still obscure. Quasars (for "quasi-stellar objects") sometimes change their luminosity within a single day and thus cannot have diameters much exceeding one light day (26,000 million kilometres, about ninety-fold the diameter of the earth's orbit); otherwise, these changes in luminosity would cancel out. Nevertheless, these quasars radiate about a hundred times as much energy as a whole galaxy of

100,000 light years diameter. Some of them exhibit the expulsion of matter in a sharply defined jet (Maeder, 1977).

With the big radio telescope at Westerbork (Netherlands)—twelve reflectors in a straight line of 1.6 kilometres length—radio galaxies of huge size have been discovered in the past few years which are evidently the result of gigantic explosions (Strom *et al.*, 1975). The largest of these radio galaxies measures 18 million light years in its longest dimension, or in other words 180 times as much as our Milky Way system. But what is most conspicuous about them is their double structure which originates in the explosion when hot gas is simultaneously expelled in two opposite directions as if through two narrow nozzles. This energy-rich gas acts as an extended radio source. If the cores of radio galaxies move at high speed, gaseous tails form the parts of which originated in different phases in the history of these objects. Periodic explosions in the core may in this way be deduced from the structure of these tails. The radio galaxy in the Perseus cluster which is classified as NGC 1265, for example, exhibits a gaseous tail from such periodic explosions the last three of which, as is clearly discernible, must have taken place in intervals of 4 to 6 million years. If one is to believe outer appearances, there may be limit-cycle behaviour in spatially as well as temporally gigantic dimensions. But we know far too little about the dynamics and the energetic processes in such objects to hypothesize a kind of self-organization dynamics which we know from autopoietic, evolving systems on earth. There is also the possibility that gravity waves or shock waves act as a trigger for such explosions.

A similar long-time rhythm may also play a role in the formation of stars within galaxies. The British cosmologist Freeman J. Dyson (1971) in Princeton suspects such a rhythm of gravity and shock waves running over galactic sectors every 100 million years like a rotating beacon. The condensation of large gaseous masses would be accelerated thereby. The stars of largest mass would shine very brightly for a few million years and then die as supernovae in gigantic explosions. From a distance this relatively short period of high luminosity would appear as a bright spiral arm. Stars of smaller mass, in contrast, would continue their evolution through thousands of million years. The explosion of a near-by supernova not only furthers by local shock waves the condensation of stars but also mixes the heavier elements into the proto-stellar cloud from which stars and planets form. In the case of our solar system, there is indeed considerable evidence for two local supernova explosions 4700 and 4600 million years ago. The first one only mixed heavy elements into the protosolar cloud, the second one triggered its condensation. The proof of plutonium fission in meteorites which date back to the origin of the solar system and isotope ratios of other elements point to a decisive supernova explosion within 60 light-years (6×10^{14} kilometres) distance and not earlier than a million years before the formation of the solar system (Schramm and Clayton, 1978). Perhaps the rotation of the planets in their orbits around the sun may also be traced back to such an event. Thus, we may even speak of a hierarchy of shock waves—a hierarchy of dynamic phenomena—which underlies a multilevel cosmic reality, or at least is instrumental in bringing it about (Cameron, 1975).

Quasars, Seyfert and other radio galaxies and normal spiral galaxies form an almost continuous spectrum of macroobjects with decreasing energy dissipation. At the same time, they represent in this order a scale of decreasing average distance. This means that the observed radiation in this scale dates back to increasingly more recent times. Therefore, quasars and radio galaxies may be interpreted as predecessors of spiral galaxies. But this would imply that the formation of galaxies and superstructures cannot be explained by the condensation model alone and that there is considerable interaction between gravity and nuclear forces, macro- and microevolution even at this stage. Electromagnetic fields, in other words, physical forces of intermediary range, also seem to play a significant role.

Galaxies and large structures apparently do not continue to form today. But galaxies interact dramatically in the gravitational maelstrom of rich and dense clusters, stripping each other of stars and generally leading to the emergence of giant galaxies in the centre (Gorenstein and Tucker, 1978). The birth of stars does continue in our days, although apparently restricted to spiral galaxies. On the day on which this is written, the discovery of a star is reported which seems to be less than 2000 years old (plus the travel time of light), as may be determined by its gas expulsion. Our sun is 4600 million years of age, which is only one-fourth of the age of the universe. The first regular stars formed about 5000 million years after the universe started its expansion.

According to the simple condensation model (Steinlin, 1977), the formation of stars is imagined in such a way that clouds of interstellar matter at a temperature between 10 and 100 degrees Kelvin condense into a multiplicity of protostellar clouds, due to the effect of gravity. Stars are generally born in clusters, especially in the spectacular spherical clusters which measure 20 to 400 light years in diameter. Besides spherical clusters, there are also open clusters with 5 to 30 light years diameter. In the case of the sun, the protostellar cloud reached beyond the orbit of Pluto. When such a protostellar cloud reaches a minimum density of 10^{-13} grams per cubic centimetre, it collapses at the speed of free fall. During this very fast contraction—it is estimated that the sun contracted within a decade from a diameter corresponding to the orbit of Pluto to one corresponding to the orbit of Mercury—pressure and temperature increase enormously. Thereby, conditions are being re-created which correspond to an early phase of the universe, but which are more favourable for the synthesis of heavier atomic nuclei. Macroscopic evolution acts as a booster for microscopic evolution which had become stuck.

Going into detail, the development of a typical star is imagined to follow a sequence of phases (Maeder, 1975). In the first phase, the transformation of hydrogen (making up 70 per cent of the star) into helium starts in a core zone at approximately 5 million degrees Kelvin. In the presence of carbon, which has found its way from the explosion of older stars into the protostellar cloud, a catalytic reaction cycle may form at 10 million degrees Kelvin which has been proposed by Hans Bethe and Carl Friedrich von Weizsäcker (Fig. 23). In

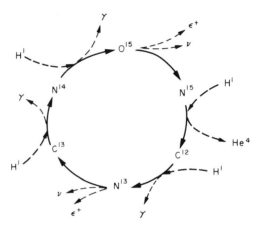

Fig. 23. The carbon cycle according to H. Bethe and C.F. von Weizsäcker which, in its overall effect, catalyzes the fusion of four hydrogen nuclei (protons) H^1 into one helium nucleus. He^4. The various isotopes of carbon (C), nitrogen (N) and oxygen (O) are always reconstituted in the cycle. Energy is dissipated in the form of γ-radiation as well as positrons (ε^+) and neutrinos (ν).

this cycle, four hydrogen nuclei (protons) are transformed into one helium nucleus, while the intermediary carbon, nitrogen and oxygen isotopes are reconstituted in the cycle. The energy dissipated from the cycle seems to be responsible for at least a significant part of the solar radiation. Since radioactive decay plays a role in the cycle, the slowly acting weak nuclear forces dictate the rhythm. As has already been mentioned, this fact is responsible for a long-lasting, steady energy liberation instead of a violent explosion. The formation and maintenance of such a cycle which runs irreversibly in one direction and reconstitutes its participants and thereby itself, is possible only far from equilibrium. It constitutes an autopoietic reaction system which, as a whole, acts as a catalyst for a specific transformatory reaction.

A macroscopic self-regulatory mechanism ensures that the reaction probability which is growing smaller with the burning of hydrogen is balanced by an increase in temperature. Eventually, when hydrogen in the core has been burned off to a large extent, the interior of the star contracts making the peripheral layers around the core sufficiently hot for a continuation of the transformation of hydrogen into helium in the peripheral zone. In the inactive core, now consisting mostly of helium, density and temperature rise until a new fusion reaction starts at about 80 to 100 million degrees Kelvin which transforms helium into carbon. Since energy is liberated faster than it can be transported to the outer layers, the temperature in the core rises to 500 million degrees Kelvin resulting in almost explosive burning of helium—the so-called "helium flash". The core regions expand vehemently which, in turn, leads to a calming down of the helium burning and renewed stability. This cycle may be repeated several times. Finally, almost all helium will be burned in the core and the transformation of helium into carbon continues only in more peripheral layers. This "onion skin" model also holds for the synthesis of heavier elements. In stars with a mass greater than that of the sun, the fusion of helium with various carbon and oxygen isotopes may yield certain elements up to iron Fe^{56}. The rest, as well as elements beyond iron are only formed under the extraordinary conditions given in the burst of a supernova, the explosion of a star with a mass greater by at least a factor of 6 than the mass of the sun.

I have described the basic mechanism of stellar evolution in some detail in order to show how the interaction of forces of macroscopic and microscopic range (essentially of gravity and the strong nuclear forces) brings about the co-evolution of macro- and microcosmos. Not only do macro- and microevolution mutually generate the conditions for their acceleration and continuation, they also result in astonishing stability of the macroscopic structure (the star), or, to be more precise, in an astonishingly long sequence of globally stable structures which transform themselves across instability phases into new structures. During this evolution, an extraordinary spectrum of conditions for existence may be realized. The density of stars varies over not less than 21 orders of magnitude, whereas the temperature may range from practically zero to 10^{12} (a million million) degrees Kelvin.

In this cosmic domain, we already find those forms of system existence which we know in principle from dissipative structures: an evolving sequence of autopoietic structures whose top criterion for self-regulation seems to be the maintenance of a dynamic structure, a specific régime of nuclear processes. The system itself, that is to say, the star evolving through many structures, seems to regulate the evolution of its own structures in such a way that the continuity is maintained of the transformation of mass into energy, with changing transformation rates and mechanisms over longer periods of time. The formation of convection layers, self-organizing co-operative structures (analogous to the hydrodynamic Bénard cells in Chapter 1) seems of importance in this self-regulation.

There is a difference between stars and the form of autopoiesis known from dissipative structures. There, energy is obtained by means of exchange with the environment, whereas in stars it is obtained from the transformation of constituents of the system itself. Like a dissipative structure, however, a star also shows true individuality; it regulates its dimensions and processes independent from the environment. Like a dissipative structure, it also oscillates in characteristic rhythms. Our sun, for example, pulsates in periods which, according to observations so far, range over at least six or seven orders of magnitude. The shortest observed period concerns the extension of the sun's corona by 1000 kilometres every 10 to 12 minutes. The longest observed periods are the well-known 11-year sun-spot cycle and another superimposed cycle of about 80 to 90 years.

The emergence of stars also clearly marks the appearance of that autonomy of a self-organizing system which implies the independence of the emergent structures from the expansion dynamics of the universe as a whole. With galaxies and their superstructures we may not be so certain in this respect. Do super clusters and clusters of galaxies, and galaxies themselves, expand with the expanding universe, or do they contract ever further? According to the already mentioned cosmological theory of Dirac, the gravitation constant decreases over time and all cosmic systems expand. This effect, however, makes a marked difference only with the largest systems. The smaller the systems considered, the more they appear like islands which, in an ocean of emptiness, become ever more distant from each other and find their own "optimal" structure according to their own laws.

Matter transfer and cosmic "phylogeny"

Old stars practically consist of only the primary matter of an undifferentiated universe, namely, hydrogen and helium. Young stars, in contrast, generally contain 2 to 4 per cent of heavier elements. Our relatively young sun consisted originally of 70 per cent hydrogen, 28 per cent helium and 2 per cent heavier elements, which remained to a large extent in the outer layers of the protostellar cloud of which the planetary system formed shortly after the birth of the sun, 4600 million years ago. Part of these heavier elements stem from the explosion of a supernova, another part may have been expelled in the instability periods of older, bigger stars. In the latter case also molecules which formed in the relatively cool outer layers of blown-up giant stars got dispersed in space. This contributed to the contents of cosmic clouds which were mainly made up of the primary matter hydrogen and helium from the early phases of the universe. One may perhaps conclude therefrom that only younger stars were able to form planetary systems with participants having firm surfaces and thereby fulfilling one requirement for more complex life. The earth on which we stand consists to a large extent of matter which has been synthesized in the womb of strange stars. Perhaps the largest part of it stems from that supernova explosion for whose role in the birth of the sun and the solar system there is mounting evidence. In addition, there is a startling variety of organic molecules which radio astronomy has recently found in interstellar matter and whose origin may at least partly point to the centre of the galaxy. This centre itself is mostly shielded from our observation by dark clouds. But it may not be implausible to assume that certain basic prerequisites for life are achieved not only locally, but perhaps also "centrally" by the galaxy.

We may almost speak of a cosmic "phylogeny" in which the products, and with them the experience, of various phases and lines of development in the evolution of the universe and in particular the galaxy join as in a phylogenetic tree. But we may also think of the recycling processes of life in which the entire matter of an organism, every single molecule, returns to the earth after the organism's death and is reused for new life. In contrast to the phylogeny of life, however, it is not information which is transferred but matter. The interaction of the physical forces reorganizes this matter in ever new systems, which dissipate energy by means of transforming matter and thereby aliment the self-organization of life's more complex systems. In the cosmic recycling of matter there is no downgrading of complexity as in biological recycling. In the latter, macromolecules are only partially reused directly (in the form of food), whereas in death, complexity becomes reduced to simple molecules and basic chemical elements. In cosmic evolution, however, matter is generally becoming more complex in organization.

Without such material, that is, phylogeny/recycling across thousands of millions of years, our solar system would have remained sterile with respect to life. After 4600 million years our sun is still busy with the conversion of hydrogen into helium and will stay with this task for another 5000 million years. Thereafter it will probably add the fusion of carbon and oxygen, but get stuck there. The multifaceted needs of life will never be met by the sun's own activities. In the contrary, its luminosity will rise about a hundred-fold during hydrogen burning and later perhaps ten thousand-fold, and its radius will increase about fifty-fold. The further evolution of the sun from a red giant will probably lead to a planetary nebula and further to a nova, before it ends as a white dwarf and ultimately as a burnt-out, black dwarf. In this tremendous increase of energy dissipation and dimensions, life on earth and possibly on other inner planets will hardly be able to continue in the forms in which we know it today. But at the summit of its self-realization, the sun will donate part of its mass to newly emerging stars and planetary systems and thus perhaps make a contribution to life on yet unborn planets. Somewhat larger stars (up to two or three solar masses) seem to end as pulsars, tremendously dense neutron stars which rotate unbelievably fast (at about 30 revolutions per second!). Stars of more than three solar masses may collapse into the already mentioned "black holes".

The phase of cosmic co-evolution comes to an end when the dimensions meet in the differentiation from both the macroscopic and the microscopic side. This is the case on cool planetary surfaces where the formation of crystals in microevolution meets with rock formation in macroevolution (Fig. 24). This closing of the gap, signalling a fully differentiated physical world, is a task for the electromagnetic forces. It becomes now clear that co-evolution is neither the formation of building blocks nor continuing differentiation of an originally homogeneous universe. It is the emergence of hierarchically ordered complexity to the full structuration of all hierarchical levels.

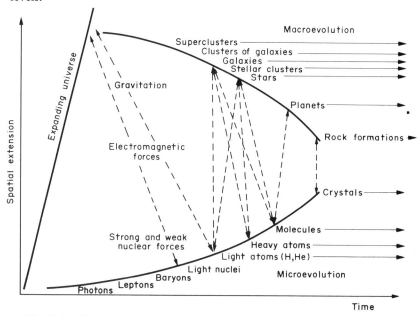

Fig. 24. Cosmic co-evolution of macro- and microstructures. The asymmetrical unfurling of the four physical forces calls into play step by step new structural levels, from the macroscopic side as well as from the microscopic. These levels mutually stimulate their evolutions.

The arrow of cosmic time

Before we continue with this co-evolution scheme by adding the bio-chemical and biospherical phase, we ought to discuss one more symmetry break, namely the one in *time*. With the big bang, time was given a direction on the macroscopic branch of cosmic evolution. The expansion of the universe is an irreversible process, either in an absolute sense (if the universe is open), or at least during a long expansion phase which will in any case last longer yet than it has already. Things are different with the microscopic branch of cosmic evolution. In the earliest phase, when matter and radiation continuously became transformed into each other, the occurring processes might have been called reversible. Matter/antimatter appeared and disappeared. Within the corresponding temperature boundaries, the matter/antimatter/radiation mixture did not show any evolution. Everything was in thermal equilibrium; all processes were reversible. Time did not yet have a preferred direction. Within the observable space, past and future were not yet separated from each other, with the possible exception of some relatively unimportant "inner friction".

The macroscopic direction of time, however, the irreversible expansion of the universe, brought about a continuous decrease in density and temperature with corresponding phase transitions which introduced, one after the other, important irreversibilities in microevolution: the production of baryons (in particular, protons and neutrons with their antiparticles) came irrepeatably to its end, then the production of leptons (electrons and positrons). The left-over matter and the photons did not continue to destroy and produce each other, but in collisions exchanged kinetic energy instead. This means that they now obeyed the laws of statistical mechanics and formed an isolated, if expanding, thermodynamic system. Thermodynamic irreversibility, the qualitative distinction in the macroscopic development of the energy contents of the universe, was thereby introduced. It has already been mentioned that in a pure condensation model an increase in entropy is compatible with an increase in structure. But the transformation of matter into energy and the dissipation of energy in stellar evolution and in the core regions of galaxies tends to complicate this picture again.

A last fundamental question may be raised here, a question which continues to preoccupy cosmology: Is the universe with its macroscopic characteristic parameters and an accordingly determined dynamics a random product, or is it, at least partially, the product of a certain degree of self-organization and self-regulation? It becomes ever more evident that a universe rich in structures was able to emerge only if a few narrowly defined boundary conditions were met. These boundary conditions concern, for example, the original high isotropy (i.e. uniformity in all directions), because otherwise the forming structures would have been torn to pieces by tidal effects. Furthermore, even the smallest local disturbances would have had catastrophic consequences and prevented the formation of galaxies and their superstructures, were the expansion velocity of the universe not close to the escape velocity (the limit velocity between an open, infinitely expanding, and a closed, pulsating universe). Connected with the latter factor is also the density of the universe so that the measure of the symmetry break between matter and antimatter and the mass of the left-over matter also enter here. That in micro-evolution some very narrow boundaries were respected was already mentioned when we discussed the boundaries between the effects of strong and weak nuclear forces which are responsible for a slow, controlled transformation of hydrogen into helium. There seem to be simply too many "accidents"—and more are becoming discovered every year

An immediate "Darwinian" explanation has, of course, been quickly offered. Many universes may form of which many stay poor in structures and sterile, until the conditions are accidentally met for the formation of structures, as it happened in the case of our universe. In less favourable cases, there is no life and therefore no observation and self-reflexion in these universes. The same argument has also been made in connection with the above-mentioned model of a vacuum fluctuation. Among innumerable smaller fluctuations, there is every now and then one which is sufficiently big and durable to permit the emergence of life; of the others, we shall never know. But in the same way in which we understand life today as self-organizing process which partially creates itself the conditions for its own continuation and complexification, and which enacts a far higher measure of self-determination than is implied by the Darwinian model of confirmation or elimination of random mutations, in the same way we may one day perhaps understand the self-organizing processes of a universe which is not determined by the blind selection of initial conditions, but has the potential of partial self-determination. We do not know whether the symmetry breaks between the physical forces which set the stage for the unfolding of evolution may occur only in the way witnessed in our universe, or also in other ways. Perhaps in a pulsating universe—but whether our universe pulsates or not, we do not know—the interactions between the physical forces and thereby the entire space-time continuum of evolution also evolves from cycle to cycle. Perhaps there has been an optimization of the variable parameters and relationships over many cycles in such a sense that the emergence of structure is increasingly favoured. Perhaps cosmic evolution shares this general theme with the evolution of life. Perhaps. . . .

But what we already know with sufficient certainty to be able to discuss, at least in its general features, is the manner in which life has itself created the conditions for its richly differentiated unfolding. This story will be told in the next chapter.

6. Biochemical and Biospherical Co-evolution

> All real living is meeting. Meeting is not in
> time and space, but space and time in
> meeting.
>
> Martin Buber

Energy flow as a trigger for chemical evolution

The age of the earth is assumed to be 4600 million years, roughly one-quarter of the age of the universe. The oceans formed 4400 to 4100 million years ago and the oldest igneous rocks appeared on the volcano-studded earth 4100 million years ago. Today we know that the largest part of the earth's history has been connected with the unfolding of life. Since the formation of sedimentary rock, the oldest of which date back 3800 million years, there are fossil traces of life. The oldest known fossils belong to single-cell organisms which lived about 3500 million years ago. Conditions on earth seem to have favoured the emergence of primitive life from the very beginning. That these conditions continued to be favourable is less surprising because, once started, life created the conditions for its continuation and further evolution to a good deal by itself, as will be shown in this chapter. Nevertheless, the examples of the other inner planets of our solar system—Mercury, Venus and Mars—seem to indicate that there is not too wide a margin for the development of higher complexity over thousands of millions of years.

For the first step to life the anorganic molecules which had formed during the condensation of the planet and during its cooling off were not sufficient. Whereas on the one hand the formation of equilibrium structures generated a sufficiently stable, energy-rich and dense environment for the unfolding of complexity, the temperature is not sufficiently high to ensure the continuation of evolution toward organic macromolecules. But electric discharges, or lightnings, during short time periods provide extraordinarily high energy penetration which results in high temperatures (recently measured at up to 30,000 degrees Kelvin) at which chemical reactions occur in which radicals and ions dominate. The very fast reactions lead to a non-trivial chemical kinetics whose equilibrium distribution includes already highly complex organic molecules. During the following fast cooling-off period, the most stable of these complex chemicals are "frozen out". They form the "ashes" of the high temperature reaction systems and find themselves at an environmental temperature at which they do not at first undergo further chemical reactions. But these ashes include some of the basic elements of life: carbohydrates, nucleic acids, amino acids (of which proteins consist), and even some smaller protein molecules. The two last-mentioned types of molecules are excellent catalysts, suitable for starting further dynamic steps in microevolution. The trigger for the beginning of the microevolution of life is thus provided by the macroscopic branch, the energy processes of a planetary environment. An important role seems to have been played by the energy-rich ultraviolet light of the sun which, in the absence of atmospheric oxygen, reached the surface of earth.

When, in the year 1953, Stanley Miller, then still a student, undertook the experiment which was to become famous—to vaporize a stimulated "primeval soup" consisting of water, methane, nitrogen, traces of ammonia and small amounts of hydrogen and to send electrical discharges through the vapour—he instantly obtained organic substances, such as sugars, bases and above all amino acids as well. The essential evolutionary step from simple anorganic to more complex organic molecules was thereby clarified in principle and found its experimental proof, but the result was not immediately understood. Instead of a very broad spectrum of compounds in small amounts, there was an extraordinarily selective spectrum of partly highly complex molecules. Today it is possible to understand why this specific selection of molecules emerges from the high-temperature reaction kinetics with subsequent quick cooling. It is interesting that the further steps in the evolution of life seem to require rather cool temperatures, perhaps in the range between 0°C and 25°C (Miller and Orgel, 1973).

Organic molecules, including amino acids, also hit the earth from outer space. Traces of organic substances of undoubtedly extraterrestrial origin were found in meteorites and also occur in comets. They have apparently formed at the same time as the solar system which adds plausibility to the assumption of a high-energy event, such as a near-by supernova explosion triggering the formation of the solar system. In this case it might be possible that the sun, the planetary system and the organic substances from which life was to emerge may have a common origin in the same giant fluctuation which seems to have forced by shock waves the protostellar cloud to its macro- and microscopic structuration. Organic molecules, in this case, did not form on earth but in the gas and dust cloud out of which emerged the planets. The further possibility to obtain organic substances from the centre of galaxy where they also occur may not have been necessary for the earth. But one may conclude that at very different initial conditions in the universe, the first steps of life are very similar.

Seeding from outer space with the products of a more advanced stage in the evolution of life, e.g. spores, is assumed by the old "panspermia" hypothesis which has recently been revived by such noted scientists as Nobel laureate Francis Crick and Leslie Orgel. Some believe even in the possibility of "directed" panspermia, planned and engineered by extra-terrestrial beings. But this implies just a shift in the locality. The question is still: How did life (anywhere) evolve by self-organization?

Prebiotic self-organization: Dissipative structures and hypercycles

The emergence and unfolding of life has often been told in more competent ways than I would be able to do. However, I should like to emphasize in this chapter the different levels of evolutionary processes, the nature of the symmetry breaks which separate them, and the co-evolution of macro- and micro-aspects of life. That these connections emerge only from recent and partly still controversial theories—the products of the metafluctuation sketched in the introduction—does not come as a surprise. Although there is no "standard model" for the evolution of life, as there is in cosmology, the ideas which I am attempting to present in a tentative logical structure, form in the eyes of many sensitive scientists a unity.

The first symmetry break in a world which, in the course of its condensation and cooling, has become structured into equilibrium systems such as crystals and rock formations, is the appearance of dissipative structures. As has been pointed out in Chapter 1, this corresponds to the break of spatial symmetry. The catalytic potential of those organic "ashes" which were left over by the high temperature reactions is here of decisive importance. The characteristic self-organization dynamics of dissipative structures may contribute importantly to the clarification of the following two steps of prebiotic evolution (Prigogine *et al.*, 1972). In the centre of interest there are autocatalytic steps which may be studied with the help of quantitative models of evolutionary feedback (see Chapter 3).

The first step concerns the formation of biopolymers from monomers, in particular the *polymerization of polynucleotides* for which Agnès Babloyantz (1972) in Brussels has developed a model. This model describes the competition between two types of polymerization, on the one hand linear chain growth as it dominates in equilibrium and leads to a stationary state with relatively low polymer concentration, and on the other, co-operative polymerization based on complementary molecular templates as proposed by Eigen (1971) and in accordance with the pairing rule of Watson-Crick. This rule, named after the discoverers of the double helix structure of genetic molecules, James Watson and Francis Crick, states that the four elementary bases in the double helix pair off in a particular way, adenine always joining thymine and guanine joining cytosine. For linear chain growth, normal clay with traces of metals or silicates may act as catalyst. But if the non-equilibrium is enhanced by a higher concentration of monomers, for example in deposits at the edge of a "prebiotic soup", the autocatalytic co-operative mode dominates and leads to a marked increase in the polymer concentration. Under certain conditions, this may result in instability and the formation of a dissipative structure. In polymerization based on template action a new type of memory plays a role, chemical stereospecificity or the capability of molecules to recognize each other in their form, or spatial structure. This may be viewed as the first step toward genetic communication which is based on information storage in conservative structures.

Dissipative structures imply an extraordinary intensification and acceleration of processes which otherwise might not lead anywhere. Simple catalysis leads to linear growth, autocatalysis to exponential growth. If in cosmic evolution the "task" was sometimes to delay the processes of energy liberation in order to ensure a fuller unfolding of evolution, it is now primarily the acceleration of processes. Another effect derives from the spatial concentration in dissipative structures and in the relatively high autonomy from the environment. Earlier theories assumed mechanical isolation as necessary in order to ensure the continuity of slow evolutionary processes. Half a century ago, when the Russian chemist Andreas Oparin (1938) formulated the first biophysically and biochemically well-founded theory of the origin of life, he postulated the inclusion of biopolymers in lipoprotein membranes and the formation of spherical coacervates ("underwater soap bubbles") which were selectively penetrable for ions, but otherwise ensured isolation of the primitive biosphere from the environment. Such coacervates of about one-hundreth of a millimetre diameter have actually been found in experiments with protein polymerization. But neither proteins nor nucleotides in this isolation can reach the degree of complexity which characterizes life. Only a common, mutually determined development seems to lead to such a complexity. Along this way, dissipative structures maintain their openness and selective exchangeability even without previous mechanical isolation. They are themselves capable of forming membranes and thus to accelerate the reaction kinetics by concentrating the catalysts on the one hand, and to separate the chemical reaction chains on the other. The much more flexible principle of the autonomy of self-organizing systems takes over from the principle of isolation. Finally, dissipative structures do not require any outside "drive" to maintain their self-organizing processes. Such an outside "drive" has to be assumed in other theories in the form of natural fluctuations of environmental factors, for example in Hans Kuhn's (1973) theory which focuses on random reproduction in the semi-isolation of rock pores.

The competition of theories may be characterized in such a way that in the processes in which life originated many things may also be explained by dull and highly unlikely accidents resulting from constellations arrived at in the slow rhythm of geophysical oscillations and chemical catalytic processes. But for every conceivable slowly acting random mechanism in an equilibrium world, there is a mechanism of highly accelerated and intensified processes in a non-equilibrium world which facilitates the formation of dissipative structures and thereby the self-organization of the microscopic world. It is not difficult to choose between these theoretical possibilities and it is probably also so that only beyond a certain critical threshold the continuity of a dynamics may be ensured which leads to higher complexity.

For the above-mentioned common evolution of proteins and nucleotides, Manfred Eigen (1971) in Göttingen has proposed a model which may be called a stroke of genius. Following the formation of sufficiently complex molecules of proteins (polypeptides) and polynucleotides, populations of both molecular species enter a phase in which they interact at many steps. In a *"self-reproducing catalytic hypercycle"* (Fig. 25), as Eigen has named it, the polynucleotides I_i carry the information both for their own autocatalytic self-

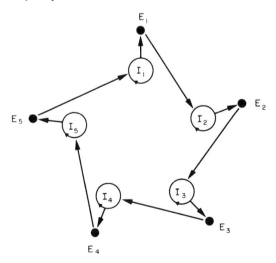

Fig. 25. A self-reproducing hypercycle of second degree, as it may have played a decisive role in precellular evolution. Each information carrier I_i (a nucleic acid molecule) carries the information for its own self-reproduction — indicated by the arrow in a closed circle — as well as for the production of an enzyme E_i (a protein molecule). The latter acts as catalyst for the formation of the next information carrier I_{i+1}. A closed hypercycle of this type is capable of a high degree of error correction in its self-reproduction and therefore of the preservation and transfer of complex information. After M. Eigen and P. Schuster (1977/78).

reproduction—making use of the catalytic capabilities of the preceding polypeptide E_{i-1}—and for the synthesis of the polypeptide E_i next in line. Because of this double function, this particular hypercycle is called a hypercycle of second degree (Eigen and Schuster, 1977/78). If the hypercycle is closed, so that the last enzyme E_n becomes the catalyst for the formation of the first polynucleotide I_1, the overall system is autocatalytic. This interweaving of the development of two molecular species may be viewed as a first expression of *symbiosis*. Each molecular species has to offer something which the other one does not have: polynucleotides are, because of their molecular structure in long strands, the best available information carriers and proteins are excellent catalysts; both capabilities together make self-reproduction possible. As Eigen and Schuster (1977/78, Part C) show for a simpler predecessor of this hypercycle—one in which only two polynucleotides (guanine and cytosine) and the two proteins for which they carry the information, participate—it can evolve only if each protein preferably catalyses the *other* polynucleotide, that is to say, if mutual enhancement prevails over self-enhancement. At this stage already, altruism appears as a basic evolutionary principle.

New substances may now arise from "copying errors" in the reproduction via templates. In this way, new non-linearities may be brought into play and drive the system across an instability into a new régime. This may be repeated many times. The entropy production may rise by several orders of magnitude in the transition phase between régimes. In the "build-up phase" of prebiotic evolution, we may therefore recognize the already mentioned principle of maximum entropy production at work. The "economic" principle of minimum entropy production holds for a stationary state. It comes into play when the evolutionary chain is interrupted, for example, when the system has attained optimal stability *vis-à-vis* fluctuations and its own errors. The history of this development is stored in the information contents of the nucleic acid molecules which may be regarded as *predecessors of the genetic code.*

Linear self-reproduction—the vertical aspect of genetic communication

Eigen's hypercycle represents in a pure form the co-evolution of nucleic acids and proteins which continues in the further development of single-cell and multicellular organisms. Fully developed, the basic process of genetic communication emerges as it will be briefly sketched in the following.

Information carriers are the DNA (desoxyribonucleic acid) molecules which occur in the cell nucleus in double helix strands. The parts (nucleotides) of the strands facing each other are composed according to the already mentioned pairing rule by Watson-Crick. Human DNA consists of no less than 2300 million nucleotides which contain the information for nearly a million genes. In replication, the two strands separate and double by acquiring the complementary nucleotides. In this process, a protein, called DNA polymerase, plays the role of the catalyst. In cell division, a complete copy of the DNA molecule may be transferred in this way.

The DNA molecule contains the information necessary for the production of proteins and therefore for the formation and continuous regeneration of the cell. This is achieved in such a way that the DNA molecule is copied by single-strand messenger-RNA (ribonucleic acid) molecules which remain stable for only a few minutes. In the so-called ribosomes within the cell, the twenty varieties of amino acids, each of which is addressed by three messenger-RNA-nucleotides ("codons"), are being joined to form proteins; in this process, another type of RNA molecules, transfer-RNA, is instrumental. In this way, a large variety of proteins is formed which, in turn, act as catalysts for tremendously complicated biochemical processes which occur simultaneously and side by side. For some of these processes it was possible to show that they organize themselves in the framework of tiny dissipative structures within the cells. The fixation of catalysts on membranes makes dense spatial packing possible as well as the intensification of the processes and thereby also the formation of complex networks of biochemical pathways.

In Eigen's hypercycle the principle of chemical autocatalysis is replaced by the principle of self-reproduction of entire, cyclically organized, process systems which introduce autocatalytic functions at a higher level. This new version of autocatalysis extends single-generation autopoiesis by copying to a sequence of generations. We may also speak of linear self-reproduction. Linear because it represents the strict replication of a given structure. What emerges are identical copies of one and the same individual. But also the fluctuations which result from copying errors are transferred along the line of generations. In contrast to cosmic evolution it is not matter which is transferred but *information* for the organization of matter. A new dimension of openness is introduced since via information the cumulative experience of many generations may be handed on. Whereas a chemical dissipative structure is merely capable of ontogeny, of the evolution of its own individuality, and its memory is limited to the experience accrued in the course of its existence, phylogeny (the history of an entire phylum) may now become effective. At first, the ancestral tree is no tree, but a single thin line. The experience of earlier generations as well as the fluctuations and evolution are transferred vertically, which here means along the axis of time. This time-binding makes the development of higher complexity possible than seems attainable by the ontogeny of material systems.

In contrast to the hypercycles underlying simple dissipative structures in which generation and degeneration are balanced, the pre-genetic type of hypercycle is geared to net increase. The participants in the hypercycle are relatively stable in comparison with the copying processes which take only minutes. Therefore, their decay may be neglected in first approximation. The result is not growth of the original system, but its multiplication. Ideally, as Eigen (1971) has shown, there will be hyperbolic growth of the numbers of active hypercycles, that is to say a type of growth which increases faster than exponential growth. Whereas in the latter the doubling time remains constant, it decreases in hyperbolic growth. For several hundred years, the human world population increased hyperbolically; each doubling required only half the time the previous doubling had taken. Hyperbolic growth of hypercycles, according to theoretical considerations by Eigen (1971), implies that in the competition between two or more basic forms of self-reproducing bio-molecules there will be a clear decision for one of the competitors. This is the basis for an explanation of the fact that all life on earth, animal as well as plant life, is based on the same kind of genetic structures. The individual molecules, however, may not be fully identical but probably formed a "quasi species", as Eigen and Schuster (1977/78) have called it, a species consisting of very similar members.

The reproduction of the participants in the hypercycle and their engage-ment in new hypercycles became obviously only possible by an equally fast-growing metabolism. This means that the function and self-reproduction of hypercycles probably depended on the formation and multiplication of dissipative structures. At the present level of life, the metabolic function is included in simpler hypercycles, such as the self-reproduction of transfer-RNA; new RNA is synthesized from energy-rich molecules and energy-deficient molecules are ejected as "waste". In complete cells, however, the metabolism is primarily taken care of by numerous auxiliary loops.

But there are also the viruses which consist only of genetic material (mostly RNA) enveloped in protein and which are incapable of metabolism. In order to multiply, they have to invade host cells and, as a kind of molecular cuckoo's egg, use their metabolic mechanism (Campbell, 1976). Development to the first cells, however, obviously depended not only on the capability of copying complex information, but also on the capability of securing the metabolism in the self-reproducing and evolving units themselves. Whether the dissipative structures go along with hyperbolic growth without any difficulty, is at best uncertain. How many products may have fallen under the table without becoming structured immediately in new hypercycles? Perhaps the develop-ment resembled short bursts of growth in the individual hypercycles, with contemplative, autopoietic periods in between. This would correspond to the theory of dissipative structures in so far as it predicts in the formation of a new structure an entropy production higher by several orders of magnitude in comparison with the autopoietic state. In an "established" structure this hectic activity decreases again.

In any case we may in the prebiotic phase, even before the emergence of the first cells, already speak of matter systems which metabolize, reproduce themselves, evolve through mutations and compete with other systems for selection. Who would have dared to make such a statement only a few years ago? And how many biologists will accept it today? The enumerated capabilities have stood for a concise definition of life. Today, however, we recognize that these capabilities are general properties arising from dissipative self-organization and bridging the gap between the realms of the animate and the inanimate.

If one looks exclusively at the microevolution of life, that is to say at its development from hypercycles to the first cells, Darwinian selection may be assumed with Eigen and Schuster (1977/78) in so far as evolution of the phenotype (the individual) in this phase falls together with genetic change. The phenotype of the precellular stage consisted of not much more than its genetic material. There was hardly a margin for the flexible utilization of this genetic material in the exchange processes with the environment, a theme which was to appear only later in the form of epigenetics. It may appear ironic that strict Darwinism—the thesis of the evolution of the genotype by environment-dependent regulation of the chances for the reproduction of the phenotype—has been formulated for highly developed organisms, but may hold fully only in the molecular domain.

According to Eigen and Schuster (1977/78), there are good arguments for a strictly Darwinian selection in relentless competition. The genetic processes and information carriers exhibit highly specific characteristics which appear to be the result of very precise optimization and are hard to explain in other ways than by selection of the fittest elements and mechanisms. But this would mean that at this early stage of life there was no co-evolution of macro- and microcosmos, or to be more precise, that both sides alternated in being dominant. After the preparation of the organic starting material by the planetary system and perhaps macrosystems of even wider scope, a phase of "pure microevolution may have followed. But as we shall see, the interweaving of macro- and microevolution becomes firmly and durably established at the latest with the emergence of the first primitive single-cell organisms.

It may nevertheless be worthwhile to keep an eye on the possibility that the reverse may be true. In the evolution of life, individualism appears only at a late stage. Primitive life is determined to a large extent by macrosystems such as colonies, societies and ecosystems. Might then the beginning of this development not have been determined by a kind of planetary, prebiotic ecosystem? The most prominent representative of such a view is the Spanish ecologist Ramón Margalef (1968). He distinguishes between three channels in which in the domain of life information is transferred. The channel of ecological information exists since the beginning of life and widens slowly. Later only, the channel of genetic information is added, at first broadening quickly and later less quickly. Last, the channel of behavioural and cultural information is added; it broadens even faster than the others.

In this scheme, the beginning of life is characterized by the dominance of ecological information. Margalef even speaks of an ecosystem consisting of the "different parts of a chemodynamic machine" which existed even before the appearance of the first self-reproducing molecules—a thought which seems to anticipate the concept of dissipative structures. In fact, the existence of dissipative structures is almost totally determined by horizontal, ecological relations with the environment. There is ontogeny, but no phylogeny, individual experience, but not its transfer along a phylum. The correspondence between structure and function which holds for a dissipative structure may also be effective in the regulation of ecosystems which consist not of organisms, but of dissipative structures. In this view, the emergence of a capability for self-reproduction would not have had to result from a fight for survival between vertical development lines, but might also have arisen from a co-evolution of many dissipative structures which exchanged their information continuously and jointly produced the nucleic acid solution. The almost total openness toward ecological information in this case implied that fluctuations of periodic or irregular type exerted great influence. The overall ecosystem, still little emancipated, followed these fluctuations. For such a co-evolution of the subsystems of an ecosystem, Thomas Ballmer and Ernst von Weizsäcker (1974) have proposed in a different context the term *ultracycle*. In such an ultracycle the evolution of higher complexity does not result from competition, as in the hypercycle, but from interdependence within a larger system. Ballmer and Weizsäcker have not formulated this idea for the precellular phase, however, but for ecosystems in later stages of evolution.

Be that as it may, both microscopic and macroscopic principles, embodied in the concepts of the hypercycle and the ultracycle, may have contributed to pave evolution's path to the first, primitive cell—and possibly an ingenuous combination of both principles. But whereas the concept of individual selection in hypercycles already boasts an impressive theoretical foundation, a precellular ultracycle is at present at best a vague idea. But the uniformity of the basic structures of life is hardly explainable at a molecular level exclusively. When in a later stage of evolution, multicellular organisms appeared, there were already millions of types of single-cell organisms; but even the most complex organism, such as the human organism, consists of not more than two hundred cell types which all originate in a single cell.

Horizontal genetic communication—the stage of prokaryotic cells

Following chemical and biochemical dissipative structures, which also include the individual steps in the evolution of self-reproducing hypercycles, the primitive cell represents the next higher step of a system stabilizing itself in autopoietic structures and being cyclically organized. It is the cell which, at a higher level, co-ordinates and separates the functions of nucleic acid and protein. Of decisive importance was the development of the membrane which permitted at the same time the separation and reinforcement of biochemical pathways and cycles. Membranes enhance the local non-equilibrium. A cell may contain many dissipative structures of microscopic dimensions. In comparison with the pure process systems of simple chemical dissipative structures, they have become highly miniaturized structures by the introduction of solid or semi-solid elements.

A cell embodies macroscopic order at a higher level of complexity and this order may establish itself in very small volumes. The smallest free-living cells, pleuropneumonia-like organisms, have a diameter of only about 0.0001 millimetre, corresponding to only a thousand-fold the diameter of the hydrogen atom (Morowitz and Tourtellotte, 1962). All they need is about 1200 biomolecules. The diameter of bacteria is in the average about ten times larger than these minimum dimensions of life, the diameter of a mammal tissue cell a hundred times and the diameter of a protozoon, such as an amoeba, a thousand times. Even more dramatic is a comparison of masses which vary in a ratio of one to a thousand million. The mass of the smallest cell is 5×10^{-16} grams. A cell which is enriched with nourishing material, such as the yolk of a bird's egg, may become much bigger and heavier than even a protozoon.

The co-ordinating function of a cell manifests itself in the development of complex multistage metabolic processes. At first, they were geared to the fermentation of material from the environment. A first degree of flexibility was achieved with the development of preliminary steps in which these materials were produced within the cell by means of enzymes and by using different starting materials. The absolute dependence from specific food materials was thereby broken. Besides these diligent and inventive autotrophs (organisms producing their own food) there appear to already have been at an early stage some heterotrophs which fed on cell fragments and smaller cells and thereby saved the metabolic costs of the production and transfer of genes representing the capability of synthesis from food. Perhaps this also meant an advantage for selection.

The microorganisms which are discussed here were nucleus-free single-cell organisms (prokaryotes) of two basic types. The direct descendants of one type are today's bacteria. But most of the prokaryotes belonged to another type, the much larger, thread-shaped blue-green algae which also occur today in many types of environment, including such types which are unfit for any other vegetation (Echlin, 1966). They are also called cyanobacteria or cyanogens to emphasize that they are closer to bacteria than to green algae. Most of them live in fresh water. In humid tropical regions, they may cover wide areas with a gelatine-like mass. In moderate latitudes they often cover rocks and tree stumps in moist valleys.

One-sided Darwinian thinking imagines their evolution in such a way that over-population occurred and a crisis in food procurement developed from which those species emerged as survivors which were capable of opening up more effective energy sources and processing methods. Such linear thinking which considers the competition between development lines (species or phyla) which are assumed as clearly separable, meets with an unexpected difficulty at least in respect of bacteria: there were no species in a strict sense.

Bacteria do not reproduce in a strictly vertical way by transferring their genetic material to the next generation which emerges from cell division. In a strict sense, cell division (binary fission) does not constitute a sequence of generations, but multiplication as we encountered it in hypercycles. The complete genetic information, which is present in several copies, is passed on. There is no natural death separating generations from each other. Each cell continues to live and in many steps of dividing becomes an extended space-time structure. Therefore, it may not come as a big surprise that genetic information is not only multiplied by dividing the identical DNA copies in a prokaryotic cell between the two daughter cells emerging from cell division. There is also a feedback mechanism which permits the exchange of genetic material within the same generation.

The horizontal exchange of genes between individual bacteria even plays a significant role. Genes may be transferred in the form of dissolved DNA molecules or on plasmids (parts of cells). Or one of the two to four identical chromosomes of a bacterium may be transferred via a temporary bridge ("conjugation"), or also by means of a special carrier, in particular, a virus. With conjugation, there is a first touch of sexuality, when a "male" cell transfers its chromosome to a "female" cell. The difference between the bacterial "sexes" lies in a specific gene which initiates this pseudo-sexual process. If it is transferred to the "female" cell, the latter becomes "male" and takes the initiative the next time (Broda, 1975). Transsexuality is nothing unusual with bacteria. The horizontal gene transfer by means of viruses which "fit" into a host cell and become their integral part also occurs in humans and may play a hitherto under-estimated role (Campbell, 1976).

This so-called parasexual gene transfer seems to be possible among all types of bacteria, even between very different types in multiple steps. Thus, bacteria employ vertical as well as horizontal gene transfer. The role of the latter is so important that it has been proposed (Hedges, 1972) that bacteria be considered as one huge, common gene pool, from which temporarily defined "species" obtain the genetic information which they need for changing situations, i.e. for their relationships with a changing environment. Perhaps here a mechanism is already anticipated which, at later stages in the development of life, is called epigenetic. This notion, introduced in 1947 by Conrad Waddington (1975), stands for the selective utilization of genetic information carried by an organism, in dependence of the ever-changing relations between the organism and its environment. Here, with the bacteria, there is basically the same flexible utilization, but the entire gene pool is available instead of an over-determined hereditary information carried by a single organism. With the bacteria, one may perhaps speak of *external epigenetics* in contrast to the internal epigenetics at later stages of evolution.

The retrieval of this information from a generally accessible "central library" which stores the totality of all mutations—and thereby also the totality of the experience made by all bacteria—is obviously less well aimed than the utilization of the genetic "private libraries" which are inherited at higher levels of life by means of meiosis and sexuality. But even a totally random exchange ensures that mutations with survival value are spreading quickly whereas "bad" mutations are still rapidly eliminated. In Chapter 16 we shall learn how the bacteria develop a tactics based on a so-called "random biased walk" to unerringly get closer to the optimum food concentration in their environment. Bacteria seem to be real masters at pursuing a purpose with random processes.

Again we are facing the question: is the image of a microevolution of life not one-sided and misleading? With the hypercycles, there was no clear answer. But here, with the first single-cell microorganisms, it becomes evident that there is a co-evolution of macro- and microsystems. The macrosystem in question comprises the totality of all bacteria. The evolution of this totality only provides the possibility for the unfolding of the individual bacteria and the mutations occurring in this unfolding keep the overall system alive and dynamic. The horizontal exchangeability of microscopic information guarantees the epigenetic flexibility which provided the best opportunity for the gradual collection of macroscopic experience perhaps even in this early period of trial and error. The fluctuations, which might have endangered single individuals, but also colonies and temporary species, act subdued in the macroscopic framework on the one hand, but stimulating for the overall system on the other.

The strategy of evolution in this early phase may perhaps be compared with the strategy of a multi-person expedition into dangerous, unknown territories. Individual participants may die of illness or the attacks of wild animals or hostile natives. But if they exchange their experience every day, not only will they enhance their capability of surviving in this environment, but only a single participant need return in order to save the results of the entire expedition—to the extent that this returning participant is a generalist. In the world of the prokaryotes, everybody is more or less a generalist, even when they act temporarily as specialists.

We already realize that in a multilevel system which is self-organizing at least at a macroscopic and microscopic level, it is no longer the same if we assume that energy organizes matter or that the reverse holds. If the emphasis is on matter systems organizing energy, a microscopic, Darwinian description seems to follow. If the aspect of an energy system organizing matter is in the foreground, a macroscopic description in terms of ultracycles imposes itself, in which entire ecosystems evolve to higher complexity.

In this first phase of life on earth the subject of evolution was ultimately the entire biosphere which, in this phase, consisted only of prokaryotes and some "fillers", the less significant archebacteria which occupied only ecological niches with extreme living conditions. This becomes even clearer with the history of the appearance of free oxygen which was the prerequisite for the evolution of more complex forms of life. This history is the history of a massive transformation of the earth's surface and the atmosphere—a history in which the creative principle of the co-evolution of the macroscopic and the microscopic manifests itself perhaps in the most splendid way.

The build-up of an oxygen-rich atmosphere—life itself creates the conditions for its further evolution

Until the middle of the 1960s life in the earliest phases of evolution was mostly a matter for speculation. Apart from some mat-like structures, later identified as fossil microbiotes, the oldest known fossils were the 500 million years old trilobites and other marine invertebrates. First microscopic plants were found in 1954. Since then, however, newly developed methods have permitted the discovery of microfossils of unicellular life forms which existed thousands of millions years ago. It is assumed today that in this early period prokaryotes existed side by side with so-called "arche-bacteria", especially "methanogens", which had an even simpler structure than the prokaryotes and transformed carbon dioxide and hydrogen into methane using hydrogen as their sole energy source. The contemporary descendants of the archebacteria live in absolutely oxygen-free environments at the bottom of the ocean, in sewers, in the hot springs of Yellowstone National Park in North America and—in cow stomachs. The hypothesis is favoured today that archebacteria and prokaryotes stem from a common predecessor of even simpler design, capable of rapid evolution. It appears even plausible that in the oxygen-free primary atmosphere they were linked by a hydrogen/methane cycle in a similar way as animals and plants are linked today by an oxygen/carbon dioxide cycle, with photosynethic pro-karyotes providing the hydrogen and breaking up the methane. Without such a cycle, the gases would have quickly become exhausted *(Neue Zürcher Zeitung,* 1978).

The oldest sedimentary rock formations apeared 3800 million years ago. The oldest microfossils are not much younger, an indication that life appeared on earth very early, perhaps 4000 million years ago. With regard to the microfossils which had been found at several places in Swaziland in Eastern Transvaal (South Africa), one was not so certain at first. A group of approximately 200 fossilized cells, however, which were recently identified by Elso Barghoorn and Andrew Knoll (1977) and which appear in different stages of cell division, are without doubt blue-green algae which lived 3400 million years ago in warm and relatively shallow waters. In these prokaryotes of 0.0025 millimetre diameter the complex pathways of photosynethesis were apparently organized which means that these prokaryotes were capable of producing carbohydrates from carbon dioxide and water by using solar energy in the form of light, while giving off oxygen. This, at least, seems to follow from the ratio of carbon isotopes in the sediments which is characteristic for photosynthesis (see below). According to recent reports, fossilized cells found in Western Australia and also exhibiting cell division may even be older (3500 million years).

Of somewhat more recent origin are the so-called stromatolites, reef-like, densely packed and finely laminated sediments, measuring centimetres to metres in length, which stem undoubtedly from mat-like aggregations of prokaryotes in which the blue-green algae dominated (Schopf, 1978). The oldest of these fossils belong to the Bulawayo group of Rhodesia and date 2900 to 3200 million years back. The Steeprock Lake stromatolites in Canada come next with 2600 million years of age. These oldest stromatolites are not different in any way from considerably younger ones which seems to indicate a certain "stagnation" in microevolution and also in particular the very early development of photosynthesis. Since about 2300 million years, stromatolites occurred in more places, especially in Africa and Australia. So far, forty-five of such places have been found (all but three since 1968) and most of them are between 2250 and 725 million years old. This quantitative boost of evolution in an age which is called the proterozoicum may well be linked to the appearance of free atmospheric oxygen. Stromatolites continue to form today in rare places, such as Shark Bay (Australia), where the high salinity prevents invertebrates from flourishing and feeding on the prokaryote communities.

Organic oxygen is produced in aerobic (oxygen-dependent) photosynthesis. With photosynthesis which has been developed very early by bluge-green algae and in a different form by bacteria, access has been found to inexhaustible energy flows and thus to significantly enhanced flexibility. Photosynthesis also marks the inclusion of a cosmic environment for the non-equilibrium system of the biosphere. Without such an inclusion, the energy-rich organic materials would soon have been exhausted, the entropy of the biosphere would have increased and life come to an end. Aerobic photosynthesis by blue-green algae was supplemented by anaerobic (oxygen-free) photosynthesis by bacteria, using hydrogen sulphide (H_2S) instead of water and giving off sulphur instead of oxygen. Aerobic photosynthesis dates back at least 2200 million years, because from this time we have fossil records of specialized thick-walled cells, so-called heterocysts, which protected enzymes from oxygen. But probably aerobic photosynthesis preceded the anaerobic version and occurred already 3800 million years ago.

In photosynthesis, one or two photons transfer part of their energy to an electron which is thereby excited and in turn invests its excess energy in several biochemical process stages. In this way, energy may be stored in the end product, in particular in the glucose molecule, for later degradation and use. Is it to be called an accident that many chemical processes—and especially the biochemical processes—require energy jumps in the order of one electronvolt* and at the same time the energy of a typical photon in the visible spectrum of the sun's light is between 2 and 3 electronvolts and thus may easily be degraded by the required amount? Short-wave ultraviolet light already has double that energy and destroys biochemical processes, whereas infrared (heat) radiation, with about half that energy, is too deficient in energy to spare one electronvolt. The most intense part of the solar radiation impinging on the earth's surface, however, suits the subtle biochemical processes to perfection. Is this the result of co-evolution in the sense that life took a specific chemical pathway? Without this favourable coincidence, would life only be possible on the basis of fermentation and thereby be trapped in its earliest steps? Or is life possible at all in the light of only a star which corresponds to the sun's spectral class? These questions can hardly be answered at present.

Whereas the fantastic development from anorganic molecules to living cells and from simple dissipative structures to photosynthesis required less than 1000 million years after the birth of the earth, not much happened in the microevolution of life over the following 2000 million years. The more active, however, were the interactions between microscopic life and the planetary macrosystem in this period. It is characterized by a tremendous transformation of the earth's surface and atmosphere. The first primitive life forms set the stage for the development of more complex life forms. They created the preconditions which, above all, included free oxygen in the atmosphere. Only oxygen-breathing cells with a nucleus are capable of forming cell tissues and multicellular organisms. The nucleus-free prokaryotes, however, went to

---

*One electronvolt is the energy gained by an electron while travelling through a potential of one volt. In metric units, one electronvolt corresponds to 1.602×10^{-19} Joule or watt-seconds.

work to produce this oxygen. This was not an easy task because it required a seemingly endless detour via the oxidation of the entire surface of the earth.

Generally speaking, there are two mechanisms of oxygen production which were possible in the early age of the earth (June, 1976). Besides aerobic photosynthesis, in which oxygen occurs as a side product, the photo-dissociation of water vapour by short-wave ultraviolet radiation in the higher layers of the atmosphere may also produce oxygen; the simultaneously produced hydrogen would in this case, due to its light atomic weight, escape from the earth's gravitation. There are good reasons, however, to assume that the latter non-biological process cannot have played an important role. For example, the atmosphere is increasingly shielded from ultraviolet radiation by the accumulating ozone and the feedback system of oxygen production and shielding aims at an equilibrium, the so-called "Urey point", named after the American chemist Harold Urey who already in 1959 drew attention to it. This point is reached at only a thousandth of the present oxygen concentration. If, however, the shielding from the same ultraviolet radiation which has played such a positive role in the prebiotic phase but is now becoming an obstacle to life's further evolution is due to biologically produced oxygen, we may recognize a particularly impressive example of the co-evolution of macro- and micro-cosmos on earth.

But even more important is the evidence of carbonous sediments dating back 3300 to 3800 million years and including formations in which some of the oldest microfossils have been found. These sediments consist at 80 per cent of biologically produced carbon originating from the transformation of carbon dioxide in photosynthesis. This may be deduced with great certainty from the ratio of the carbon isotopes C^{13} and C^{12}. It is for biologically produced carbon 2.5 per cent smaller than for non-biologically produced carbonates. However, it is not possible to decide in this way whether the photosynthesis was aerobic (oxygen-based) or anaerobic (sulphur-based).

On the other hand oxidized iron minerals found in the earliest sedimentary rocks in Greenland and between 3800 and 1900 million years old, already indicate the presence of oxygen in very early phases of evolution. However, this oxygen can hardly have occurred in the form of free atmospheric oxygen because much younger deposits of uraninite (UO_2) in the form of smooth, round grains have been found on various continental shelves which would quickly have been oxidized to U_3O_8 and decomposed in the presence of oxygen. These deposits indicate that in a period starting about 3200 million years ago and ending about 2000 million years ago, there was essentially no free atmospheric oxygen.

These apparent contradictions may be resolved by assuming the origin of the very old oxidized iron minerals in an early ocean containing iron in the ferrous state with a valence of 2 (Junge, 1976). This soluble iron was furnished in the ocean's microorganisms with the oxygen required to form indissoluble ferric iron oxide, Fe_2O_3, with a valence of 3, and subsequently expelled. A further argument for the biogenous origin of these oldest sediments is provided by the identification of fragments of the chlorophyll molecule.

Oxygen is a dangerous substance for cells which are not capable of using it in a controlled way in their biochemical processes. It burns the cell tissue which is based on carbon. However, this circumstance did not play a role as long as the oxygen produced in these first cells by photosynthesis is used up immediately for the oxidation of the sediments. Only 5 per cent of all oxygen that was ever formed in the history of the earth occurs freely today in the atmosphere—where it makes up 21 per cent of air—or is dissolved in water (a small amount only, in the order of 1 per cent of free atmospheric oxygen). The rest has been bound in minerals by oxidation, approximately 56 per cent in sulphur (SO_4) and 39 per cent in iron (Fe_2O_3). No wonder that this tremendous work took so long. The prokaryotes, however, are no slow workers. It has been calculated that one gram of them, under ideal conditions of unlimited growth, could have produced in only 40 years all the oxygen present in today's atmosphere.

Only when the oxidation of the sediments was approaching saturation was a partial pressure of oxygen able to build up in the atmosphere. The increasing appearance of prokaryote aggregations since about 2300 million years ago seems to indicate that oxygen became increasingly available at this time. The massive depositing of banded iron oxides in this era lasted until 1800 million years ago. There is indeed much evidence of the occurrence of significant amounts of oxygen in the atmosphere since approximately 2000 million years. From this time date reddish minerals whose colour is due to the oxides of the ferric iron of valence 3. About 1800 to 1900 million years ago the oxygen concentration of air seems to have been around 1 per cent of today's concentration. This already resulted in considerable shielding from the sun's ultraviolet radiation. About 1500 million years ago, today's oxygen concentration was reached and remained remarkably stable ever since. However, this state was hardly reached without periods of major fluctuations. It seems significant that the biochemistry of contemporary prokaryotes is optimized for an oxygen concentration of 10 per cent, or half of what it is today (Schopf, 1978).

With the liberation of oxygen, its utilization commenced. The function of respiration serves the downgrading of energy stored in the glucose molecule. Without oxygen, this energy was recouped only at less than 5 per cent. With oxygen, however, the efficiency rises to 65 per cent—a fifteen-fold increase. However, the biochemical pathway is not designed in a totally new way. The oxygen-burning step replaces the end of the old pathway which used to end with alcoholic fermentation or the formation of lactic acid. This formation of lactic acid is still experienced as tiredness when an oxygen deficiency leads to the reactivation of the old pathway end.

Among the first prokaryotes adapting to oxygen were the same blue-green algae which produced most of it. Respiration was probably introduced by photosynthesizing bacteria (Broda, 1975). The flow of energy-transferring electrons and the biochemical pathway are very similar in photosynthesis and in aerobic, oxygen-breathing processes. In both cases, they are separated by membranes into various steps. This raises the possibilities that an entire class of photosynthesizing bacteria has undergone this development in an "obligatory" way or at least in many parallel cases. However, even today there are many prokaryotes which can only live in oxygen-deficient environments—in the earth, in mud, or in the bodies of organisms.

As with every "invention" of a new principle by life we are again facing the question: How did this innovation come about and how did it break through? A strictly Darwinian explanation places the fluctuation within a single individual which, due to better adaptability to the environment, has more offspring which in turn are a little bit better off than their contemporaries in each generation, and so forth, until after many generations only descendants of the mutants are left. Darwinism is based on vertical transfer of genetic information. With bacteria, however, horizontal genetic and ecological mechanisms play an important role as well. Without having a detailed idea of what really happened, we may perhaps assume that fluctuations were very fast in spreading horizontally and thereby in their self-reinforcement and that in such a way large groups of individuals in local aggregations were forced into new relations with their environment. When such a profound change occurred as the appearance of oxygen in the atmosphere, suitable bio-chemical-biophysical fluctuations were probably quick to follow. Once they occurred, their breakthrough was facilitated and accelerated by horizontal gene transfer.

Gaia—the planetary self-organizing system of bio- and atmosphere

It is photosynthesis which transforms the biosphere into a system which is open with respect to its energy exchange, if generally not with respect to matter and information if we neglect meteorites and the already-mentioned "panspermia" hypothesis of spore transfer across cosmic space. This openness makes it possible to obtain free energy from solar radiation and to export the accruing entropy with waste-heat radiation into space. The atmosphere acts like a buffer system which stores heat and which regulates the waste-heat radiation in such a way that there are no extremal temperature differences. Might it then not be also possible that the entire biosphere together with the atmosphere acts as an autopoietic system which organizes and regulates itself? This is indeed the basic idea of the American microbiologist Lynn Margulis and the British chemist and professional inventor James Lovelock (Margulis and Lovelock, 1974) which may be called another stroke of genius. William Golding, author of *Lord of the Flies,* has found a suitable name to express the weight and dignity of an idea which touches the totality of life on earth: *Gaia hypothesis,* in honour of the earth mother in Greek mythology.

The Gaia hypothesis proposes a dynamic view instead of a static one; in the latter, as everybody knows, the atmosphere consists of 79 per cent nitrogen, 21 per cent oxygen and traces of other gases. In a dynamic view it turns out that a whole spectrum of biogeneous gases show the same through-flow through the atmosphere expressed in moles (or numbers of molecules) per year (see Table 3). This through-flow is about 1 per cent of the through-flow characterizing the primary oxygen/carbon dioxide cycle of life. Only their residence time is very different. This may be made clear by imagining a bathtub into whicn water is running at the same rate as it is drained from the bottom. The through-flow is always the same whether the tub is full or practically empty. The atmosphere is also in high chemical non-equilibrium. Of some gases, up to 10^{30} the amount is present in the atmosphere as would be "permissible" in an equilibrium system with given oxygen content. We may remember here that non-equilibrium is one of the basic prerequisites for self-organizing and autopoietic behaviour of dissipative structures.

The autocatalytic units in this system which make possible the formation of a dissipative structure far from equilibrium and maintain the through-flow of the various gases are none else but the prokaryotes. It seems that after the profound transformation of the earth's surface by the oxidation of sediments and the accumulation of free oxygen, they have been instrumental in bringing the overall system bio- plus atmosphere into global, autopoietic stability, reigning now for 1500 million years. Figure 26 shows a comparison of the development which the earth's atmosphere would have taken without life and the development which the Gaia system actually took. The already mentioned possibility of a methane/hydrogen cycle mediated by archebacteria and prokaryotes in the early oxygen-free atmosphere points to a possible evolution of earlier structures.

It seems that we are dealing here not only with the largest, but also with the most durable of all autopoietic structures on our planet. It maintains its energy exchange with the cosmos but its matter exchange with the anorganic world of the earth's surface by means of biochemical processes which are organized in highly efficient microscopic subsystems. In addition, there are matter-exchange processes among the subsystems which are obviously of

Table 3

The most important gases in the atmosphere. A dynamic perspective (throughflow per year) renders a picture which is very different from that obtained in a static perspective. Oxygen and carbon dioxide play the major role in the recycling of gases of organic origin; six other gases, however, form another group with a throughflow of about 1 per cent of the oxygen/carbon dioxide cycle, when measured in moles. (One mole is the mass of a substance which measures as many grams as the atom or molecular weight indicates. One mole corresponds therefore to the mass of 6.023×10^{23} atoms or molecules of the substance; this number is known as Avogadro's number). Note the high chemical non-equilibrium, calculated with reference to the oxygen present. Modified from Margulis and Lovelock (1974), with additional information from Margulis and Lovelock, *CoEvolution Quarterly*, No. 6, Summer 1975.

Gas	Static concentration in parts per million (ppm)	Mean residence time (years)	Chemical disequilibrium factor	Throughflow per year		Inorganic share in throughflow	
				in 10^{13} moles	in 10^9 tons	Nature	Human technology
Nitrogen (N_2)	790,000	10^6–10^7	10^{10}	3.6	1	0.001	—
Oxygen (O_2)	210,000	1000		344	110	0.00016	—
Carbon dioxide (CO_2)	320	2–5		354	156	0.01	0.10
Methane (CH_4)	1.5	7	10^{30}	6.0	1	—	—
Hydrogen (H_2)	0.5	2	10^{30}	4.4	0.09	0.00016	—
Nitrous oxide (N_2O)	0.3	10	10^{13}	1.4	0.6	<0.01	—
Carbon monoxide (CO)	0.08	0.3	10^{30}	2.7	0.75	<0.001	0.20
Ammonia (NH_3)	0.006	0.01	10^{30}	8.8	1.5	—	—
Hydrocarbons $(CH_2)_n$	0.001	0.003			0.4	—	0.50

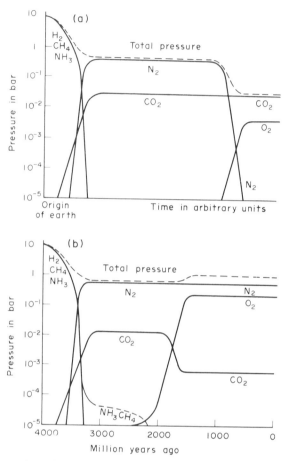

Fig. 26. Comparison of two development patterns for the earth's atmosphere: (a) under the assumption of no life on earth; (b) actual development in co-evolution with life on earth (Gaia system). After L. Margulis and J. Lovelock (1974).

decisive importance for the self-renewal and self-regulation of the Gaia system. The most conspicuous example is the closed cycle between oxiding processes in animals and plants (oxygen is transformed into carbon dioxide) and reducing processes in plants (carbon dioxide is reduced to oxygen). Are the prokaryotes no longer exclusively at work here? Yes and no. In a certain sense it is still this earliest life form which manages the entire energy household of the biosphere even when it appears in disguise. But this is a story for the next chapter. It may only be anticipated here that the descendants of the prokaryotes, in the form of parts of more highly developed cells, are still at their old job. Besides these, contemporary free-living prokaryotes, especially soil bacteria, play an important role in the management of the Gaia system.

As may be seen in Table 3, the abiological contribution of human technology has little effect, except for the increase in the carbon dioxide concentration of the atmosphere. It may be significant that the corresponding warming trend due to the "greenhouse" effect of carbon dioxide has been offset by a cooling trend in the Northern hemisphere which makes itself felt since 1945. Perhaps we are worrying too much about the distortion of equilibrium due to the impact of technology (Stumm, ed., 1977). We should worry about not distorting the maintenance of non-equilibrium. But which factors in the Gaia system easily readjust themselves and which fluctuations touch sensitive points and might lead to a new autopoietic structure—which will imply a new structure of the biosphere as well—cannot be estimated at the basis of equilibrium concepts.

It seems that Gaia has not only stabilized the composition of the atmosphere and the through-flows of gases essential for life, but also the average temperature over thousands of millions of years. Already before the appearance of free oxygen, ammonia seems to have assumed this regulatory function. Today the main role in temperature control is assigned to the infrared radiation due to carbon dioxide whose atmospheric concentration is only 0.03 per cent. According to conservative estimates, the solar radiation has increased by at least one-fourth since the birth of the earth, perhaps by even more. Correspondingly, it may be estimated that the surface of the earth should have been frozen more than 2000 million years ago. However, it was not, as is proven by the sedimentary rocks. A concentration of one hundred-thousandth ammonia in the atmosphere—kept at this level by biological action—would have been sufficient to keep the temperature above freezing point. The importance of temperature control for the evolution of life on earth is obvious. Complex life occurs almost exclusively in the narrow temperature range between 0°C and 50°C. Variations within these boundaries as they occurred, for example, in the glacial periods in higher latitudes may one day be understood as pulsations in the dynamic behaviour of the Gaia system.

In any case, however, the Gaia hypothesis makes clear what is meant by the co-evolution of macroscopic and microscopic aspects of life on earth. There are also other factors which play a significant role such as the shielding from ultraviolet radiation, inimical to life, with the liberation of oxygen—after the same ultraviolet radiation has rendered valuable services to the formation of macromolecules in the prebiotic phase. Life to a large extent indeed creates its own conditions. With this insight it is already possible to overcome Darwinism which views life one-sidely as adaptation to a given environment without feedback effects.

With the appearance of free atmospheric oxygen the microevolution of life started rolling again after 2000 million years. An entirely new type of cell emerged which made it possible to restart in a radically new way. However, the achievements of the prokaryotes were not rendered invalid, but served as a platform for the next start. The physical tasks of energy and matter exchange were not taken away from these masters or slaves of life on earth—whether master or slave depends on the angle of view. The new task of life, however, was to generate the functions for the management of higher complexity.

7. The Inventions of the Microevolution of Life

Time is invention or it is nothing at all

Henri Bergson, *L'Evolution créatrice*

Emergence of the eukaryotic cell from symbiosis

The next important evolutionary step which probably followed soon after the appearance of free oxygen was the emergence of the eukaryotic cell, the cell with a real nucleus in which the genetic material, organized into chromosomes, is aggregated. All eukaryotes depend on oxygen and are air breathing. It is assumed today that the first free eukaryotes appeared approximately 1500 million years ago, just at the time when the concentration of atmospheric oxygen reached its present value. The oldest microfossils with a cell structure resembling today's eukaryotes have been found in Northern Australia and are 1500 and 1400 million years old. These fossils stem from free-floating algae which, in contrast to the prokaryotes, did not form mat-like structures and became part of the sediments in deep-water shale. Their eukaryotic nature, however, is still subject to controversy.

The classic theory of the origin of eukaryotes is captive to a model which has been gathering dust for more than a hundred years. This model is Ernst Haeckel's theory of separately developing "kingdoms" of plants and of animals. In accordance with this theory, the same path of cell differentiation was followed in parallel developments. Although this view still represents academic conventional wisdom, it is becoming untenable. It is replaced by the still controversial theory of the endosymbiotic origin of the eukaryotic cell, developed by the same Lynn Margulis (1970) whom we already know as co-author of the Gaia hypothesis. Endosymbiosis may be viewed as fusion without total loss of the participants' identity.

The participants in this fusion were various prokaryotes, the result is the eukaryotic cell which includes the former prokaryotes, now called organelles. An important argument for this joining together of originally free-living prokaryotes may be seen in the fact that the organelles carry their own genetic material and rudimentary mechanisms for their own protein production; that is to say, their own DNA, RNA and ribosomes. In the same way as the newly appearing cell nucleus, the organelles are separated from the rest of the cell contents by a double membrane. This relatively far-reaching autonomy leads to two semantic levels, the levels of the endosymbiotic organelles and the level of the cell as a whole which co-ordinates the activity of the organelles. This maintenance of individuality and partial autonomy in a multilevel semantics is characteristic for the organization and management of complexity in life.

Endosymbiosis occurred in several steps. According to Margulis it started with the swallowing of an oxygen-breathing prokaryotic cell by a fermenting prokaryotic cell. The integrated oxygen-breathing cell is now called mitochondrion. The improved energy supply led in the next step to the incorporation of another prokaryote celle, a spyrochaeta, which had a motility system out of which were to come propulsion systems such as flagella and cilia which are all constructed after the same 9 + 2 formula (two central tubes are surrounded by a ring of nine peripheral microtubules). The same motility system (cytoplasma) also facilitated the formation of a real nucleus in which the genetic material is arranged neatly and wrapped in a membrane. With asexual cell division (mitosis) the nucleus also divides. In such a division, the motility and contraction apparatus separates the twenty-three pairs of chromosomes with great precision and assigns one-half each, representing the complete genetic information, to the newly emerging nuclei. This process needs oxygen. Finally, a third participant in this endosymbiosis, a photosynthesizing blue-green alga, becomes integrated which is then called chloroplast. Chloroplasts carry their original genetic material much more completely than mitochondria which speaks for a later integration. The mitochondria, which were probably first integrated, are already partly dependent on protein which is produced not with the help of their own DNA, but with the DNA of the cell nucleus and is "centrally" distributed. It has been shown that in the cells of the same organism, the DNA of all mitochondria is alike, a fact which also holds for the cell nuclei and their DNA. This seems to point to an adaptation over very long spans of time.

Today, the majority of all living species—green algae, higher plants, fungi, protozoa (single-cell organisms) and animals—consists of eukaryotic cells. Only this new cell type is capable of forming cell tissues and giving rise to multicellular organisms. The number of mitochondria in a eukaryotic cell runs between one and several thousand (in vertebrates), the number of chloroplasts between one (in green algae) and several hundred. The decisive point, however, is that the organelles do not simply sum up their capabilities. The eukaryotic cell represents a newly emerged level of co-ordination, a new autopoietic system level.

The presence of the organelles leads to an entirely new organization of cell functions at a much more complex level. The most important difference seems to lie in the mode of regulation (Stebbins, 1973). In the prokaryotes, groups of genes are activated or deactivated by the products of specific inhibitor producers in the cell, a process which is often disturbed by strange molecules. In the eukaryotes, in contrast, an always available basic activity is normally suppressed and becomes activated in a specific way by cancelling the inhibition for the functions in question. The difference may be compared to the difference in regulating the illumination of a house (Stebbins, 1973): Whereas candles or petroleum lamps have to be switched on and off individually, electricity is always and everywhere available in the complex wiring of a modern house. Normally, electricity remains suppressed as long as it is not switched on for a single lamp or centrally for whole groups of lamps. The brightest light of activated genes, it has recently been found, shines in the early phases of embryonic development; thereafter, genes are increasingly deactivated.

With this central control, important functions are taken away from the mitochondria and carried out by the cell as a whole. It is similar with the chloroplasts. It is the strict separation of process pathways serving photosynthesis and respiration which imparts enhanced flexibility to the green (eukaryotic) plant cells. In prokaryotes these pathways overlap.

Again the question may be raised: How did the fluctuations leading to this stepwise endosymbiosis originate and break through? The first step in the integration of a mitochondrion is relatively easy to explain. The emerging cell is as open toward horizontal genetic information transfer as a prokaryotic cell and the decisive energetic advantage of oxygen breathing has a marked impact on horizontal ecological relations. At the next step, however, the formation of a divisible cell nucleus, things get more complicated. The horizontal genetic openness of the prokaryotes comes abruptly to an end. The genetic material becomes separated by a membrane. With the asexual division of such a cell, the genetic vector is switched to practically exclusive vertical information transfer. The important factor of isolation of a development enters for the first time. But this does not mean falling back to a strictly Darwinian gene selection because in the eukaryotic cell the chromosome is no longer rigid but, like the other parts of the cell, continuously built and degraded. The chromosome field theory by Lima-de-Faria (1976), which elaborates the self-renewal of the genetic material, will be discussed in more detail later in this book. Here it may suffice to indicate that epigenetic development—the selective utilization of genetic information in feedback with the environment—enters the stage with the eukaryotic cell and shares this stage henceforth with purely genetic development.

With the orderly division of the cell nucleus, the vertical genetic vector nevertheless acquired great effectiveness. But in the early phases of epigenetics, verticality still dominated and prevented significant genetic innovations. Besides random mutations (for example caused by cosmic radiation), errors in self-reproduction and first epigenetic changes the cell plasma may have played a role in favouring mutations. But obviously not much happened. It is therefore no wonder that again a longer time interval, about 500 million years, passed before the next decisive step was taken with the introduction of sexuality. An exception is made by the integration of the chloroplasts which became the basis for the evolution of higher plants. Although here, in contrast to the integration of the mitochondria, there was no longer horizontal genetic openness, it may be assumed that horizontal genetic processes greatly favoured the emergence of a photosynthesizing match for the oxygen-breathing eukaryotes.

The more or less rigid lines of genetic evolution must have caused considerable tension in the horizontal processes of self-organizing and evolving ecosystems. These tensions contributed to the selection pressure and favoured diversifications. In this way it may be possible to explain not only the emergence of true species, but also the separation of groups of development lines into the plant, animal and fungi "kingdoms". Plants transfer four separate sets of genes—those of the cell nucleus, the mitochondria, the chloroplasts and the motility apparatus. Green plants need photosynthesizing chloroplasts besides mitochondria because they degrade the energy stored in photosynthesis in the glucose molecule in the same way as animals with the help of breathed oxygen. But they can use the resulting carbon dioxide immediately for photosynthesis, whereas animals breathe it out. Between day and night, the emphasis on the individual pathways changes. Animals transfer only three sets of genes since they are lacking the chloroplasts. The development line of the non-photosynthesizing fungi which live off organic material seems to have split from the development line of the animals at a later point in time. Figure 27 shows the scheme of evolution as it corresponds to the endosymbiotic theory of Lynn Margulis.

In spite of their partial independence, the functioning of the organelles may be interpreted in such a way that they continue the activity of the prokaryotes in managing the Gaia system. They are still playing the principal role in the maintenance of the most important of all macroscopic biochemical circular processes which links oxidation as an energy source for animals and humans, and reduction processes as vital for plants. As within the microscopic cell, it is now becoming clear in a macroscopic perspective as well that in an unfolding multilevel world the individual levels function semi-autonomously. In its atmospheric aspect, the Gaia system is primarily managed at the level of the prokaryotes. Whereas the prokaryotes became vertically organized in more complex cells and the latter in turn in multicellular organisms, they have not lost their capability to maintain exchange relationships with the environment according to their own laws. That this exchange creates the conditions for the formation of higher cells and organisms is a manifestation of systemic co-evolution.

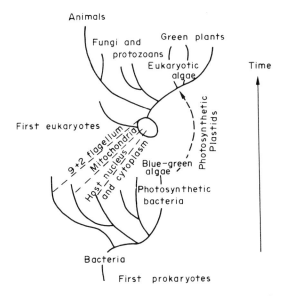

Fig. 27. Endosymbiotic origin of the eukaryotic (nucleus-possessing) cell from the joining of various prokaryotic (nucleus-free) cells. After L. Margulis (1970).

Sexuality

It is assumed that sexuality was already well developed 1000 million years ago. This, at least, is the age of spore-like cells which appear to be of meiotic origin, or in other words originated in the division of a fertilized cell, a zygote. Sexuality implies the fusion of two eukaryotic cells whereby the complete genetic material of both parent cells (one-half of each of the duplicated chromosome pairs) is united in the nucleus of the new cell. There it is decided which genes of each parent cell will dominate. The genetic material which becomes effective is not to have more genes as each parent cell—with doubling, the limits of complexity would quickly be reached. The subsequent division of the new cell transfers information which stems from two separate development lines in the past. Viewed toward the past, this scheme of unification of past experience appears like a tree and indeed we speak of an ancestral tree.

The result is extraordinary genetic variety. It is so vast that only a part of it is used in a lifetime—the rest is the "reserve" for epigenetic flexibility. The balance between horizontal and vertical information transfer now makes true phylogeny possible for the first time in evolution. Out of the *ad hoc* genetics of the bacteria in which horizontal and vertical vectors mixed in a haphazard way, and via dominant verticality (in asexual cell division or mitosis), emerged the orderly interaction between horizontal and vertical genetic vectors. If horizontal information transfer is characterized primarily by novelty, vertical transfer provides confirmation. The evolution of genetic information transfer thus brings these two aspects of pragmatic information (see Chapter 3) into a specific balance. This means a near-optimization of the effectiveness of pragmatic information. We may also put it that evolution runs in the direction of enhanced genetic autonomy, or in other words, enhanced individuality—not referring to the individual organism but to the dynamic process in which generations follow after generations.

Viewed from this angle it now becomes clear that sexuality can only represent one side of a principle the other side of which is *death* (of the individual organism) or devolution in ontogeny. It is death which forces the gene pairing in each generation. In purely vertical reproduction by asexual cell division, there is no natural death, only forced death. Amoebae do not die. The dividing cells do not age and continue to divide if the environmental conditions are favourable. The prokaryotes living today are still the same which populated the earth in the early phases of life—but they have branched into a nearly infinite multiplicity of individuals.

It is precisely environmental conditions, however, which control the population growth of the amoebae. With unicellular organisms capable of partly sexual, partly asexual reproduction, such as the green alga *Paramecium*, a cell colony ages even when it reproduces over longer periods of time in an asexual way. In the human organism there are only a few cell types left which divide. With the exception of the liver and spleen, our organs are incapable of regeneration. Those cells, however, which still divide are limited in this capability. Blast cells divide only 40 to 60 times (the Hayflick number named after its discoverer).

The invention of sexuality may, in principle, be assumed to have known primarily vertical spreading and breakthrough of the fluctuation it represented first. But in the beginning there must have been at least a whole group of cells capable of sexuality in order to ensure some variation in the genetic material. The initial pool must not have been too small if the variation is not to be left to random mutations over long periods of time.

Sexuality was one of two essential factors which resulted in an extraordinary acceleration of evolution and the emergence of a great variety of life forms. The other factor is heterotrophy or the capability of feeding on other life. It already played a certain role with early, fermenting life forms. In connection with aerobic single-cell organisms it becomes the source of a multifaceted, multilevel unfolding.

Heterotrophy—life feeds on life

Already the first prokaryotes occurred in aggregations characterized by different species. But in the higher precambrian (in the early period of the prokaryotes) these aggregations were dominated by a few types of photosynthesizing autotrophs (organisms feeding themselves), especially by blue-green algae. At the higher level of eukaryotes, however, evolution became accelerated by sexuality which introduced systematically enhanced variety. This was about 1000 million years ago.

A little later (800 million years ago) a second boost of evolution occurred, this time due to the massive and orderly appearance of heterotrophy. Heterotrophy signifies the capability of an organism to live off other organisms—be they plants or animals, or unicellular organisms. At today's stage of evolution, practically only plants are autotrophic, except for fermenting bacteria (such as yeasts), whereas all animals are either herbivores or carnivores or (as most humans) both. The secret of autotrophic organisms, of course, is primarily photosynthesis which permits the direct conversion of solar energy and thereby the synthesis of organic molecules from inorganic materials (lithotrophy). It was already the secret of the types dominating the early microbiotes. Photosynthesis is so superior to fermentation of organic materials from the environment that the photosynthesizing species spread more easily than others. But preferred quantitative spreading was all. The photosynthesizing blue-green algae apparently did not have that horizontal gene transfer which resulted in the *ad hoc* appearance and disappearance of millions of pseudospecies of bacteria.

It seems that with variation by sexuality some species lost their capability of photosynthesis. This, at least, is the way in which the appearance of heterotrophic organisms is explained which, at first, probably were feeding on everything in sight, especially on organic waste and smaller living cells. Later, the heterotrophs specialized and split into herbivores and carnivores. With the appearance of heterotrophs, the monopoly of photosynthesizing autotrophs for the best places in microbiotes was broken. The heterotrophs made room. In this way only higher complexity in the ecosystem became possible.

Sexuality and heterotrophy go hand in hand in the development of variety. Ecology, which has really started only with the appearance of the heterotrophs, gets into motion and develops in the direction of greater complexity. A large variety of multicellular organisms appeared. This development started probably more than 750 million years ago, not long after the entry of the heterotrophs, with small worms. The beginning of the fossil-rich cambrian age, 580 million years ago, is characterized by the appearance of numerous species of invertebrates and their prey, multicellular algae, within the relatively short time span of 10 to 20 million years—a spectacular evolutionary explosion. Life had finally created the macroscopic and microscopic conditions for the unfolding of that wealth of forms and relations which surrounds us today and of which we ourselves are part. These conditions were first realized in water which probably to some degree had to do with the shielding from ultraviolet radiation. Only about 450 million years ago, the colonization of land by plants started which was made possible by the ultraviolet shielding effect of atmospheric oxygen. But this can hardly have been the triggering factor since free oxygen occurred much earlier already. A little later, about 400 million years ago, animals followed the plants on to the land.

The urge toward multicellularity

The formation of tissue, which is possible only with eukaryotic cells, is obviously closely linked with the aggregation of unicellular organisms. The origin of multicellular organisms is still very obscure. There are theories which hold that the first multicellular organisms arose from the differentiation of one and the same cell. This is not so far-fetched since the embryo also develops from the division and differentiation of a single cell, the zygote. But even today there are so many intermediary steps on the scale of societal organization from unicellular to multicellular organisms that a horizontal joining together—an endosymbiosis as it characterized the origin of the eukaryotic cell—may have played an important, if not unique, role. In this connection, Eigen and Schuster (1977/78) suspect another level of evolving hypercycles with eukaryotic cells as participants.

The communication mechanisms between cells which are joined to form colonies seem to be in principle the same as the mechanisms guaranteeing the co-operation of cells in tissue. They transfer metabolic information and are based on chemical and electric (ionic flow) processes. Cells recognize each other and associate with their own kind.

If sponges of different kinds, for example yellow and orange ones, together with water are put into a blender and transformed into a homogeneous suspension which is then left standing, yellow and orange sponges will have formed again after a few hours. Even more interesting is an experiment with small fresh-water polyps, which are 5 millimetres long. If they are mechanically dissected into single cells, of which they have about 100,000 belonging to a dozen cell types, they try to rearrange themselves, first into cell clumps, and later to monster formations in which head, gut and foot regions grow in greatly disordered ways. Eventually, normal animals will emerge from such tissue formations. Apparently, certain substances which exchange morphogenetic information in the cellular tissue first act in disordered ways. It was possible to isolate such a substance of which very low concentrations activate the formation of tentacles in the head region (Gierer, 1974). If this substance is put in contact with cells of the gut tract, the formation of tentacles starts there also in contrast to normal development.

In a mixed gruel of living cells, liver cells join with liver cells and retina cells with retina cells and they will try to reconstitute the organs from which they were taken. In certain cases, even specialized cells co-operate which fulfil the same function in different kinds of organisms. Like biomolecules which recognize each other's spatial form through stereospecificity, cells, too, have at their level a recognizing memory which, however, seems not to be geared to spatial form only, but also to exchange processes, in other words to the dynamic process structure of the system. A new type of communication, *metabolic communication,* comes into play. It underlies the formation of auto-poietic units which consist of cell populations—whether these units are multi-cellular organisms or cell aggregations.

The process nature of the criteria for recognition becomes clear with the example of the slime mold (see Chapter 4). Such a slime mold is a temporary aggregation of amoebae, bacteria-eating eukaryotic protists. In periods in which it is easy to obtain food, they function as independent individuals. In scarce periods, however, the steady secretion of cheomotactic substances occurs in characteristic pulses which originate in dissipative structures within the cells. In this way they attract each other irresistibly. The resulting slime mold moves along the earth in search of more favourable feeding places and disintegrates after 2 or 3 days into individual amoebae which reproduce by cell division.

There is an interesting parallel to the behaviour of a specific human society (Friedman, 1975). In good times, the Burmese mountain people of the Kachin live in several separate tribes which maintain trade relations with each other, but are politically independent. In scarce period, however, when the harvests are poor, they spontaneously form a hierarchic order in which the chieftain of one tribe rules as king over the entire Kachin people. Each of these alter-nating phases usually lasts for decades. Such an alternating societal system seems to have characterized the Kachin for many hundred years. It represents a more or less historyless dynamics which orients itself according to the hori-zontal relations and shows little vertical development.

But back to the efforts of unicellular organisms to form societies. Apparently only photosynthesizing eukaryotes succeed in forming durable societies. They may link up with other cells by fine protoplasma threads and thereby achieve a communication system which acts much faster and is much more dependable than, for example, mere chemotaxis as exhibited by the amoebae. Perhaps it is at this point where the beginnings of multicellular organisms may be found. Among the species existing today, many steps may be studied in this development, from totally uniform behaviour of all indivi-dual cells within a colony to societies with marked leadership functions and even further differentiation.

In her extremely interesting book, Larison Cudmore (1977) describes as the crowning of this development the green alga *Volvox*. This unicellular organism forms spherical colonies with 500 to 500,000 cells each which move along with the co-ordinated beat of their flagella. Each cell is connected with six others by protoplasma threads. In the Southern hemisphere of this spherical colony, a few cells (between 2 and 50) are chosen to form new colonies. They may reproduce by simple cell division or sexually. In the former case, a new sphere forms within the original sphere with cells which do not yet have flagella and the "heads" of which are turned inward. At the proper time something happens which looks like birth. The inner sphere turns inside out and at the same time leaves the mother colony by a pore in its wall. Arriving outside, with heads now turned outward, flagella grow on the young cells and the newly born colony swims away to its own life. But the new colonies cannot be born until the time has come for the mother colony to die. In this way it may happen that the daughter colonies in turn develop daughter colonies even before they are "born". In such a case, three generations exist simultaneously, but only the oldest one lives in freedom. The reproduction cells may also form male and female sexual cells (gametes). The male cell may break loose from the colony and seek a female one. Only after fertilization does the female cell break loose, too, and develop a hard and spiny protective coat. The colony which forms beneath this coat remains immobile on the bottom of the fresh-water pond until spring when it unfolds its life to the fullest extent.

In the formation of multicellular systems, just as in the formation of dissipative structures, the dimensions of the system seem to play a decisive role. Cells of an organism differentiate into cell cultures only when they find themselves in an aggregation of a certain minimal size. In the normal development of the embryo, the differentiating cell mass grows to a certain size. Then polarization sets in (for example, the formation of a head) and from then onward there is a "positional field" which determines the further development along the already mentioned chreods (Nicolis and Prigogine, 1977). With polarization a new level of self-organization seems to be reached which is visibly characterized by a break of spatial symmetry. The complementarity of stochastic and deterministic factors, of novelty and confirmation, reappears at a new level. The development along chreods is regulated by regulator genes which, it is assumed, represent up to 95 per cent of the genes in complex animals. Their effect resembles the neural mind which will be discussed in Chapter 9. There is a continuous supply of concepts which are tested as to their compatibility with other concepts, mutations are reversed or fitted and so forth. In short, a learning process unfolds which is partly open and partly heuristic. As Wolfgang Stegmüller (1975) emphasizes, it is especially the capability of anticipating not yet realized processes which plays a role in this development.

The difficult balance between novelty and confirmation

Recapitulating the most important stages of the microevolution of life, we obtain a picture according to Table 4. In the more than 3000 million years before the appearance of the first multicellular organisms, three main levels of autopoietic existence appear: dissipative structures, prokaryotes and eukaryotes. In macroevolution, however, the identification of autopoietic levels is more difficult. Nevertheless it seems that the prokaryotes are matched on the macroscopic branch by the autopoietic Gaia system which embraces the entire biosphere of the prokaryotes together with the atmosphere and has stabilized itself for 1500 million years. The appearance of the eukaryotes finds its macroscopic correspondence in the formation of ecosystems which become more complex with multilevel heterotrophy. With multicellular organisms, finally, complex societies with a division of labour appear on the macroscopic branch. Figure 28 gives a scheme of the co-evolution of macro- and micro-cosmos in the evolution of life on earth.

Table 4

Steps in the unfolding of life on earth

Time past (in million years)	Event/Evidence
4600	Origin of the earth
∿4000	Origin of life; common ancestors of prokaryotes and archebacteria?
3800–3300	Oldest biogenous carbon deposits (from photosynthesis)
3500–3100	Oldest microfossils; prokaryotes (bacteria and blue-green algae) dominate; anaerobic photosynthesis; environmental evidence for aerobic (oxygen-producing) photosynthesis
3200–2900	Oldest stromatolites (deposits from prokaryote aggregations)
2350– 725	Frequent appearance of stromatolites, proterozoicum
2200	Microfossil evidence for aerobic photosynthesis (heterocysts)
2000	Appearance of free atmospheric oxygen
2100–1900?	Respiration
1500?	Atmosphere (Gaia system) stabilized
1500?	Eukaryotes, mitosis (cell division)
1200?	Sexual reproduction, meiosis
800	Widespread heterotrophy
750	Microfossil evidence for multicellular organisms
580	Beginning of the cambrian period with numerous fossils of invertebrates
500	First vertebrates
450	Colonization of land by plants
400	Colonization of land by animals
370	First amphibians and winged insects
330	First trees and reptiles
200	Beginning of the dinosaur period
165	First mammals
130	First flowers
64	Dinosaurs become abruptly extinct
50	First real primates (predecessors were contemporaries of dinosaurs)
14	*Ramapithecus*, first erect ancestor of man

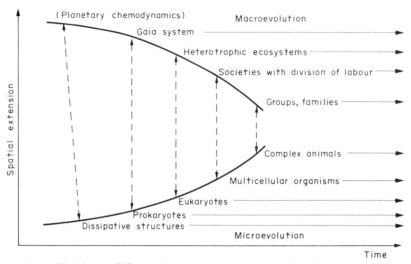

Fig. 28. The history of life on earth expresses the co-evolution of self-organizing macro- and microsystems in ever higher degrees of differentiation.

If we now ask what primarily characterizes the three microscopic auto-poietic levels—dissipative structures, prokaryotes and eukaryotes—we may recognize important symmetry breaks in the transitions between them. As they did in cosmic evolution, symmetry breaks unfurl space and time for a specific system dynamics. Dissipative structures mark a break in spatial symmetry (see Chapter 1). With prokaryotes, or even with their predecessors, the hypercycles according to Eigen, self-reproduction appears for the first time, implying vertical information transfer. In this way the time symmetry of experience is broken. Experience is no longer symmetrical to the acquisition of this experience in the present and in the exchange processes with the environment. The experience from the past may now become equally effective in the present. Thus, a special kind of time-binding is the correlate of this symmetry break. However, vertical genetic information transfer is mixed at this level with direct horizontal transfer. Whereas the time symmetry of genetic information transfer is broken, the spatial symmetry continues to be maintained at first. It is broken only by the eukaryotes which, on the basis of sexuality, develop the systematic inclusion of experience from the past which has accrued in all branches of an ancestral tree. Information cannot stem from the total gene pool as it may with the horizontal gene transfer of the bacteria, but is derived from a specific structure of past experience formation—the ancestral tree. With this, a two-step spatial symmetry break is completed at the level of the eukaryotes which has led through the linear principle of more or less identical generations in mitotic cell division to the principle of the ancestral tree in sexuality.

It is conspicuous in this development that important inventions of micro-evolution, such as respiration with the prokaryotes and sexuality with the eukaryotes, have to be "worked out" in steps. Symmetry breaks may become modified. Evolutionary mechanisms first put forward a one-sided principle "for discussion", which eventually becomes refined and is brought into contact with other principles. One may express this in such a way that the completeness of the complementary process principle is not always attained at the first attempt and has to be put straight by somewhat clumsy dialectics.

The principle of self-reproduction, which is of decisive importance for life, is not based on matter transfer, but on information transfer. To be precise, it is process programmes which are transferred and which provide guidance for the formation of structures—not only material structures but also structures of relations and processes, in other words dynamic space-time structures. This seems self-evident. But in cosmic evolution it was the direct transfer of matter which had the effect of a certain space- and time-binding in evolution. In parti-cular, as has been discussed in Chapter 5, the matter of which our planet and the life forms which have developed on it consist does not stem from the homogeneous primary matter and also not from the sun. In comparison with cosmic evolution, life has ensured for its evolution an extraordinary measure of flexibility. It hands on process programmes which locally may be used in creative ways in the exchange of an autopoietic structure with its environ-ment. This openness manifesting in the exchange processes with the environ-ment is even the prerequisite for replacing matter by information in phylogeny. If an architect sends the plan of a house to a far-away building place he must first be sure that the local conditions and the relations to the near and far environment make it possible, for example to find clay and produce bricks, to engage skilled masons, to import marble from Italy and to have sufficient water available for the garden.

In the evolution of this information transfer the task is now obviously to find the right balance between novelty and confirmation, the complementary aspects of pragmatic information. With the prokaryotes this balance is still found in an *ad hoc* mixing of vertical and horizontal vectors—a truly anarchistic solution which, in principle, makes every combination possible. Only with the eukaryotes did some order enter, first in the form of excessive emphasis on confirmation and linear self-reproduction by cell division. The beginnings of epigenetic feedback with the environment and the remainders of horizontal gene transfer (perhaps only by means of viruses) introduced probably little novelty. The right balance was found only with sexuality. And indeed, an extraordinary acceleration of evolution followed and led to an explosion in the variety of life forms.

Relatively soon after the introduction of sexuality, an intricate web of ecological processes started with multilevel heterotrophy. This web is basically not of a material nature, but also represents energy and information transfer. There is indeed a material event when a predator eats his prey. But the matter which is quickly ejected really does not count; what counts is the energy in the form of food. When insects transfer plant pollen, this process also is basically not a material one but represents the transfer of genetic information. The domain of ecological processes—we may also say, of ecological information—at first shows a strong horizontal orientation and thus emphasizes novelty before confirmation. This, too, leads to quickly expanding variety. With the increasing importance of epigenetic development in multicellular organisms, however, the horizontal vector combines with a vertical one, novelty combines with confirmation. Organisms may adapt better to the environment on the one hand and drive the evolution of that same environment on the other.

If the microevolution of life starts with an emphasis on confirmation, the macroevolution starts from the other end, from novelty. From both sides, the balance increases. The resulting optimization of balanced pragmatic information may be called the real triumph of the principle of co-evolution of macro- and microcosmos in the realm of life.

8. Sociobiology and Ecology:
Organism and Environment

> Father: (Life is) a game whose purpose is to discover the
> rules, which rules are always changing and always
> undiscoverable,
> Daughter: But I don't call that a *game,* Daddy.
> Father: Perhaps not. I would call it a game, or at any rate
> "play". But it certainly is not like chess or canasta.
> It's more like what kittens and puppies do. Perhaps.
> I don't know.
>
> *   *   *
>
> Daughter: Daddy, why do kittens and puppies play?
> Father: I don't know—I don't know.
>
> Gregory Bateson, *Metalogue:*
> *About Games and Being Serious*

A clarification concerning the terminology

In the middle of the 1960s the notion of ecology entered the consciousness of a broad public, and a decade later the narrower notion of sociobiology followed. Both terms were originally conceived in the framework of stationary steady-state thinking. Both sociobiology and ecology, as they were recently understood, are dealing primarily with horizontal relations among biological organisms and groups. Both, however, lead to the principle of co-evolution if applied in a dynamic context.

The term sociobiology is often used in such a way that it only addresses the genetic origin of behaviour (Wilson, 1975; Barash, 1977). Such an unrealistic limitation is not meant in this book. I understand the notion of sociobiology in such a way that it covers all co-operative material processes among biological systems of the same kind whether these processes take part between individual organisms or between groups and systems in which they participate. The restriction to processes of a material nature is essential; it marks the boundary between sociobiological and sociocultural processes which will be

135

discussed in the following chapter. The much wider notion of ecology, in contrast, includes all co-operative processes which take part in a self-organizing system which in turn is composed of biological self-organizing systems. The term "co-operative" is viewed here from the angle of the overall system and includes competition and predator-prey relationships although it would be difficult for the affected individuals to recognize the co-operative aspect. Both sociobiology and ecology are thus characterized by a minimum of two semantic levels, the level of the individual organisms on the one hand and the level of the macrosystem on the other. The number of semantic levels may, of course, be higher than two if the macrosystem differentiates further.

Sociobiology in a real sense starts with the division of labour among the participants of an aggregation. In spite of horizontal gene transfer, the world of bacteria is not yet truly sociobiological. The first approaches to specialization and division of labour may be found with colonies formed by eukaryotic cells. But only multicellular organisms are sociobiological to a full extent.

Ecology in a broad sense starts with the relations within the Gaia system of bio- plus atmosphere which are carried by the prokaryotes. It becomes quickly more complex with the addition of symbiosis and heterotrophy. In symbiosis, the advantage of co-operation between two organisms lies in the improved viability of the emergent total system which represents a higher level. Heterotrophy builds upon a first autotrophic level which obtains energy from photosynthesis. The herbivores already represent a second level, whereas carnivores bring a third and possibly further levels into play.

Optimal utilization of energy

Nothing demonstrates the process nature of life more impressively than the energy cycles in an ecosystem. The primary energy is practically exclusively derived from solar radiation. It enters as visible light and leaves as long-wave heat radiation of low photon energy. The boundaries of an ecosystem may be meaningfully defined in such a way that they enclose all energetic processes from entry to exit of the photons. On the average, only 1 per cent of the involved solar energy is transformed by plants in photosynthesis and stored. The highest efficiency is approximately 3 per cent. The rest is either reflected back or rejected as heat in the autopoietic exchange processes of the plant with its environment.

In an ecosystem with several trophic levels (levels of food hierarchy) the 1 per cent of stored solar energy is used by the herbivores to aliment two different processes. One process serves the build-up and growth of the animal body, or in other words, the storage of energy, which again is subject to a relatively low efficiency, again of the order of 1 per cent. Almost all the energy, however, is used in metabolism, the direct exchange processes in autopoiesis. An adult human takes in an average of 2700 calories per day, which corresponds to 3 kilowatt-hours, but rejects almost all of it in a steady heat flow to the environment measuring about 100 to 150 watts. Taking the lower heat production in the infant stage into account, during the first 20 years of human life 15,000 kilowatt-hours of heat have found their way from the body to the environment and ultimately into cosmic space. But the 70 kilograms of flesh, fat and bones for the benefit of which all this effort has been undertaken, represent hardly more than 100 kilowatt-hours which may be available to a predator animal. In the later years of life, the growth of mammals (in contrast to fishes, for example) stops altogether; all energy is used up in the autopoietic metabolism.

The hierarchy of trophic levels thus forms a pyramid which from level to level transfers only about 1 per cent of the energy. One square mile (about 2.5 square kilometres) of grassland feeds a herd of 100 gazelles which, in turn, feed a single lion. The lion's share of the originally involved solar energy is only 1 per cent of 1 per cent, or a hundreth of a per cent. The rest serves the autopoiesis of the grassland and the gazelles.

Before the invention of agriculture, when man lived as a hunter and food collector, he needed approximately 10 square kilometres *per capita* for his own nourishment. Today's agriculture has lowered this measure to the order of 1 hectare (one-hundreth of a square kilometre) and is thus a thousand-times more efficient. In cases of most intensive agriculture only one-third of a hectare *per capita* is sufficient. Accordingly, the world population, which had become stable at about 5 million around 10,000 B.C., grew with the introduction of agriculture to 100 million by 3000 B.C. and to 500 million by the end of the eighteenth century. The industrial era with its intensified agriculture increased this figure by a further order of magnitude. Thus, the earth is carrying and nourishing today a thousand-fold the human population which existed before agriculture.

Plants are capable of producing all twenty amino acids which are necessary for the formation of proteins, heterotrophic animals only about half of them. The loss of the capability to synthesize all essential intermediary products is suggested to have resulted from the selective advantage of energy saving which may be used for other functions of the body. The acquisition of ready-made products turned out to be cheaper—as it also is in the later industrial era of evolution. The loss of autonomy, however, tends to create problems which become bothersome only at much later stages. This is the case with ascorbic acid (vitamin C). It is assumed that the capability for its synthesis emerged in amphibians more than 300 million years ago, but that a mutation about 25 million years ago caused animal species to lose an enzyme which is needed as catalyst in the last step of the transformation chain from glucose into ascorbic acid (Scrimshaw and Young, 1976). The economized glucose benefits the body directly, but humans, primates, guinea pigs, bats and also some species of birds have become dependent on a diet which contains ready-made vitamin C. In a tropical environment there is no problem. But with the migration into moderate climatic zones, and especially with the switch from raw to cooked food, a situation of chronic deficiency arose which made the body much more illness-prone (Pauling, 1977).

Ecosystems may be stratified in more than three trophic levels when longer food chains develop. This is the case in the oceans where the average length of the food chain is 3.5 to 4, not counting the plankton which utilizes primary solar energy. Fish at the end of the food chain, such as tuna, thus utilize less than a millionth, or even a thousandth of a millionth, of the original solar energy. It is significant, however, that terrestrial ecosystems rarely transcend four trophic levels—plants, herbivores, primary and secondary carnivores—with multiple overlaps such as by omnivores.

Only a part of the energy stored in organisms, however, finds its way to higher trophic levels. In forests, those richly structured ecosystems, only about one-tenth of the primary production reaches the herbivores. Ninety per cent is utilized directly by the "decayers", i.e. bacteria and fungi, and is broken down into the original chemical elements in order to serve new energy cycles. In one square metre of fertile earth, there is approximately 1 kilogram of soil bacteria in which we may recognize the contemporary version of the earliest life on earth, the prokaryotes. They also become active at several points in the energy cycle at which problems arise for various reasons. Plants, for example, need nitrogen for the synthesis of proteins, but cannot use the gaseous form of nitrogen which is so abundant in the air. They require nitrogen in bound form, such as nitrates and nitrites in the soil. Leguminous plants, however, which include pulses, clover, alfalfa and lupines, make use of symbiosis with soil bacteria owning the necessary genes to build nitrogen-fixating enzymes. These genes commence action only when the plants have fulfilled their side of the symbiotic pact, namely, to ensure an oxygen-free environment in their roots. The way in which the leguminous plants fulfil this obligation anticipates a development which is only to again play a role with vertebrates. They produce haemoglobin, the basis for blood and an effective means for the binding of oxygen. A major task of blood is oxygen transport.

The largest part of carbon which plants use for their energy storage is bound in the form of cellulose. Only a few types of bacteria are capable of producing the enzymes which are needed to break down cellulose. Humans have such bacteria at the rear end of the digestive tract where they are of little use. Cattle, sheep, goats and deer, however, have large colonies of these bacteria at the beginning of their digestive tract and keep them busy with the transformation of cellulose to proteins and nucleic acids.

The relationships described here are of a symbiotic nature. Symbiosis may be understood as the intensification of environmental relations by process links between two or more organisms. As with the differentiation into several trophic levels, here also the result is a multilevel semantics. The individual organisms do not lose their identity and yet the symbiotic relationship establishes an autopoietic unit of higher order. The most frequent and successful example for symbiosis are the lichens which are the most widespread form of life. Lichens are nothing but symbiotically linked single-cell algae and fungi which, under favourable conditions, would also be capable of existing on their own. Whereas the alga takes care of the energy supply by means of photosynthesis, the fungus contributes water, carbon dioxide and a firm grip. But perhaps we may recognize here the continuation of an old passion of unicellular microorganisms, namely the massive transformation of the earth's surface in order to make it more viable. Lichens are an important factor in the transformation of rock into earth, of dead equilibrium structures into a substrate for life.

Macrodynamics of life

Following the entire biosphere, the highest macrolevel is the level of so-called *biotic* or *biogeographical provinces,* of which appproximately two hundred have been identified (Dassman, 1976). They are primarily determined by climatic and geographic factors. An example is the Central European Forest which stretches from West Germany to the Black Sea and in a small tongue further to the Urals. Adjacent biogeographical provinces in many cases share only little of their fauna and flora. I should like to think that they qualify for a separate level of autopoietic macrostructures. This is also suggested by the intereaction of human civilizations with such provinces.

Biogeographical provinces are often settled by different civilizations in succession and transformed in totally different directions. The grassland, for example, which today gives the American Middle West its character, has replaced the original forest. The drought in the African Sahel, which already lasts for several years, has been traced back to changes in traditional behavioural patterns by the advent of technology. Such biogeographically interpreted cultural zones are called *bioregions.* As long as everything goes well, they may be regarded as evolving structures which are forced into new dynamic régimes by ethnic fluctuations. Sometimes, however, they are ruined. This may happen especially when culturally fixated relationships with the environment are transplanted from one biogeographical province into another, totally different, one. In this respect, nature and culture still meet as opponents and one-sided confirmation of culture may turn against life. Here, too, the aim must be to strike a balance between novelty and confirmation which furthers the co-evolution of both sides.

Ecosystems function undoubtedly as autopoietic structures. They clearly show the three characteristics which already hold for chemical dissipative structures. There is openness with respect to the energy exchange by means of solar and heat radiation; autocatalytic steps characterize processes at all levels of life from intracellular biochemistry to the reproduction of organisms; and rich differentiation, and thus high non-equilibrium, corresponds to a basic trend exhibited by all ecosystems within the limits of the possible. The maintenance of non-equilibrium may be observed even with very simple ecosystems. In plankton ecosystems, for example, structuration into a honeycomb pattern is the eventual result, with a region of low diversity surrounded by regions of high diversity (Margalef, 1968). The movements which lead to this structuration are partly caused by winds and partly by the activity of the microorganisms themselves. A pattern of hexagonal cells—how could one not think of the Bénard cells in fluid dynamics which have been mentioned in Chapter 1!

Of special interest are ecosystems whose evolution may be observed within long periods of time. The biomass, i.e. the mass of all life in the system, increases, usually also the primary production from direct photosynthesis (plants) as well. But these two factors do not increase at the same ratio. With the formation of a complex system of trophic levels the energy derived from the same primary production may be handed on from level to level, even if only to a relatively small extent. The total system's efficiency of energy utilization increases and so does the total energy stored in the system. Diversification also increases in typical cases. In African biotopes the biomass of diversified, free-living hoove animals reaches 15 to 28 tons per square kilometre, fifteen-fold the biomass of deer in medium Alpine ranges and five- to ten-fold the biomass attained by African cattle raising. Only European cattle raising, at 20 to 22 tons per square kilometre, comes close to this natural efficiency, due to the intense use of industrial fertilizers (Zwahlen, 1978).

Young ecosystems or ecosystems of low "maturity" are characterized by species which produce a large number of offspring in each generation, of which, however, only a small part survives and reproduces. In "mature" ecosystems, in contrast, those species which produce only a small number of offspring (such as the mammals), dominate but they take good care of these offspring and ensure their survival to a high degree. Therefore, the selection pressure acting from the environment is much more effective in immature systems and accelerates their evolution. In mature systems, the fluctuations originating in the system itself play an accelerating role in evolution. This may be interpreted as a continuation of that development toward higher autonomy which has started with the appearance of death as a correlate to sexuality. It is interesting that the complexity of the mature system, expressed by the number of participating species, generally remains the same, even if the species themselves come and go due to emigration, extinction and other factors.

In this dynamics the principle which has already been formulated for chemical dissipative structures may be recognized. At first, high energy penetration and maximum entropy production act as stabilization criteria whereas after the establishment of a basic structure there is a gradual shift toward a criterion of minimum entropy production per unit of mass.

If we consider the ecosystem as a matter system its first concern is with the present in view of the organization of energy flows, whereas the evolutionary aspect emerges at a later stage. The more complex the system becomes, the larger the share of the energy throughflow which, at a given moment, is stored in the system (Morowitz, 1968). It is the structures which are capable of influencing their own future with the least energy that have the least difficulties in evolving.

If, however, we consider the ecosystem as an energy system which manifests itself in the organization of matter, maximum "engagement" in matter (i.e. energy storage) and maximum process intensity (i.e. entropy production) are the criteria for optimal stability. This may explain to some extent why the most differentiated and mature ecosystems occur at high temperatures. According to the equation for the total energy $E = F + TS$ (see Chapter 1), the entropy S is multiplied with the absolute temperature T, resulting in a higher share of entropic energy at higher temperatures. The most mature ecosystems occur in the tropics, namely, coral reefs and tropical rain forests. But there are also particularly mature ecosystems at low temperatures in the deep sea and in caves.

In a sociogenetic view, finally, the ecosystem appears as an information system which, through suitable interactions and corresponding behaviour of the participating organisms, ensures the continuity of the genetic information potential. Its thrust is expressed by the slogan "the selfish gene" (Dawkins, 1976).

The angle of view from which an ecosystem appears as a matter-organizing energy system emphasizes the aspect of novelty, the reverse angle of view (an energy-organizing matter system) emphasizes the aspect of confirmation, which also dominates the information system. The flora seems to correspond more clearly to the energy and information system aspect than the fauna. Plants do not need the same material continuity as animals. Trees hibernate without leaves and with greatly reduced metabolism. In periods of drought many plants retreat from the exchange with the environment but maintain their chances of a quick come-back with changed environmental conditions by forming extremely resistant information-storage facilities (seeds, pollens, spores). The information stored in them may be activated after decades. Deserts start to bloom with the first rain fall and the "golden" (i.e. dry) hills of California immediately turn green with the first autumn rains.

Animal populations do not follow equally closely the seasonal and other environmental oscillations. They have a higher degree of autonomy. Accordingly, such notions as autonomy, matter-emphasizing system and confirmation—but also higher vulnerability—would be linked in some way. Life which has become emancipated from the environment to a higher degree may generally not retreat from the environmental dynamics (with exceptions such as hibernation and larval stages) but has to continuously accept the challenge of physical unfolding.

Margalef (1968) has described the evolution of an ecosystem as process of information accumulation. Information is not only generated by the differentiation of the participating species and the structuration of their life processes, but also in the establishment of paths, burrows, signals and other physical structures which result from multiple confirmation of life processes. The environmental information gained by the system is subsequently applied to the acquisition of higher autonomy and thus, paradoxically, to the partial blocking of further information intake from the environment.

Novelty is continuously transformed into confirmation, as in every life process. Novelty alway enters with fluctuations which break through. However, it is decreasingly environmental influences which dominate but the evolutionary self-organization dynamics of the system itself. We may recognize again the urge toward higher autonomy which may be interpreted as an urge toward higher consciousness.

In mature ecosystems the fluctuations arriving from the outside, such as climatic oscillations, become increasingly damped. Instead of a rhythm dictated by reactions to environmental events, the endogenous rhythm of the system unfolds to an increasing extent. A mature ecosystem, even in the tropics, is not a confused tangle of wild growth, but incorporates a very fine order. The richly orchestrated vibrations of this order may be felt strongly and clearly before the outbreak of a tropical thunderstorm. The "law of the jungle" is a very high law.

The Canadian ecologist Holling (1976) has vigorously emphasized that healthy, resilient ecosystems are those which live with high local fluctuations. This may already be recognized in the complicated, wave-shaped boundaries of mature ecosystems. In Chapter 4 examples were presented for the ways in which a management geared to equilibrium may ruin ecosystems.

Nature, however, always has fluctuations in store which may shake up from the outside even the most autonomous ecosystem. According to recent concepts, the sun is not nearly a steady-state, but a globally oscillating system (Eddy, ed., 1978). But these fluctuations do not only affect the earth's climate (e.g. through glacial periods). The magnetic field of the earth, too, is subject to oscillations. This is known because sedimentary rock keeps the magnetization impregnated by the magnetic field of the earth at the origin of this rock (Cox, 1973). Already in 1906 the French physicist Brunhes pointed out that certain minerals are reversely polarized. Around 1960 further effects of the same type were discovered in volcanic rock formations the precise age of which is known. The world-wide agreement among these data permits us to conclude with a high degree of certainty that the magnetic field of the earth has been repeatedly reversed at least over the past 5 million years. The present polarity has existed only since 690,000 years (Lowrie, 1976). The reversal of the magnetic North and South poles does not occur in an instant. The magnetic field becomes weaker over a period of perhaps 5000 years and rebuilds equally slowly with reversed poles. This, however, implies that the earth is subjected during this period of a weakened or absent magnetic field to the hard solar radiation (the "solar wind" remembered from lunar experiments) and cosmic radiation to a much higher extent. Within each epoch of polarization there seem to be shorter intervals in which the magnetic fields becomes reversed. Apparently very small shifts in the outer earth core are capable of reversing the dynamo properties of the rotating earth.

In evolution, many things must have come into motion with such fluctuations. A possible correlation between the reversal of the magnetic field and the extinction of certain species of the ocean fauna has been found. But extinction is hardly the only effect. Sometimes I cannot help imagining a powerful Mother Evolution stirring the soup in the pot of life with no other purpose in mind than to keep things moving and thereby stimulate innovation. Novelty may break into life at many macro- and microlevels. On a grand scale, the galactic shock-wave triggering every 100 million years nearby supernova explosions (see Chapter 5) is bound to leave its mark on evolution. Some scientists make it responsible for the sudden extinction of the dinosaurs 64 millions years ago—and thus for a decisive stimulation in the development of the mammals.

Evolution is certainly not the "game with sticky cards" as the British philosopher Jacob Bronowski (1970) has described it. According to his scheme, with every shuffling of the cards more of them would stick together and ultimately all cards would form a solid block. This is pure equilibrium thinking. If something happens in nature, for example in a mature ecosystem, there is soon a new dimension of openness or an entirely new level of evolutionary mechanisms at which true autopoiesis in high non-equilibrium may become re-established. Perhaps the natural macrofluctuations serve basically the maintenance of openness. Life as a total phenomenon has never been seriously threatened by them.

The feedback loop between organism and environment— epigenetics and macroevolution

With the advent of sexuality, the genetic information transfer is ensured by vertical and horizontal vectors. The horizontal genetic vector takes care of mixing the genes from an entire ancestral tree from the past. Phylogeny, however, is not only the result of genetic information transfer, but also, and very importantly, of ecological conditions. Each generation of a phylogeny becomes involved in horizontal processes which play in the ecosystem. Microevolution in phylogeny and macroevolution in the history of ecosystems and the entire biosphere enter in a mutual relationship. In this relationship there is never full adaptation, never perfect equilibrium. Just as non-equilibrium forms a prerequisite for the self-organization of dissipative structures, it does so also for the co-evolution of living systems. The tension which acts between a multiplicity of phylogenies and the history of the ecosystem becomes, in a long-range perspective, the creative drive of co-evolution.

The principle which has already been pointed out in the history of the prokaryotes acts also at other levels. Life itself creates the conditions for its further evolution. Four hundred and fifty million years ago, plants started to colonize land and preceded animals by about 50 million years—they had to be first since only they understand how to harvest solar energy. Today, plants also act as pioneers in the settling of new land; newly emerging islands, for example. Sometimes, they even create this new land in an unselfish way. On the west coast of Florida, for example, huge colonies of red mangroves growing in shallow water lead to the formation of new islands. Thus, they generate the conditions of life for several species of pioneer plants including another type of mangrove. But while the young ecosystem differentiates quickly, the red mangroves become extinct; they only grow in water. A similar role in other biotopes is played by papyrus plants. Again, the ultimate principle of evolution does not seem to be adaptation, but transformation and the creative diversification of evolution. Similar to the death of individuals, the death of whole species in ecosystems, too, furthers evolution.

In the equilibrium thinking of Darwinism the interaction between macro- and microworld focuses on the genetic adaptation of microevolution in small steps to an environment the origin and changes of which remain outside this consideration. There is no feedback link between organism and environment. The subject of Darwinian selection is the phenotype (the individual organism) which, by virtue of its morphology—a spatial structural concept!— attains advantages in the prolongation of its life and thereby in the production of a larger number of offspring. Neodarwinism uses the same scheme, applied to groups. All Darwinism belongs to structure-oriented thinking. Basically, it considers the stabilization of certain organismic structures by adaptation to a given environmental structure. A dynamic factor only enters with the competition for scarce resources.

However, the processes which determine the evolution of a species within an ecosystem are not only of a genetic nature but also result from the ways in which the organism copes with its environment. True, genes determine form and to some extent also behaviour. But it is far more important which relationships with the environment have led in phylogeny to the development of particular physiological and behavioural patterns. It is not successful adaptation to a given environment which is the foremost formative factor in life, but the web of ecological processes in an environmental system which shape physiological and behavioural patterns which subsequently may become genetically anchored.

Relying on a scheme proposed by the British biologist Conrad Waddington (1975) we may distinguish at least five subsystems of the biological evolutionary process system. They are sketched in Figure 29. Each subsystem introduces a new dimensions of basic indeterminacy. But they are all systemically interlinked.

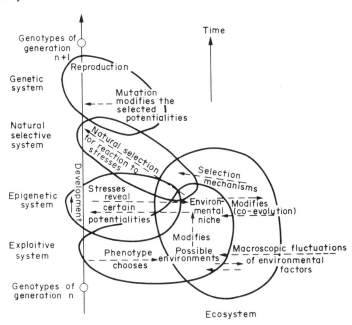

Fig. 29. The systems of biological evolution which, in their interaction, determine the direction taken by evolution. After C.H. Waddington (1975), with the addition of the ecosystem and several feedback loops.

The choice of habitat and thereby of the exploitive system may, of course, be heavily determined in a genetic way. For functions which are closely linked with reproduction, ranging from courtship and pairing to birth, many animals choose not only a particular type of environment, but actually the same locations. Fish populations stay faithful to their breeding grounds even under great difficulties and birds have their particular trees for courtship.

The Monarch butterfly alternates between the same deposits for its eggs in North and South America, even if usually each generation of caterpillars/butterflies makes the trip only once, in one or the other direction. How such a precise geographical memory functions genetically is one of the great mysteries of nature. But the Monarch, as we have seen in Chapter 4, is a champion of co-evolution in other respects, too. Another factor in the determination of the exploitive system, which is obviously genetically influenced, is the choice of the sexual partner which, with evolution proceeding to more complex organisms, becomes ever more specific.

The environment which is modified by the organism forms an ecological niche. This niche, however, undergoes further changes in co-evolution with other niches in the ecosystem. This mechanism of the ultracycle has already been mentioned in earlier chapters. Each niche, in turn, exhibits certain elements to which the organism is not well adapted and which give rise to a state of tension or "stress". With this stress, certain genetically available potentials are brought into play which, in turn, may contribute to the modification of the niche. The phenotype thus is formed in the living-out of relationships. This mechanism forms the core of the *epigenetic subsystem,* as Waddington has called it. In this subsystem it is determined which part of the total genetic potential is activated.

It is evident that the choice and the modification of the exploitive system as well as the processes of the epigenetic system influence natural selection to a large extent. With the habitat, for example, the natural enemies and their relative importance vary. It is also of importance whether an animal feeds during the day or at night, where it sleeps and under which circumstances its offspring are born. In the same way, natural selection will be strongly influenced by the activation of genetic potentials in reaction to stress in the epigenetic system. Thus, the individual may itself determine within relatively wide boundaries to which natural selection it subjects itself. It may often be observed that the principle of natural selection itself is not questioned. In the animal-rich regions of East Africa, for example, one may often observe herds of gnus or gazelles grazing in the immediate neighbourhood of groups of lions. Each evening one of the animals will lose its life—but the life of the herd continues.

We may recognize here at a higher level of evolution the same principle of partial self-determination which was already evoked in connection with the choice of autopoietic structures in the evolution of chemical dissipative structures. Again, we may speak of the self-finding of optimal stability although multiple criteria may be in play here. The principle of the co-evolution of macro- and microworld acts also on this level. As the individual organism contributes to the shaping of the ecosystem, it determines itself the type of natural selection which will influence the genes of generations to follow. And as the ecosystem provides selection mechanisms which influence the evolution of its members, it changes itself the structure of the relationships by which it is constituted.

Epigenetics and microevolution

Epigenetic development eventually reflects in genetic changes. Such changes also affect tested behavioural patterns which, in their genetic fixation, manifest themselves as instinct. There is no detailed description of these processes in terms of molecular biology. Certainly, the "wiring" of the information units arranged in the DNA molecule plays an important role (Riedl, 1976). Such a wiring-up may, for example, work with the help of "operons" which block certain parts of the DNA strand by sending out molecules of specific shape; or, on the other hand, such operons may also activate information. Such processes have actually been observed in bacteria, i.e. in prokaryotic cells.

In eukaryotic cells the processes are far more complicated. The chromosome field theory by Antonio Lima-de-Faria (1975, 1976) is an interesting approach to clarifying matters. It emphasizes the co-ordinated and integral character of the chromosome as a holistic system which organizes itself on the basis of a mechanism of molecular messages. In other words, an ordering principle is active at the level of the chromosome which makes self-organization possible. This would contradict the conventional view that chromosomes develop in accordance with random laws and natural selection. Whereas the prokaryote chromosome has a fixed sequence and a fixed number of operons, the eukaryote chromosome, according to Lima-de-Faria, is a dynamic self-organizing system capable of dismantling and rebuilding its operational units depending on the physiological functions required by its own genes and by the cell environment. Structure and function are not fixed but evolve jointly. But the chromosome does not constitute a complete autopoietic system since, for this dynamics, it requires the metabolic functions of the cell. Self-organization dynamics is co-ordinated at the level of the cell. With this approach, it may become possible to elaborate the "behavioural epigenetics" with which Gunther Stent (1975) wants to replace the conventional "behavioural genetics" which, in a strict sense, would only hold for prokaryotes.

The chromosome field becomes observable not only in the positioning of partial developments in the embryo but also in the fusion of plant and human cells (Dudits *et al.,* 1976) in which a new order spontaneously establishes itself. Perhaps the chromosome field also cancels out many of those deleterious mutations which seem to run against evolution. At a molecular level 99 per cent of all mutations are believed to be deleterious—and yet evolution has continued to proceed toward ever higher complexity and new levels of functioning. If this was exclusively the result of Darwinian selection, such a correcting factor would have been eliminated from the human world with its health and nutritional technology and the cumulative evolutionary consequences would show up eventually in the form of biological decadence. A new ordering principle at a supramolecular level, however, as is implicit in the idea of the chromosome field, would be able to correct errors in the same way in which the hypercycle of the Eigen type has corrected errors at the molecular level.

In the 1950s, when molecular biology started with great elan to study the grammar and syntax of the genetic lanaguage, it neglected semantics, or the context of meaning in a specific situation. Such a semantic reference is expressed, for example, in the three-dimensional spatial structure of the protein molecules built on the basis of genetic information. When a protein molecule has been built in a one-dimensional way by joining amino acids together in a specific sequence, it folds into a three-dimensional structure. Only then is it capable of certain catalytic functions. Gunther Stent (1975) speaks of the "context hierarchy" of genetic information, the first level of which is represented by the protein-folding process, and the second level by the principles of chemical catalysis, both of which are not given with the DNA nucleotide base sequence. There is an implicit semantic content of genetic information.

Conrad Waddington (1975) has graphically described the development of an embryo from a fertilized egg cell in terms of an "epigenetic landscape", an image which has stimulated the development of catastrophe theory based on differential topology. While the embryo crosses the topology of the epigenetic landscape along the time axis, it activates functional relationships which describe both the organism and its environment and which become manifest in chemical and physical processes giving shape to the embryo. Biological gestalt is not built from the joining of spatial structures, but emerges from the interaction of processes in a self-organizing space-time structure. The organs do not develop independently of one another, but shape one another to a certain extent so that their later integration in the functions of the organism as a whole poses no problem. The muscle tone partly determines the shape of the bones with which they are connected. In animal experiments it has been shown that different body regions have their own epigenetic landscapes. If a part of the embryo is transplanted into another part, everything becomes disordered.

However, as Waddington puts it, there is a "subtle balance between flexibility and lack of flexibility". Only certain types of cells develop and no intermediary types. Development is canalized into specific process chains which Waddington calls chreods. These chreods represent at the level of the cell the evolutionary paths of autopoietic structures as they already become observable in chemical dissipative structures. In biological development, however, there is not the same measure of indeterminacy since the individual evolutionary paths are embedded in a system of innumerable chreods and thus, in a macroscopic perspective, themselves become a more or less unambiguous chreod. Along such a path (called histogenesis) a cell may exercise different related functions in the embryo and repeatedly participate in the formation of tissue of a new kind until the stability which corresponds to the adult organism has been reached. In this process, a positional field is always established first, whereupon the cells interpret their integral genetic material in specific ways according to their position. If the epigenetic process in unicellular organisms is characterized by the selection of individual genes and the synthesis of certain chemical substances with their help, interactive chreods are the subject of selection in multicellular organisms and they control the growth of whole body systems (such as the digestive or the reproductive system). This means that no single genes are activated or inhibited, but entire gene groups which belong together in their developmental functions. It becomes clear now why only eukaryotic cells are capable of forming tissue and multicellular organisms. With their chromosome field, they mark the beginning of true epigenetics.

Of particular interest in this connection, of course, is the change from the caterpillar to a butterfly. In the transition only the brain, the hind guts and the heart of the larva remain functionally intact. The rest of the body is broken down into its molecular parts and reused by cells standing by in baglike annexes called imaginal discs. At an early stage, when the larva consists only of about 6000 cells, the cells are divided into two development lines, one for the body of the larva and the other one for the imaginal discs (Dübendorfer, 1977). The latter come into play only after the metamorphosis. The caterpillar/butterfly system thus evolves through two entirely different structures, characterized by two sets of chreods which exist from the beginning.

A much faster acting process which is physiologically anchored in epigenetic development may be found in the central nervous system. The full number of nerve cells which the organisms will ever possess are apparently there at birth. However, these nerve cells may form many new filaments (dendrites) and, through them, become "wired" in many new ways with adjacent cells. Brain growth, especially in certain phases of childhood and adolescence, is not growth in the number of cells, but growth of dendrites of which each cell may have as many as thousands or tens of thousands.

The same flexibility which is expressed in the frequently mentioned complementarity of structure and function also becomes evident in the development of whole organisms. With social insects, such as bees or ants, the further development of an egg produced by the queen as a "central mother" is not prejudiced. It depends on the treatment by worker insects (acting as nurses) whether soldiers or workers or even a new queen will come out of such an egg. The type of treatment required is regulated in a chemotactic way by an enzyme diffusing through the colony and inhibiting the formation of a new queen. When the colony becomes too big in size, this chemotactic control starts to fail, a new queen is produced and the colony divides.

Manipulation of history in long-range evolutionary strategies

Another principle from the last century besides Darwinism, Ernst Haeckel's biogenetic law—"ontogeny recapitulates phylogeny"—has long dominated the image formed of biological evolution. This law implies that evolution acts in an additive way, so that new achievements are added at the end of a long liner development. Each organism then carries the entire history of the phylum within itself. This is not altogether wrong, but the law does not act in a rigid and structurally oriented way, but in the sense of the flexible modification of a space-time structure.

In the same way as the individual chreods in the development of the embryo, the "macrochreod" of phylogeny also makes its insistence felt with a certain stubborness. Mammals and humans, too, in their embryonic phase first develop those gills which their evolutionary ancestors in the oceans used for breathing. With the conquest of land the vertebrates developed bone structures and lung breathing. But those marine mammals which about 60 million years ago returned to the sea, for example whales and dolphins, kept these features. Whereas fins grew over the bone structures of the hand and the feet, lung breathing still forces the animals to surface periodically (the blowing of the whales)—certainly no great adaptation to the wet element. But the interesting aspect is that in the whale embryo there are gills still which grow at first and vanish eventually. Ontogeny cannot branch off from this early phylogenetic advantage. It has to pass through the "land stages" to develop the later, and somewhat cumbersome, marine features. The flexibility which facilitated the double change of environment emerges in another way.

The genetically transferred experience and the physiological programmes connected with it are indeed transferred in phylogeny. Their application in ontogeny, however, by no means has to be rigid. On the one hand, as has been discussed, the genetically available, rich information is only partially utilized, depending on environmental conditions. It is estimated that humans and higher mammals utilize only between 2 and 50 per cent of their genes, with 15 per cent as a frequently cited average value. The rest is epigenetic reserve. In the 300 generations of humans which have lived since the Stone Age, the genetic heritage has been changed only insignificantly. But the environmental relations which we master have changed tremendously.

On the other hand, the activation of genes in ontogeny may shift in the time dimensions (heterochrony). In particular, certain bodily features and puberty may either become accelerated or delayed. This may happen by a simple shift in the balance of endocrinal secretions. Heterochrony apparently plays an important role in the interactions of phylogeny with the evolution of ecosystems, that is to say, in the co-evolution of macro- and microsystems (Gould, 1977). Both effects together, selective utilization of genes and heterochrony, form the most important elements of epigenetic development known today. They are the sources of increasing flexibility in microevolution. It has even been suggested that evolution consists primarily of reusing already formed genomes (E. Zuckerkandl, quoted in Gould, 1977). According to Antonio Lima-de-Faria, most of the genes evolved in unicellular organisms. Higher stages of evolution focus on their combinatorial organization.

By simple prolongation of certain features of the phylum—a mechanism corresponding more or less to the idea of linear recapitulation expressed in the biogenetic law—bigger and more complex organisms may develop (hypermorphosis). Often, such a hypermorphosis leads into an evolutionary *cul-de-sac*. Puberty is delayed in relation to the development of the body. An example is the development of huge and heavy antlers (such as in the Irish Elk) which, ultimately, become a useless burden. The elephant has survived whereas its over-sized cousin, the mammoth, has become extinct. Viewed from the angle of phylogeny, it is the characteristics of the *adult* organism which are added on here in a chain of indefinite length. We may say, that confirmation is added to confirmation. This often leads to a bad end.

But it is also possible that the characteristics of the *juvenile* organism enter phylogeny (paedomorphosis). This may happen in two different ways. On the one hand, puberty may be accelerated (progenesis). In this case, organisms reproduce which are still young in their bodily characteristics and whose epigenetic development phase reflects this youth. Progenesis is characteristic for ecosystems which are relatively immature. As has already been mentioned, such ecosystems are dominated by species which produce offspring in a quick rhythm but abandon them in a relatively helpless state to the dangers of the environment. In these phases of ecosystems, the chief criterion of evolutionary fitness is not so much physiological form and functioning, but primarily the dynamic quality of being able to colonize an open, largely unexploited system as quickly as possible. The subject of selection is the capability to reproduce quickly a dynamic instead of a morphological quality. Progenesis may even be regarded as a laboratory for morphological experiments in evolution. It is characteristic for insects, for example, which usually find themselves in a wide open ecosystem in which there are still many green leaves to feed from. With few exceptions (such as those which lead to locust migration) the relations between the insects and the ecosystem remain open since the insects fall prey to their natural enemies before they reach the limits of food availability. This is one more reason for them to produce offspring early in life.

The other cause of paedomorphosis may be found in a delay of the appearance of bodily characteristics (neoteny). Not the characteristics of the adult organism are inherited, but those of the juvenile organism. In this way, considerable openness and flexibility are imparted to ontogeny. Neoteny is characteristic for species in mature ecosystems which are capable of learning. Only few offspring are produced, but they are carefully educated. Man is a prime example for neoteny. His body is getting taller, but puberty has become delayed and the formation of bodily features takes a longer time than it used to with his distant ancestors.

Man, as also some of the primates, attains his puberty at about 60 per cent of his stabilized body weight. Almost all the other mammals attain puberty at only 30 per cent of their stabilized weight. The distinguished Swiss biologist Adolf Portmann (quoted in Gould, 1977) suggests that the development of the human foetus should last 21 months instead of 9 months, in comparison with the development of the body in other mammal species. The human baby is indeed so helpless that its first stage of life is more like a continuation of the foetal stage outside the womb. According to Portmann, this is not so much due to the difficulty of giving birth to a larger baby but to the necessity of stimulating the development by sensory impressions from outer reality. Both arguments may be valid.

It is of particular interest here that selection acts less on biological forms and functions of the organism, but primarily on *long-range evolutionary strategies*. Not structures are ultimately decisive in each phase of phylogeny, but the *dynamics* of this phylogeny. We may even speak of a self-selecting dynamics. The preferred dynamics corresponds to an evolutionary strategy of openness. This conclusion falls far from the traditional dogmas of Darwinism and Neodarwinism.

The open strategies of paedomorphosis or evolutionary rejuvenation do not yet imply that novelty enters by way of this openness. The general result of progenesis in immature ecosystems is a loss in flexibility by the simplification of the body structure and the loss of genes which correspond to the adult phase. The general result of selection in mature ecosystems, on the other hand, is hypermorphosis which also is coupled with a loss in flexibility. Neoteny, however, preserves juvenile flexibility in mature ecosystems; but its prerequisite is the capability to learn. Both strategies of evolutionary rejuvenation may in principle save developments from over-specialization, but both are not all too frequently creative in the sense that they bring novelty into play.

With earlier puberty in progenesis, new morphological and functional combinations emerge because an entire complex of functions may now become active at an earlier stage. Other bodily features, too, are realized at an earlier stage. This acts as a source of variety. But perhaps the duplication of genes and, connected with it, the possibility of "liberating" functionally undetermined genes for other tasks plays an important role here. It is assumed that progenesis only rarely leads to evolutionary innovations. Generally, it leads into high-grade specialization and gets stuck there. Perhaps this furnishes an explanation for the stabilization of the insect world which has not evolved significantly since 100 million years.

Neoteny may be expected to introduce novelty more often, chiefly because the manifold relationships in a mature ecosystem favour the co-evolution of learning species as it is expressed in the already-mentioned idea of the ultracycle. As mentioned in Chapter 4, it may be shown that in a co-evolving predator-prey relationship both species, the prey included, benefit. Adolf Portmann assigns a decisive role in the development of the brain to neoteny. In this way, sociocultural evolution is prepared, which will be discussed in the following chapter.

With both evolutionary strategies of rejuvenation—progenesis and neoteny—the possibility of relatively large and quickly spreading evolutionary jumps is essential. Openness is always possible and it is not necessary to first overcome the rigidification of old age. Epigenetic development also implies that very different features and behavioural patterns may emerge from the same genetic structure. Conversely, the same features and behavioural patterns may be based on a different genetic inheritance. In the same way, acceleration or delay in the ontogeny of an organism may entail very different morphological and evolutionary consequences.

But what is the meaning of the conspicuous acceleration of puberty in the contemporary Western world, holding at present at an age of only 11 years? Is it only a short-range oscillation, or is it a true evolutionary fluctuation? I do not know the answer.

Man as a product of epigenetic evolution

Epigenetic development emphasizes a broad evolutionary stream across a wide ecological spectrum, in contrast to the idea of relatively uninfluenced development lines in purely vertical, genetic evolution. Until the second half of the 1960s the origin of man has been viewed as the result of such a single development line from *Australopithecus* to *Homo erectus* and further to *Homo sapiens*. New relics found in East Africa as well as the development of molecular anthropology which investigates the degree of relatedness between the genetic material of humans and other primates, however, have changed this picture thoroughly (Fig. 30).

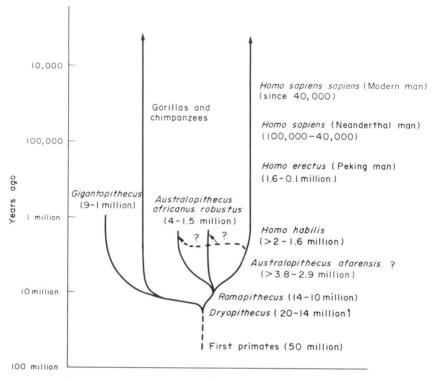

Fig. 30. The evolution of man. For several million years, man's ancestors shared their living space with similar, erect species of *Australopithecus*.

Today it may be assumed that the development lines branched several times and that the direct ancestors of man lived at the same time as other erect creatures (the two kinds of *Australopithecus*) and, at least in East Africa, also in the same regions as well (see, for example: Leaky and Lewin, 1978). Their common ancestor may have been an ape-like creature, *Ramapithecus*, whose relics have been found in many areas, including East Africa, India, Pakistan, the Middle East and Central Europe. *Ramapithecus* seems to have left the forests and entered the grassland, thereby initiating an accelerated development. This daring step seems to have been taken in the context of major geological changes, such as the emergence of mountain ranges. Again, macroevolution is the trigger for a new development in microevolution. *Ramapithecus* apparently was a hunter and carnivore. In response to the new conditions of his life he developed an erect posture which is of great advantage in the grassland.

The development line of man proper seems to have branched off 6 million years ago from the two *Australopithecus* lines, according to an interpretation of recently found remains (Johanson and White, 1979) perhaps only 2.6 million years ago. In the latter case, a common ancestor is assumed, *Australopithecus afarensis*, whose fossils date back 2.9 to 3.8 million years. The oldest stone tools which have been found date from 2.6 million years ago. However, they have been found near the same places where the *Australopithecus* fossils have been found and it may not be entirely certain that only man's direct ancestors used tools. It is considered certain, however, that *Homo habilis* used hunting tools.

Another alleged advantage of the hominides may be tested more directly. Whereas the brain volume of *Australopithecus* measured only about 0.45 to 0.55 litre, a brain volume of 0.7 litre was found in a skull of *Homo habilis* which is 1.8 million years old, and a brain volume of 0.9 litre in a skull of *Homo erectus* which is 1.5 million years old. A 2 million-year-old skull of *Homo habilis* which has been found at the borders of Lake Turkana in Kenya falls somewhat outside this scheme; its brain volume measures 0.8 litre. With the appearance of *Homo sapiens*, the brain volume has reached an average of 1.4 litres, a value which remained practically constant ever since. However, Peking man, a version of *Homo erectus*, had already tamed the fire half a million years ago—the Promethean adventure of human evolution was about to begin.

Sociobiological evolution in the direction of individuation

With earlier phases of evolution, the bond with the collective is stronger and more uncompromising. It is generated and maintained primarily—perhaps not exclusively—by material exchange processes, in particular by chemotaxis, the diffusion of chemical substances which generate in a compelling way a certain type of behaviour. This type of bond may be already observed with the prokaryotes which even in our days are linked by chemical exchange processes to their macrosystem, the Gaia system. As has been discussed, bacteria utilize a common gene pool and in this respect act as one big system.

The sociobiological type of binding is also very strong with those multi-cellular organisms which appeared early in evolution, the invertebrates and insects. It becomes increasingly weaker in vertebrates. Sexuality is still regulated to a large extent by the diffusion of chemical substances and the activation of the smelling sense—in rudimentary form even at the human stage in which sexual advertisement is enhanced by perfumes, a kind of artificial chemotaxis. But individual behaviour is no longer dictated by the collective. There are mammals living solitary most of the time, for example, male elephants and lions. The leading insect specialist Edward O. Wilson (1975) has drawn the conclusion that evolution has gone astray. Such utterances of hybris are, alas, characteristic for that reductionist branch of sociobiology which should better be called sociogenetics. The mistake is due to a one-sided microevolutionary view which considers a hierarchy of life building from the bottom up and leading to crowning sociobiological super-structures.

If, however, we consider evolution as a holistic phenomenon in which micro- and macroevolution are interdependent and which runs in the direction of increasing complexity, the increasing emphasis on the individual in sociobiology appears quite natural. With the autopoietic level of the complex multicellular organism, the phase of ecological determination, of the supremacy of sociobiological insurance of community and species, of genetic and later increasingly epigenetic development in the co-evolution of macro- and microsystems, draws to its end. The organism has matured and, in the case of the human organism, practically does not evolve any further—with one important exception, the brain. What follows is sociocultural development in which sociobiology is turned upside down. Man enters into co-evolution with himself. He carries the macrosystems of his life *within* himself and in relative freedom determines their organization. But this task requires a communication mechanism which is capable of acting much faster than metabolic communication. The emergence of such a mechanism forms the core of the following chapter which completes that part of the book which is devoted to sketching the history of evolution.

9. Sociocultural Evolution

I am in a world which is in me.

Paul Valéry

The dynamic unfurling of biological communication

The exchange of energy and matter with the environment is one of the basic prerequisites for self-organization in energy/matter systems. The energetic and the material points of view appear as complementary and express the interaction of novelty and confirmation. Novelty and confirmation, however, are aspects of pragmatic information. Energetic/material systems may be co-ordinated and steered by suitable information; on the other hand, the self-organization of information generates new information.

The experiencing and utilization of environmental information is already linked with the exchange of energy and matter. No nervous system or brain is needed for that. If a unicellular alga is continuously disturbed, for example by touching with a needle, it will first try to protect itself by contracting and eventually will swim away. One of my house ferns, from which I am learning a lot, recently developed three new sprouts at the same place, one after the other. The first two headed straight forward, became entangled with an already fully developed leaf system, which apparently is "not permitted", stopped and died. The third sprout, however, circumvented the leaf system in a bold and almost incredible curve, found its way to the light and was able to unfold. Only then did it straighten and cut unconcerned through the obstacle whose turn it now was to die. In principle the learning capability does not require a nervous system or brain.

In the biological domain of life, three types of communication play a role, according to Gunther Stent (1972). *Genetic communication* acts in time intervals which are long compared to the lifetime of an individual. It makes phylogeny and coherent evolution across many generations possible. *Metabolic communication* which in the organism is transmitted by special messenger molecules, the hormones, fulfils two tasks within an organism.

One task is the regulation of the development of multicellular organisms, in plants as well as in animals. The other task concerns the damping of the consequences of environmental fluctuations for the organism, or in other words, the enhancement of the organism's autonomy. Metabolic communication based on hormones acts relatively slowly, corresponding to the transport in the organism which may take time spans from seconds to minutes. In superorganisms, such as amoebae aggregations or insect societies, chemotaxis, which is a form of metabolic communication, may act in a matter of minutes. Quite generally, we may call metabolic communication any exchange process of a self-organizing energy/matter system. In ecosystems, it then includes the complex relationships among organisms of different species; in societies, the material and energetic processes among members. The third type of biological communication, finally, is transmitted by the nervous system; it may therefore be called *neural communication.* This type acts in organisms typically within a hundredth to a tenth of a second and is thus in the average about a thousand times faster than metabolic communication. To this triple scheme may perhaps be added *biomolecular communication* which acts in small volumes in milliseconds.

Genetic communication already is capable of forming and projecting models and even of anticipation. The same holds for metabolic communication. Metabolizing systems, starting from chemical dissipative structures, develop their structures autonomously in interdependence with their functions. The principle of the complementarity of structure and function leads to such an active role even in the simplest domains of self-organization. Perhaps there is also metabolic anticipation. The organism seems to indeed know its life-span and even to determine it partially itself—perhaps by means of the cell-division rhythm and the limitation in the number of cell divisions by the Hayflick number, already mentioned in Chapter 7. Indian yogis know with great precision when they are going to die and long before summon their disciples for a farewell on the day of their death.

Communication of all types seems to obey certain common basic principles. Neural communication, because of its speed, accentuates these principles in particular ways so that a superficial view may recognize them only in this domain. This chapter is not concerned with sharp definitions and separations but with the recognition of principles which are of importance for evolution. In this respect, neural communication has significant innovations to offer.

Neurons, the specialists of fast communication

The development of animals led to the emergence of a particular cell type, the nerve cell, or neuron. Its predecessors in unicellular protozoa were membranes which were capable of storing and liberating energy and thus anticipated the basic functions of the neuron. Neurons seem to have no other task than to communicate. With this capability, they first contributed to the co-ordination of complex body functions and the behaviour of the animals in quickly changing situations. In subsequent development, neurons and the systems which they form—the central nervous system and in particular the brain—became the managers of information relations between the organism and the environment.

Neurons may carry out non-linear transformations in fast and efficient ways. From their cell bodies, filaments stretch out which are surrounded by membranes. One of them, a lengthy body surrounded by a coat of fat, is called axon. Whereas an axon measures about 1 millimetre in the brain, it may reach a length of 1 metre in nervous strands embedded in muscles. Other filaments, which are thinner and more numerous, are called dendrites; they may branch further. The dendrites contact the axon tips of other neurons. Usually no cell plasma bridge forms, however. This contact "at a distance" is called synapse and guarantees high flexibility in the reorganization of connections. In this way the dendrites of one neuron may be in contact with up to 100,000 other neurons; on the average, there are 10,000 connections. The growth of the brain volume in certain periods in the life of the organism, as has already been mentioned in the preceding chapter, is mainly a growth of dendrites and also of new axons. With a given number of neurons which slowly diminishes from birth (in the human cortex 10,000 million) there is practically no limitation in the formation of new communication channels and networks. Here structure remains in essential aspects malleable to such a degree that there are no narrow limits to the evolution of functions. The central nervous system, and above all the brain, are far better suited for evolution even during the onotogeny than is the epigenetic system.

Axons and dendrites supplement each other in their ways of operating. Axons transmit electrical pulse trains each of which has a specific amplitude and duration. The dendrites integrate these pulse trains by weighted summation, transform them into an electric wave—with the electric current across the synapse consisting of ions, i.e. electrically charged atoms, and not of electrons—smoothen the wave on its way along the dendrite stem and transmit it to another axon. With increasing stimulation neurons change their dynamic behaviour from constant membrane potential ("zero equilibrium") through a constant pulse rate ("non-zero equilibrium") to a constant rate of bursts of pulses (limit-cycle behaviour) (Freeman, 1975). Thus, even single neurons exhibit in their communication behaviour phenomena of self-organization which resemble the dynamics of dissipative structures.

There seems to exist a further level of circuitry which is formed by networks of thousands of millions of very small nerve cells (microneurons) and are sometimes called nerve felt (Marthaler, 1976). These circuits are built similarly to the bigger ones and also show synaptic structures, but often include only fractions of two neurons and range over micrometers (thousandths of a millimetre) only, whereas their bigger brothers range over millimetres and up to metres. Whereas the bigger circuits selectively transmit currents of 0.01 to 0.1 volt and build up an action potential, the nerve felt works with voltages of only some ten-thousandth volt. The basic potential to overcome is not here, as in the bigger network which thereby assumes the character of a binary "yes/no" decision element. Densely packed, the microcircuits of the nerve felt may quickly form systems of high complexity.

The nerve felt seems to exhibit a whole spectrum of finely tuned reactions which, besides electrical excitation also include a multifaceted exchange of chemical substances. Small proteins cross the synapses in both directions. The importance of the microcircuits of the nerve felt is underlined by the fact that its share in the forebrain increases in more highly evolved animals and reaches a maximum in the human brain. The microneurons are the last to stop their increase in the development of the organism. Like the dendrites of the big neurons, also the filaments of the mcironeurons continue growing after birth so that the effects of the exchange with the environment reflect to some extent in the microcircuits.

Single neurons do not bring much what is new. They process information and transmit it. Only larger groups of neurons lead to a phenomenon which may be identified as the *self-organization of information.* We have already seen in Chapter 3 that pragmatic information acts autocatalytically, i.e. produces more pragmatic information. But here is added a new element of gestalt, which may be viewed as a dissipative structure of information. In the domain of metabolic communication, this structure falls together with the structure of autopoietic units of life, such as cells, organisms and ecosystems. In the domain of neural communication, it finds its energetic-material correlate in the space-time structure of the dynamics of the neuron systems. But a higher level of co-ordination is active here. The gestalt quality of thought, dreams and visions is not the result of a self-organization dynamics which would be comparable to that of other matter/energy systems. It must be structured at many levels each of which brings specific ordering and co-ordination principles into play.

What is known of this at present, corresponds only to the roughest level of electrochemical self-organization. As has already been mentioned in Chapter 4, densely packed groups of neurons characterized by numerous intermeshing feedback loops may exhibit in their chemical and electrical activity typical self-organizing behaviour. The hierarchy of the macrodynamics resembles the hierarchy enumerated above for single neurons. In particular, limit-cycle behaviour may be observed (Freeman, 1975). The electrical brain waves recorded in the electroencephalogram (EEG) represent limit-cycle behaviour in extracellular potential fields the amplitude of which oscillates in a certain rhythm.

The brain is a communication mechanism which is used and directed by the self-organization of information. It has no more to do with this information than does the computer with the information it processes. Although the comparison between brain and computer should not be carried too far since, to some extent, they represent very different principles, it may be useful to also distinguish between "hardware" and "software" in the brain. The network of neurons, then, represents the "hardware" and its possibly multilevel self-organization dynamics the "software".

The essential factor is the gestalt quality emerging at several levels. Limit-cycle behaviour is a primitive gestalt and is far from representing a thought. The analysis of frequencies and intervals of mechanical oscillations as well as the rhythm of their changes says little about the gestalt quality which, in a piece of music, touches us. There, gestalt emerges from a much higher level of organization of the acoustical communication patterns. Gestalt has its own life—we may also say, its own autopoietic existence. The sensory impact plays a minor role. Many people are capable of reading music, of listening to it with their "inner ear" and of recognizing and experiencing its gestalt in this way also. The electrical impulses in the telephone wire and their acoustical translation are not identical with a beloved voice, but their gestalt matches the gestalt of this voice.

Once, at a scientific conference that I attended, an old Nobel laureate in physics exclaimed in comical despair: "The Schrödinger equation describes the motions of the electrons in my brain. But I can tell you that I have very different feelings when I think of a prime number or when I think of a pretty girl. Who can give me the supplement to the Schröndinger equation which expresses this difference?" Of course, he will never get it at the level of quantum mechanics.

The self-organization of information is an aspect of the self-organization of life and the gestalts it produces are the gestalts of life. They are autonomous, as are the gestalts of other autopoietic system dynamics. They form their own world of symbolic representation of reality and are capable of emancipating themselves from this reality. Thus they can change and redesign reality. Self-organizing pragmatic information may interfere with and co-ordinate energetic and material processes outside of the system in which this information becomes structured. Usually, this is expressed in the phrase: Mind over matter. But this is true only to the extent that the matter/energy system to which this kind of mind belongs, namely the brain, remains excluded. The mind of the human organism controls the inanimate and certain aspects of the animate world, but the mind of an ecosystem does not dominate its members—their dynamics *is* the mind of the ecosystem, just as the co-ordinated dynamics of ants is the mind of the ant colony. Control and domination are dualistic notions—there is always a controller and a controlled. But mind is a non-dualistic notion which is inseparable from the matter in whose dynamics it expresses itself.

Mind as dynamic principle

In this perspective, mind is self-organization dynamics proper. It appears wherever there is dissipative self-organization, especially in all domains and at all levels of life as it becomes expressed in macrosystems as well as in microsystems. We ought to at least distinguish between a metabolic and a neural mind and we shall subdivide the latter further in accordance with its dynamics.

At scientific conferences with brain scientists participating there are always controversies over structure-oriented concepts. One part searches stubbornly for a control hierarchy in the brain in which each level of neurons is supposed to be controlled and co-ordinated by a higher level. This leads to the assumption of an ultimate co-ordinator or dictator neuron which some call the soul while others (especially in Germany) call it "grandmother neuron" or "Aunt Emma". There is no indication that such a neuron exists. Therefore, those brain scientists who can afford to accept mind as a topic of science at all—for example, the Australian Nobel laureate Sir John Eccles—turn toward a dualistic view of mind and matter as separate entities interacting with each other (Popper and Eccles, 1977). As in many areas of quick advances in empirical research, theory-building is still amiss and finds it difficult to break loose from the old stereotypes. This in itself has to do with the structure of thought dynamics and the brain, as will be shown below.

In a dynamic view, the old dispute of structure-oriented thinking, whether mind (or the spirit) is immanent or transcendent, finds an entirely new answer: Mind is immanent, not in a solid spatial structure, but in the processes in which the system organizes and renews itself and evolves. An equilibrium structure has no mind. The British anthropologist Gregory Bateson (1979) came very close to this understanding when he equated mind with the cybernetic system, with self-regulation which also includes dynamic relationships with the environment. Thus, mind reaches beyond the autopoietic structure proper and embraces also the interactions with other systems and generally with the environment. By identifying the unit of mind with the unit of evolution—both are represented by the same cybernetic system—Bateson (1972) indicated that he was thinking of self-organization. Buddhism, too, recognizes the process nature of mind which reaches beyond the system in question and establishes a "primordial continuum of experience" (Longchenpa, 1976, Vol. 3, p. 49).

Metabolic and neural mind meet at a level which I should like to call the level of the *organismic mind.* Neural communication at first served the co-ordination of metabolic processes in organisms which were becoming ever more complex in the course of evolution. Organismic mind does not reflect, it is pure self-expression—self-representation or, better perhaps, self-presentation. A large part of what is commonly called "behaviour" is not so much the automatic play of genetically instructed functions than it is self-expression of the organism as a whole. Art, a function which we regard as a very high form of expression of human life, certainly finds its source partially in the organismic mind.

The *reflexive mind* acts differently. It mirrors an outer reality which it rebuilds in the inner world. This mirror image does not simply enter from the outside but emerges from exchange processes between a mosaic of sensory impressions and tentative models which the reflexive mind projects outward. The most significant characteristic of the reflexive mind is *apperception,* the capability of forming alternative models of reality. In this exchange between the organism and its environment which somehow resembles the epigenetic process, the model becomes more realistic and includes learned aspects such as seeing in perspective (which we learn only after birth) or the contiguity of areas and spaces. On the other hand, the emerging image of reality includes also spontaneous and creative features such as the emphasis on certain forms and colours, the gestalt association with other forms (the "man in the moon", mandrakes, or specific cloud shapes) and the suppression of certain details which might disturb the model.

This interaction between an actively produced model of the environment and fragmented sensory impressions of the same environment may already be observed with animals, at least at a later stage of evolution. Walter Freeman in Berkeley, who is carrying out the experiments with large populations of neurons already mentioned, tells of experiments with cats aimed at observing their olfactory system. When the attention of the cat is focused on searching for a new smell, there is a particular pattern of increased brainwave activity. When for some reason the expectation changes to a different odour, the model of the situation changes abruptly and there is great commotion in the electrical brain-wave pattern until a new pattern wins over and eventually becomes stabilized.

The *self-reflexive mind* acts differently again. It designs actively a model of the environment in which the original system itself is represented. Thus, the original system, which we may also call self, becomes involved in the creative interpretation and evolution of the image. The relationships with the environment become totally plastic, subject to creative design:

I am awake! Oh, let them reign,
the incomparable figures
sent there by my own eye.

(Goethe, *Faust II*, "At the lower Peneios")

With the self-reflexive mind, a new and very essential element is called into play, *anticipation*—in a passive sense as expectation and anticipated experience, in an active or creative (goal-setting) sense as creative design of the future. Walter Freeman has already observed at least the passive aspect of anticipation with rabbits. When a rabbit is shocked in connection with a specific smell, the brain shows certain reactions to this shock which, after all, implies a strong sensory impact. However, the same type of brain response occurs when the poor rabbit is no longer shocked and only senses the same smell. The brain again processes the sensory impact of a shock which has not taken place at all and was only anticipated. For the organism the experience is the same. Passive anticipative effects have even been observed in the neurons of snails and may be ubiquitous in life. Active or creative anticipation, in contrast, seems to appear only with highly evolved animals.

It is of great importance here that the processing and organization of information become independent not only of metabolic processes, but also of direct sensory impact. The self-reflexive mind may now become totally emancipated and set out on its own course of evolution. It is not "we" who think, but "it" thinks in us. Mind becomes a creative factor not only in image-forming, but also in the active transformation of outer reality. This role of the self-reflexive mind blossoms fully in the human world.

The mode in which thinking forms structures and lets them evolve is associative rather than sequential. A forgotten name or the solution of a problem come to us as a surprise, as things found accidentally. But it is probably the overlapping of many associative systems which permits such a short-cut. The self-reflexive mind also seems to be a prerequisite for dreams which are frequent and important contributions by a self-organizing consciousness and which constitute another form of mental emancipation from the outer world.

I have already pointed out that in such a dynamic view the macrosystems of life—societies, ecosystems and even the world-wide Gaia system—have mind, too. Insect societies, for example, are organized by metabolic mind. However, this mind acts much more slowly than the neural mind. The gestalts produced by the metabolic mind belong to different levels than the gestalts of the neural mind. It is precisely the profound thought that the gestalts of the human mind may meet with those of natural macrosystems which makes such a deep impression in Stanislaw Lem's science fiction novel *Solaris*. An ocean on a far planet which, in Tarkovski's film version, evolves in extraordinary dynamic structures, becomes the mirror of human mind and returns the secret desires and fears in material form.

But the electronic communication processes in modern human society have accelerated the transmission of information in extensive systems to such a degree that it becomes comparable with the speed of the neural mind of the organism. Electronic data processing, too, results in novel information structures. A new level of gestalt may be in the process of emerging if it does not exist already in certain aspects. We shall come back to this point later in the chapter.

First, however, we shall take a look at the evolution of brain stuctures which mirrors the functional evolution of the neural mind. It corresponds to the division into organismic, reflexive and self-reflexive mind.

The evolution of the "triune brain"

The American neurophysiologist Paul D. MacLean, head of a laboratory for the study of brain evolution and behaviour, calls the forebrain of the more highly evolved mammals and of man a "triune brain" (MacLean, 1973). This term is meant to express that three brains act in one. The fully developed forebrain consists of three distinct parts (see Fig. 31) each of which has its own intelligence, its own subjectivity, its own sense of space and time, its own memory and its motor and other activities. The three parts are chemically and structurally different. However, they are intimately linked with each other and normally co-ordinate their activity. MacLean has proposed the graphical image of a "neural chassis" (mainly the lower brain stem and spinal cord, later also the midbrain), which would resemble an empty and unguided vehicle if it were not driven by three "drivers". The neural chassis itself is older than the three drivers and may be traced back to an early phase of the

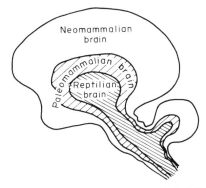

Fig. 31. The "triune brain", which expresses the evolution of the brain from reptiles and mammals to man. After P.D. MacLean (1973).

appearance of multicellular organisms. It co-ordinates aspects of self-preservation, such as breathing, blood circulation, blood pressure, digestion and motility. The midbrain is responsible for the selection of environmental stimuli which are essential for self-preservation.

The three brains are: first, the "reptilian brain" (which, in mammals, is represented by a group of large ganglia including the olfactory system and other parts); second the "paleomammalian brain" or "limbic system", sometimes also called metaphorically the "horse" (the primitive cortex and its further development in the "limbic lobe" around the brain stem—limbic means "forming a border around"); and third, the "neomammalian brain" (chiefly the neocortex). The older two brains lack the neural mechanism for verbal communication. But they are not in the "unconscious". In the contrary, they express themselves very visibly in a rich body language which ranges from chemical signals to complex rituals of expression by gestures. Like the verbal language, the body language also has its syntax (specific arrangement) and its semantics (context of meaning in a specific situation).

The *"reptilian brain"* originated approximately 250 to 280 million years ago. It serves the co-ordination of an already very rich spectrum of behavioural patterns, ranging from territoriality, ritual fights and intimidation of the opponent to the formation of social hierarchies, greeting, ritual courtship, orderly migration and hoarding. These are all behavioural expressions already exhibited by lizards and other reptiles. However, the reptilian brain seems to be less efficient in coping with new situations. In other words, it does not have good learning capability. On the other side, it provides a possibility for the development of a rich repertoire of self-expression which is only inadequately covered by the term "behaviour". Self-expression must not be understood in a strictly functional way as the securing of survival, but as true symbolic self-representation. It establishes that type of autonomy which is due to this level of self-organizing systems. Generally speaking, the reptilian brain may therefore be viewed as the material system which manages the processes of the organismic mind. Its flexibility, however, is limited and the mental processes in the reptilian brain are held responsible for irresistible drives, impulses, compulsive behaviour and possessedness of all kinds.

The largest of the dinosaurs, whose time on earth started aproximately 200 million years ago and came to an abrupt end 64 million years ago, was the *Brachiosaurus,* measuring 30 metres in length and living in swamps. It had a brain which was hardly bigger than a chicken egg (Halstead, 1975). But the mammals of the dinosaur period were hardly superior in intelligence. These small creatures, resembling mice and hedgehogs, had to hide during the day and look for food during the night. They can hardly have been responsible for the extinction of the dinosaurs which suddenly disappeared from all continents at the same time. During their long evolution, the dinosaurs had been capable of adapting to considerable climatic fluctuations and changes in their environment. The mystery of their sudden extinction suggests the possible occurrence of a macrofluctuation such as a prolonged period in which the earth was without a magnetic field or a nearby supernova explosion as it is assumed by the Russian astronomer I. S. Shklovskii (quoted in Sagan, 1977). If, by such an event, the ozone shield of the earth's atmosphere had suffered due to hard particle radiation, the penetrating ultraviolet light would have done more damage to the dinosaurs as daylight animals than to the mammals as night animals. It is perhaps significant that no animal heavier than 10 kilograms survived the cretaceous era which ended with the dinosaurs (Valentine, 1978).

The *paleomammalian brain* or *limbic system* probably originated in the oldest mammals about 165 million years ago. Reptiles have only a rudimentary cortex and it is assumed that this primitive cortex has evolved further in the intermediary steps between reptiles and mammals, the so-called mammal-like reptiles of which, however, hardly any fossils have been found. The limbic system looks alike in all vertebrates. Mammals, in particular, have a well-developed limbic system which represents the larger part of the cortex. It is structurally much simpler than the neocortex which developed in a later phase. The limbic system receives information both from the inner world of its own organism and from the outer world. Thus, it contributes significantly to the formation of a personal identity. Out of its three subsystems, two are closely linked with the olfactory system which already exists in the reptilian brain and plays a role in oral and genital functions such as feeding, pairing and aggression. In contrast, a third, younger subsystem circumvents the olfactory complex and assumes an important role in visual and other functions in social and sexual behaviour. This third subsystem is the only one which has evolved further and now dominates in man. In consequence, the socio-sexual frame of reference in the human world is primarily of a visual nature. MacLean compares the limbic system to a simple radar screen which improves the orientation in the environment and enhances the flexibility of action. In short, the limbic system generally serves those mental processes which have been identified with the reflexive mind.

The limbic system processes information in such a way that it becomes experienced as feelings and emotions which become guiding forces for behaviour. In itself totally integrated, the limbic system seems to spread its discharges of electrical or chemical nature only within its own boundaries without including the neocortex. This would help to explain that feelings and intellect are so frequently not of the same opinion. On the other hand, according to MacLean, the limbic system is also capable of narrowing down the flexibility of thinking by strong convictions and to fix it on few tracks only. Elsewhere (Jantsch, 1975) I have discussed the fatal blocking of this feedback loop in which we project models and visions into the outer world which then return as powerful, rigid myths, such as in the growth myth or in the article of faith that we must not get into the path of progress.

The limbic system, however, also seems to be the location of many of those processes which lead to so-called altered states of consciousness. Many of the psychotherapeutic drugs act selectively on the limbic system. Hallucinations, "oceanic" feelings, mystic rapture and new space-time relationships under the influence of hallucinogenic drugs are explained primarily in terms of processes in the limbic system. But the limbic system also seems to be the stage for action by the recently discovered, sensational hormones called endorphins. These endorphins consist chiefly of three polypeptide substances which normally are joined in a very large molecule (beta-lipotropin) and may induce very different types of behaviour. Roger Guillemin of the Salk Institute near San Diego, who has received the Nobel Prize for this endorphin research, has recently suggested (*Brain/Mind Bulletin,* 1977a) that production and circulation of these hormones form a hitherto unknown neuro-endocrinal system which functions relatively slowly and the effects of which last for hours or days. Such a system would establish a link between brain functions and behaviour, or in more general terms, between neural and metabolic mind.

Between the reptilian brain and the limbic system there are strong connecting pathways which pass through the hypothalamus and the subthalamic region. Here, in the exchange between the two older parts of the triune brain, the autonomous personality forms essentially along with its rich repertoire of non-verbal expression. If these connecting pathways are cut, the organism is only capable of some of the simplest and most vital functions.

The *neomammalian brain,* finally, which consists essentially of the neocortex and the structures of the brain stem connected with it, originated perhaps 50 million years ago, in the earliest phase of primates. The explosive growth of the neocortex in a later phase of evolution is one of the most dramatic events in the history of life on earth. The neocortex plays an important role with the more highly evolved mammals and dominates in primates and in man. MacLean compares it to an immense neural screen on which the symbolic images of language and logics (including mathematics) appear. With the capability of abstraction the emancipation from the reality of the outer world becomes possible. On the other hand, the neocortex receives sensory data from the outer world. Thus, there is not only the symmetry break between the outer world and its symbolic abstraction. The abstraction—we may also say, the idea or the vision—superimposes itself over the existing reality and starts the creative process of the transformation of the outer world. In this way, the function which now dominates in the technological age developed. The neocortex is the location at which information is processed in the ways characteristic of the self-reflexive mind.

MacLean has made his complex concept graphical by comparing the roles of the three brains in one with the structure of literature. In this image, the reptilian brain stands for the archetypic figures and roles which underlie all literatures. The limbic system brings emotional preferences, selection and development of the scenarios into play. And the neocortex, finally, produces on this substrate as many different poems, tales, novels and plays as there are authors. In this metaphor we recognize again the theme of evolution in the direction of individuation which has been evoked in the preceding chapter.

Autopoietic levels of mentation

If we call the activities of the neural mind mentations, we may subdivide the latter into the three levels of organismic, reflexive and self-reflexive mentations. As we have seen, they correspond to the evolution of the triune brain.

Organismic mentation belongs to the autopoietic level of the complex organism and meets there with the processes which represent the metabolic mind of the organism as a whole. We may therefore also say that organismic mentation is an integral aspect of the organism in its holistic self-expression and its environmental relations. As has already been mentioned, the reptilian brain serves to a large extent the co-ordination of the functions which characterize the organism as a whole, including symbolic self-representation.

With its focus on self-representation, organismic mentation may be assumed to play a major role in the arts. Perhaps it also transmits those primordial images which, as Jungian archetypes, underlie the images we make of our world from fairy-tales to the highest expressions of art. We do not know whether these archetypes are transmitted by genetic mechanisms or originate partly in common onotogenetic experiences, such as birth. From serial LSD experiments which the Czech psychiatrist Stanislav Grof (1975) has carried out on many patients, it becomes evident that, for example, the formation of the matrix of sexual feeling and experience is closely connected with the sliding through the birth canal. Elsewhere (Jantsch, 1976), I have proposed an ontogenetic model of human consciousness which includes such common ontogenetic experiences and describes the unfurling of the dimensional experience of the inner and outer reality.

The American astronomer and leader in the search for extraterrestrial intelligence, Carl Sagan (1977), has ventured the thesis that the archetypal dragon figures—which adolescents from St. George to Siegfried have to slay in order to become initiated as men—represent a memory of the time in which dinosaurs and mammals were engaged in bitter fight. Half jokingly, half seriously, he also interprets the Anglo-American ritual of eating two eggs for breakfast as an archetypal relic from this time in which the mammals stole and ate the eggs of the dinosaurs which were the direct predecessors of today's birds.

With the organismic mind a new level of genealogical information transfer across many generations is called into play. It is based on learning by imitation and therefore on direct communication between organisms. With the self-expression of an organism (for example, the flying of bird parents) the cognitive domain of another organism (the young bird) is oriented toward relations which are also open to him for direct experience. After epigenetics, a new level of learning enters the challenge-response relationship with the environment by which the evolution of behaviour is determined. This new level may be called the level of social and sociobiological evolution.

In contrast to organismic mentation, *reflexive mentation* opens up an entirely new autopoietic level which does not overlap with the level of the organism. Here, the symmetry between the inner and the outer world is broken which is still present in the ecological relations of the organism with its environment. The active outward projection of a model introduces a non-equilibrium factor. There is always some bias with which the image of the outer world is formed. This subjective attitude is already evident in the way in which sensory data are registered and processed. The brain destroys in several steps of abstraction part of the information—that part which cannot be expressed in the mental situation model (Stent, 1972). We may also say that confirmation is increased at the cost of novelty if novelty cannot be coped with.

We see solid objects where, according to modern physics, there is nothing solid, only ephemeral structures of energetic exchange processes. This kind of bias in seeing has to do with our perceiving the world primarily in the effects of its electromagnetic fields and manifestations. It has been shown by experiments that the human brain registers chiefly frequencies, optical and acoustical frequencies as well as those mechanical frequences which register as smells. This has led Karl Pribram (1971) to develop a holographic model of brain functions. In analogy to optical holography, first theoretically explored by Dennis Gabor, this hypothesis assumes that the frequencies of the outer world generate a kind of interference pattern which the brain—apparently the limbic system primarily in MacLean's scheme—interprets even on the basis of small parts only which still give a holistic image of the outer world. In optical holography it is possible to cut small chips from the photographic plate on which the interference pattern has been recorded which was generated by an object illuminated by coherent light. When penetrated by a beam of coherent light of the same quality, each chip of the plate gives the whole image; only the resolution of details gets worse with smaller chips. The image itself gives a three-dimensional impression and seems to be suspended in air behind the plate—an ephemeral structure of light which emerges from the interactions of optical processes. In a similar way, a world consisting of processes and vibrations and reaching us in the form of waves and specific frequencies may be translated into the image of a contiguous, solid world.

In this process of perception and transmission, the subjective aspect plays an equally important role with the objective one. The emanations from the environment are not accepted passively, but we meet them with a subjective model which is meant to order them. This model is comparable to the coherent light in optical holography (which, however, is necessary only in a particular version of holography). Only with the active appearance of the model do the frequency patterns make sense and generate the image of an orderly world. The structures which we perceive in the outer world are essentially based on the structures of our inner world—the structures of our multilevel mental organization. With the basic possibility of ordering the world with the help of alternative perception models, *apperception* comes into play. Nevertheless, it may be mentioned here that apparently bees also with only a few hundred thousand neurons are capable of apperception (Thorpe, 1976). This is at least the interpretation of their group decision-making process with which they seem to anticipate the parliamentary process.

The iterative feedback process between inner and outer world may lead to the creative evolution of the mental structure. We can learn to see a situation "with new eyes" and we become aware of it through a change in our emotional attitude. Depending on this attitude, we also influence the environment in different ways, especially the animate environment. If we feel well and are happy, this happiness will be contagious. Reality may be changed by our mental image of the situation, as it may also strongly influence this image from its side. A direct grasp of reality in the way we grasp a solid object is not possible—it emerges only in the experience of mutual, cyclically organized processes. Learning is no longer mere imitation, as it is in organismic mentation, but becomes creative experimentation, "learning by doing". We may also speak of a co-evolution of the inner and outer world.

Self-reflexive mentation, finally, again forms a new autopoietic level. The symmetry broken at its entry concerns the temporal order of experience. With dissipative structures, and still in organismic mentation, experience is bound to the processes of the matter/energy system; cognition and metabolism fall together. But in reflexive, and to an even larger extent in self-reflexive mentation, experience becomes emancipated. Not only can the experience of the past become effective in the present, as it also does in biological evolution—the new capability of *anticipation* makes the future also effective in the present. With the inclusion of the past and the future in the direct, alive experience of the present the relationships in the present become tremendously enriched. Not only the evolution of a world which represents thousands of millions of years of experience, is lived in the present, but also the vision of a multivariate, indeterminate future. Apperception, the consideration of alternative models, finds, in principle, full freedom in a future where no reality exists yet. The choice is, of course, pragmatically narrowed down. Not all imaginable futures are realizable. But in any case the concentration of evolutionary processes in mental structures enhances the intensity of life to an extraordinary degree.

Self-reflexive mentation in its higher stages of development is capable of the symbolic representation not only of the inner world (as already in the self-expression of the organismic mind), but also of the outer world. In this way, the outer world becomes manipulable—first in thoughts, ideas, plans and finally in the direct physical and social reality. With this potential power the technology emerges for enacting this power, physical technology as well as social technology. The newly gained flexibility in the symbolic representation of reality gives the anticipation of the future its full importance. Out of dreams and visions grow plans, out of wishes goals and out of hope creative action. In the pre-steps to self-reflexive mentation this was possible at best in rudimentary form.

Language

Rudimentary applies also to the language which preceded human language. The non-verbal language of the two older parts of the triune brain is sufficient for the establishment of the basic sociobiological relations. The communication of the organismic mind is limited to the orientation of another organism to specific processes which play in the cognitive domain of that other organism or are at least accessible to it. A kind of resonance is generated. The language of the bees based on their characteristic waggle dances provides a good example here. It transmits certain relationships which the reporting bee has experienced in the environment in relation to the sun's position and which may be re-experienced by all other bees. If a decision has to be made between two or more suggestions, the "enthusiasm" of the report is often the determining factor. It leads to the reinforcement of a fluctuation which finally breaks through.

Organismic communication, to put it this way, is symmetrically addressed to all sides. Reflexive mentation enters into a dialogue with other organisms, a dialogue which is emotionally coloured, shows preferences, weighs different opinions and is thus no longer symmetrical to all sides. But only self-reflexive mentation is capable of an intelligent discourse in which the outer world appears abstracted in verbal language. The simple sounds of animals, whether in self-presentation, for example in pairing, or in bird songs or in intimidating an opponent, or even in aimed signals, can hardly qualify as verbal language in the human sense. Experiments with primates have shown that they are capable of learning a simple logical language, but generally only for the enhancement of their self-expression, the satisfaction of their own needs and emotions, not for the symbolic representation of the outer world. The development of verbal human language is believed to have occurred in a period dating 100,000 to 10,000 years back. It was obviously reponsible for the extraordinary acceleration of human evolution which started 40,000 years ago with the development of complex tools and hunting weapons, shelter and boats fit even for travelling the open seas.

In the same way as autopoietic cell systems, the self-reflexive mind is also capable of storing in conservative structures the experiences from its exchange processes with the environment and in particular with other self-reflexive systems (for example, with other humans of the same intellectual capability). Works of art are one of the oldest examples. The most common form of such conservative storage, however, is *writing*. With it, a new version of genealogical information transfer comes into play which releases learning from the bonds to direct experience by observation, living or imitation.

The French ethnologist Claude Lévi-Strauss has distinguished between two basic forms of human societies, the "clockworks" and the "steam engines" (Charbonnier, 1969). The clockwork societies live practically historyless in a sociocultural equilibrium without evolution of structures. The steam-engine societies, in contrast, undergo vivid evolution such as our society does. The distinction between them, according to Lévi-Strauss, is due to writing, the invention of neolithic man roughly 10,000 years ago. Mechanized printing technology, introduced by Gutenberg in the fifteenth century, provided another dramatic "evolutionary push", as does modern copying technology in our days.

With verbal language, an extraordinary intensification of sociocultural "ontogeny" started, with writing an equally extraordinary acceleration of sociocultural "phylogeny". Even before the beginnings of that last glacial period, which started 30,000 years ago, there were distinct cultural traditions, magical rituals, burial offerings (which date back at least to a time 60,000 years ago) and social structures which were preserved over generations. With the development of fixed settlements 10,000–15,000 years ago the dialogue intensified, thus generating a need for its recording. This does not mean, however, that non-verbal symbolic self-representation stands outside of culture. On the contrary, it is, not only with animals, but also with humans, a very essential aspect of art. Art is one of the oldest forms of expressing human culture. And nobody who has ever watched the incredibly beautiful wedding dances of the strictly monogamous Bolger birds of Northern Australia, the play of dolphins or the elegance of wild horses, will doubt that this type of self-expression includes the dimension of culture as well. It is even much closer to human art than a verbal treatise in logic.

Verbal language, too, roots in the non-verbal, emotionally tainted experience of the organization of human life, as Suzanne Langer (1967, 1972) has emphasized so strongly. It is itself a product of evolution. This may explain the commonalities in the grammar of all known languages which Noam Chomsky (1976) has found. Language apparently has to do with the structure of a genetically anchored neural apparatus. It is not imprinted on the child from the outside as a "socialization theory" in the wake of Freud claims, but is learned in the interaction between genetic structure and environmental relations—in homologous correspondence to the epigenetic process which builds the physiology of the organism and to the process of perception/apperception. American newspapers recently carried reports of two children who grew up together and who were not keenly interested in their environment and consequently did not learn its language. They were classified as "mentally retarded"—until it was discovered that in their intensive bilateral communication they had created their own, highly complex language. Language, too, is a self-organizing system, a point which is also strongly made by Walter Pankow (1976).

The sociocultural re-creation of the world

Under the heading of sociobiology, I have described the development of social structures which was guided by metabolic mind. With sociocultural development, I am addressing now not only the development of culture (which belongs to multiple levels), but generally the development of a human world which is guided by the neural mind and which may already be found in rudimentary forms in the animal world. With respect to its dynamics, the human world belongs to both phases of evolution, sociobiological as well as sociocultural. It is sociobiological to the extent that it is shaped by material processes and thus by metabolic processes in the broadest sense. In the macrosystems of the human world, production and distribution processes as well as the movement of persons with and without means of transportation belong to the sociobiological aspect. It appears now as quite natural that it is this aspect of the human world which (as mentioned in Chapter 4) lends itself to modelling approaches of the same type as applied to the metabolism of dissipative structures and the dynamics of insect societies. It appears equally understandable that in Marxist theory—which embodies a model of human society which is essentially limited to processes of production, distribution and consumption—the emphasis would be placed on the collective. As has been pointed out in the preceding chapter, the sociobiological domain is characterized by a progression from rigorous collectivism toward increasing individualism. But this development is not simply continued in sociocultural development. Something new happens.

Sociocultural evolution turns sociobiological evolution virtually upside down. If sociobiological micro- and macroevolution were autonomous and connected by long-range co-evolution, sociocultural macroevolution now acts as a continuation of sociobiological microevolution. With this, I want to say that sociocultural macroevolution is now following the self-reflexive mind which unfolds in the organism of the human individual (see Fig. 32). We

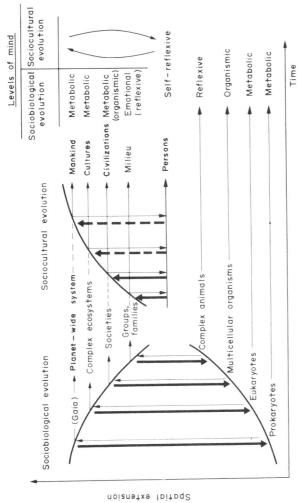

Fig. 32. The transition from the sociobiological to the sociocultural phase of evolution turns things upside down, as far as the dominant relationships in the co-evolution of macro- and microsystems are concerned. Self-reflexive mind, characterizing the individual at the level of consciousness of a person, sets out to re-create the macroworld. The dotted arrows indicate that, at the levels of culture and mankind-at-large, this is still a partially conscious process only.

ourselves shape our society to a good deal with our neural mind. We may plan many aspects of our social life as long as we follow the guiding images which are valid at a given moment. But these guiding images, too, are manifestations of our own individual mentations. We determine the "inner limits of mankind", as Ervin Laszlo (1978) calls them, to an even higher extent than we do the outer limits with the help of technology.

In the ecosystems of the metabolic and the organismic mind, niches do not emerge from the random exploitation of an offer of relations, but from the active design of a relation space which is created *ab novo* and which enhances the complexity of the overall system. In human ecology it is no different. Here, the niches are called institutions and they serve the living-out of mentally conceived values. Values, however, are irreducible to material conditions, to survival and to the satisfaction of needs and wishes—in spite of the efforts of behavioural science to prove the opposite. In other words, values are not one-sidedly confirmation. On the contrary, they develop their own life which originates in the dynamics of the neural mind and which changes, within limits, the material world.

The life of values maintains the balance between novelty and confirmation which is the sign of all life. This becomes obvious at the level of the reflexive mind at which we confront the world with our feelings. We value the world in a subjective way. Thus, it never stays the same and becomes confirmed to a certain extent. A landscape in full sunshine arouses different feelings in us than the same landscape under a cold drizzle. When we are in love with a person, he or she appears in unfathomable profundity and variety. I would even go so far as to claim that it is the inexhaustibility of a person—or, perhaps better, the inexhaustibility of our relations to this person—or in other words the source of novelty, which to a good deal constitutes the nature of love. If a person appears as totally predictable, as pure confirmation of our expectations, love has died.

The reflexive mind does away with a pseudo-objective world. It starts to align the world consciously with its values, in particular at the level of a community within the same biological species. The place of material processes in sociobiological binding is now, in sociocultural self-organization, taken over by bonds which are essentially of an emotional nature. The world assumes a new dimension of warmth. Altruism, in the sociobiological phase essentially furthered by group selection, is now consciously projected by the individual on to the level of the community. If, in the sociobiological phase, autonomy *vis-à-vis* the collective developed in steps, autonomy now broadens to the possibility of contributing actively to the design of the community.

This autonomy to redesign the outer world becomes tremendously enhanced with the advent of the self-reflexive mind. The ecology of feelings is now matched by an "ecology of ideas", as Sir Geoffrey Vickers (1968) has called it. Ideas, plans, world views and ideologies include values or whole value systems which may now enter exchange processes. They bring many elements of novelty into play. Sociocultural development has become emancipated from the dictate of the environment to a very high degree. For example, very different types of civilizations developed in the Mediterranean region under similar environmental conditions. At the level of self-reflexion, values are frozen into their structures to a much lesser extent than they are at the level of feelings. They enter a phase of accelerated co-evolution.

Co-evolution, however, appears in a totally new light at this level. The auto-poietic structures of the self-reflexive mind—the single ideas, plans and visions as well as the macrostructures of religions and ideologies—regulate the life and evolution of our societal macrosystems from families and groups through communities and nations to the systems of world-wide co-operation. As individuals we live, so to speak, in *co-evolution with ourselves*, with our own mental products. At the moment in which the evolutionary process with all its temporal span from the past into the future has become concentrated within us—with past biological and anticipated neural information—it also concentrates spatially within us. The self-reflexive mind not only relates the whole world to the individual, it also relates the individual to the whole world. From now on, everyone of us assumes *reponsibility* for the macrosystems. Not only for our societal systems, but also for the whole planet with its ecological order, and soon perhaps for space transcending our planet. . . .

Complementarity of subjectivity and objectivity

We become aware of this responsibility for the physical and social systems when, besides the relations of a material and energetic nature, we also establish relations among our emancipated mental structures. Compared with the continuing material processes through which we depend on the collective to a high degree, our mental structures may, in principle, participate in much greater freedom in this ecology of ideas. It would be difficult today to envisage the manufacturing of complex technological products in other ways than by processes at an industrial scale. In the economic domain, only those fluctuations have a chance to break through which are capable of mobilizing considerable social structures—which, in turn, may originate in an individual mental fluctuation, for example, in an invention. In the domain of mentations, however, individual fluctuations may result in a large impact in the sense of social restructuring—were there not innumerable built-in elements of stabilization.

This stabilization of specific mental structures already acts in the interplay of the three brains in the triune brain. The self-reflexive mind seems made to experience and test a maximum variety of possible views, concepts, plans and ideas. What fascinating, never-boring life this would be! But the older brains put strong reins on it. The limbic system quickly decides which ideas it wishes to support emotionally and curtails deviations, and if necessary, is supported by the reptilian brain which permits only the presentation of a single *"idée fixe"*.

It is, of course, possible to counter this to some extent. One may proceed systematically in such a way that the neocortex is confronted with new variants. The Swiss astronomer Fritz Zwicky (1966; see also Jantsch, 1967) developed his morphological analysis in America during the last World War. It consists of analysing a problem or a situation and dissecting it into its basic parameters. Subsequently, all conceivable (and not all "meaningful") variations of these parameters are laid out. Finally, all possible combinations are formed of one variation each of all parameters. The internally inconsistent combinations are eliminated and the rest is taken seriously. An analysis of chemical jet propulsion which Zwicky carried out, yielded no less than 25,344 possible configurations which also included several versions of a "hydro-jet" and of a "terra-jet". Under the parameter "medium in which the jet is operating", water and earth appeared as variations besides air and vacuum. The opposite to this eye-opening way of thinking was provided at about the same time by Churchill's scientific adviser, Lindemann. When he saw the photographs of the V-2 taken during reconnaisance flights, he declared apodictically that these things would never be able to fly. Since he was an expert in solid rocket propulsion, he excluded *a priori* the possibility of liquid propulsion.

Science itself, a field of activity *par excellence* of the neural mind, has fallen into the traps of the limbic brain and in some stubborn cases even into the traps of the reptilian brain by narrowing down to specific teachings and by claiming to represent knowledge exclusively and absolutely. There is no small irony in the fact that "objectivity"-claiming science originates in the most subjective aspect of evolution, namely, the self-reflexive mind. The narrowing down to a single view which subsequently is felt to be objective is a result of those "sub-threshold" processes which originate in the non-analytic parts of the brain and by which Western science is most horrified—and which even are connected with "unspeakable" oral and genital functions.

The same phenomenon recurs at the level of society. There, too, entrepreneurs, offices, institutions and the whole state seek to defend their structures in emotionally tainted ways. In more than two-thirds of the 150 nation states of the world, the utterances of the mind are severely restricted. Even in the nations of the so-called "free world" it is difficult for the individual to be heard—and even more difficult to present his thoughts to an audience. It is less the material processes which resist restructuring than the mental guiding images. The profound connoisseur of the mass society, Ortega y Gasset (1943), pointed decades ago to the importance of these guiding images. He has characterized their change correctly in such a way that the process takes place in the individual but becomes effective only when the fluctuation breaks through at the level of society. Guiding images are individual mental structures on the one hand and self-organizing mental macrostructures on the other.

The philosopher Nicolai Hartmann has characterized this process as interaction between the subjective and the objective mind. But what holds for science also holds for the societal domain, namely that the design of the seemingly "objective" social reality is due to the most subjective of all human functions, the self-reflexive mind. We may speak of objective mind outside the human organism only with respect to the sociobiological aspects of the human world which are based on the metabolic processes of production and distribution.

But perhaps things become clearer here also by applying the Weizsäcker notions of novelty and confirmation. Then, it may be stated in a general way that the structures which are old in evolution—and also the parts of the triune brain which have evolved first—tend toward confirmation whereas the structures which are found in evolution tend toward novelty. Confirmation may easily be confused with objectivity, novelty with subjectivity. I believe that the interplay of these complementary notions in pragmatic information expresses much more than the other pair of notions, subjectivity and objectivity. The latter pair changes its meaning depending on the angle of view and tends to confuse us in our "co-evolution with ourselves".

Evolutionary opening by creative mind

The development trend of the neural mind points in the direction of increasing novelty, as any "spearhead" in evolution does. Or perhaps we merely find ourselves in a phase corresponding to the restructuring of a level of life, or its original activation. Already in the dynamics of chemical dissipative structures the first phase after the crossing of an instability is devoted to the securing of a new structure by maximum entropy production, and thus primarily to novelty. But later, economizing, that is to say, confirmation, moves into the foreground.

According to recent research results, it seems that the neocortex itself develops the mechanisms which may bring more novelty into play. This possibility may be inherent in the much-discussed hemispherical differentiation of the neocortex which is usually characterized by the juxtaposition of notions such as "analytic, digital, verbal" (left half of the brain) and "holistic, analog, non-verbal, musical" (right half of the brain). There is some evidence for the suspicion that such a scheme mixes neocortical and limbic functions, and perhaps even functions of the reptilian brain. The co-ordination of the three hierarchically arranged brains is one of the most important tasks which man has yet to learn. One of the most important issues is the redemption of the limbic system from its rigidity. It may open to us a phantastic world of "different realities"—as it already does in dreams, hallucinations and under the influence of hallucinogenic drugs—and thus also stimulate the neocortex to let its imagination roam freely on this rich substrate instead of limiting itself to a single world view.

Results of a group of researchers at the University of British Columbia in Canada seem to indicate that the left half of the brain serves primarily the recognition of relations and the association of former experience, whereas the right half of the brain acts in a non-referential and integrative way *(Brain/ Mind Bulletin,* 1977c). In other words, the right half of the brain furthers novelty, the left half confirmation. Whereas the usually addressed complementarities verbal/non-verbal and analytic/holistic appear to be more vertically organized, with the holistic functions in the subcortical domain and the analytic functions in a part of the neocortex, the neocortex itself (or perhaps each of the three brains?) organizes itself according to the basic complementarity in pragmatic information. Logically, this ought to be precisely so—or is it only the emotions of my limbic system which makes me write this?

In this context, a first result of a research programme which Herbert Simon reported in a lecture in Berkeley in October 1977, is of interest. This research programme is devoted to problem-solving. If there are no well-tested logical-abstract operations at disposition, usually the way of a graphical imagination of the task constellation is taken, of a "physical intuition", as Simon calls it. Confirmation and logical analysis, novelty and holistic imagination seem to go hand in hand.

A new hypothesis which has been developed with great intellectual effort by the American psychologist Julian Jaynes (1976), would thereby become reversed. According to Jaynes, man learned only at a time no longer than 3000 years ago—long after the development of writing—to fully co-ordinate the functions of the two halves of his neocortex with the help of the 200 million connecting fibres in the *corpus callosum.* Since man (according to the prevalent American view) chiefly identifies with the left half of his brain, which science has promptly named the "major half"—a version of "more equal than equal"?—he experienced until then the functions of the right half of his brain as belonging to the outer world. Voices which seemed to come from there were ascribed to a god which thereby became part of a physically perceivable reality. Schizophrenia, in this view, would have been the norm at that time. Paul Feyerabend (1975), too, points out that in Homer's epic dreams, sudden fits of anger or strength were experienced as divine intervention: "Zeus builds up and Zeus diminishes strength in man the way he pleases, since his power is beyond all others" (*Iliad,* 20.241). The measure for Homer is not intensity but quantity.

Jaynes as well as Feyerabend bring this out in connection with the ways of thinking of ancient people which were very different from ours. But may it not be that these thinking functions were not the consequence of a development of brain functions, but on the contrary, characteristic of people which found themselves in a historyless phase of dominant confirmation? In such a phase, the intrusion of novelty may indeed be understood as a divine call. Sometimes, we experience such a call in our individual lives. Paul Claudel, in his play, *"Partage de Midi"* which deals with a man in the middle of his life, has given such a divine call the form of an inimitable word melody: "Mésa, je suis Isé, c'est moi!" Three times, the woman issues this call in an almost "non-verbal" sentence, the ship's clock rings noon—and Mésa rises from the emptiness of his life and is literally redeemed from it.

Out of the autopoiesis and evolution of mental, especially neural, structures emerges the magnificent wealth of human creativity. On the one hand it creates with technology a world of equilibrium structures, on the other with art and social institutions and organizations as well as in science and in the great religions and ideologies autopoietic/evolving systems of a symbolic as well as a real kind. It enters into new forms of symbiosis with other forms of life, in particular in agriculture and in the management of ecological resources.

Reaching far beyond the sociobiological bonds by material production and distribution processes, the self-reflexive, emancipated mind brings its own self-organization to bear upon the real sociocultural structures of mankind. Only a reductionist view may preach a materialism which interprets the history of mankind in terms of material processes only. In the human domain, as also at other levels in the domain of life generally, history becomes history of the mind in the true sense of the word.

In the last part of this book some of the aspects concerning the human world will be singled out and briefly discussed. However, I shall first try to give a more generally valid form to the basic principles of evolution which have been demonstrated by following the history of their emergence. In this way, the connectedness over space and time becomes visible which characterizes all aspects of evolution. This connectedness—and this is the principal message of this book—includes man and his history as integral aspect of evolution.

PART III

Self-transcendence:
Toward a System Theory of Evolution

The Hidden Law does not deny
Our laws of probability,
But takes the atom and the star
And human beings as they are,
And answers nothing when we lie.

It is the only reason why
No government can codify,
And legal definitions mar
　　The Hidden Law.

Its utter patience will not try
To stop us if we want to die:
When we escape It in a car,
When we forget It in a bar,
These are the ways we're punished by
　　The Hidden Law.

W. H. Auden, *The Hidden Law.*

Self-transcendence means reaching out beyond the boundaries of one's own existence. When a system, in its self-organization, reaches beyond the boundaries of its identity, it becomes creative. In the self-organization paradigm, evolution is the result of self-transcendence at all levels. Symmetry breaks unfurl space and time for the unfolding of self-organizing system dynamics, ranging from dissipative structures and the microstructures of life to the macrostructures on earth and in the whole universe. But these space-time structures do not remain stable. They evolve to ever new structures and at each threshold of self-transcendence a new dimension of freedom is called into play for the shaping of the future. Complexity unfolds in time and mirrors experience lived in the past as well as the creative reaching-out into the future. With the structures, the mechanisms of their evolution also evolve further. Neither the old concept of a teleological (goal-seeking) evolution, nor its contemporary modification in the sense of a teleonomic evolution (goal-seeking via a systemic network of possible processes), correspond to the new paradigm of self-organization. Evolution is basically *open*. It determines its own dynamics and direction. This dynamics unfolds in a systemic web which, in particular, is characterized by the co-evolution of macro- and microsystems. By way of this dynamic interconnectedness, evolution also determines its own *meaning*.

10. The Circular Processes of Life

Each cause is the effect of its own effect.

Ibn' Arabi

Cyclical organization—the system logic of dissipative self-organization

The notion of the organization of a dynamic system has already been introduced in Chapter 2. It refers to the *logical* arrangement of the processes playing in the system, not to their structure in space and time. The examples which have been mentioned for the system logic of self-organizing systems have always shown cyclical organization. This is to be understood in such a way that a closed process circle exists in the system which, however, is linked to the environment through other exchange processes. A system which is isolated toward its environment can only devolve in the direction of its equilibrium and when the latter is reached, the processes come to a standstill. Dissipative self-organization always involves exchange with the environment. We may speak of cyclical or closed organization in a system which in its totality, however, is not isolated.

Cyclical organization is characteristic for dissipative self-organization and especially for the system of life. In cosmic evolution, the Bethe-Weizsäcker cycle which in stars transforms hydrogen into helium, is cyclically organized; in chemical evolution it is dissipative reaction systems and in precellular evolution the hypercycles, according to Eigen. However, cyclical organization also characterizes the Gaia system, all types of epigenetic and epigenealogic development as well as the activity of the neural mind, especially in its reflexive and self-reflexive versions. Heinz von Foerster (1973) points out that already in very simple unicellular organism perception always requires the evolution of a closed process circle between changes in the sensory impact on the one hand and changes in the form of "effectors" on the cell surface, which respond to the sensory impact, on the other.

A hierarchical typology of self-organizing systems

In the introduction to their ground-breaking trilogy on hypercycles and pre-cellular evolution, Manfred Eigen and Peter Schuster (1977/78) sketch a simple hierarchy of cyclical reaction systems which are of importance in natural dynamics: A cycle of transformatory reactions, as a whole, acts as a catalyst; a cycle of catalytic reactions, as a whole, acts as an autocatalyst; and a catalytic cycle of autocatalysts, as a whole, acts as a hypercycle. This simple scheme may be extended and generalized as shown in Fig. 33. In this extended form, it provides a unifying perspective for a wide range of self-organizing systems.

Both cycles of transformatory reactions and catalytic reaction cycles use and dissipate energy, either directly by converting free energy into heat (such as is the case in photosynthesis) or by transforming energy-rich starting products into energy-deficient end products. Cycles of transformatory reactions may be aided by catalysts which do not form part of the cycle. Catalytic cycles may be aided by auxiliary cycles of transformatory reactions. Biological systems usually comprise many such cycles in a complicated network which, in turn, forms a big cycle and which makes use of universally applicable intermediary products, such as ATP (adenosinetriphosphate), the "energy coin" of the cell.

The reactions in a cycle of transformatory reactions may be of various types, especially of a chemical and nuclear nature. Catalytic action may be either of the chemical type (i.e. supportive of a transformatory reaction) or use templates in which a positive instructs the production of a negative from which a positive may be obtained again. Chemical catalysis is usually of the heterogeneous type, which is to say it represents the global effect of a small transformatory reaction cycle in which the original reaction participants are adsorbed on to the surface of the catalyst and are transformed there, whereupon the end product is released and the catalyst is recycled. In the case of template action, complex information may be transmitted; in biological systems, nucleotides with their easily recognizable molecular shape act as templates. Autocatalysis at the basis of template action is self-instruction of self-reproduction and always includes a catalytic cycle in which positives and negatives alternate. Such a microscopic view confirms the hierarchy described by Eigen and Schuster.

For hypercycles to attain their characteristic dynamics it is not necessary that all participants in the cycle act as autocatalysts. Generally it is sufficient if one link in the cycle is an autocatalyst.

The same basic types of organization appear in Fig. 33 at different hierarchical levels. The *growth characteristic* of a cycle is not only determined by its logic, but also by the built-in degeneration and diffusion mechanisms. These mechanisms are characterized by the level of participants in the cycle which remain quasi-stationary (their concentration may oscillate somewhat), the reaction participants R_i in a transformatory reaction cycle or the catalysts E_i, the autocatalytic units I_i, or the hypercycle H itself. Natural phenomena with higher than hyperbolic growth are rare. Therefore, it is of little interest to continue the hierarchy of cycles although this is possible in principle.

Since all cycles are dissipative and maintain exchange with their environment, they may be viewed as the mediators of global reactions which transform starting products into end products (waste). There is always a net

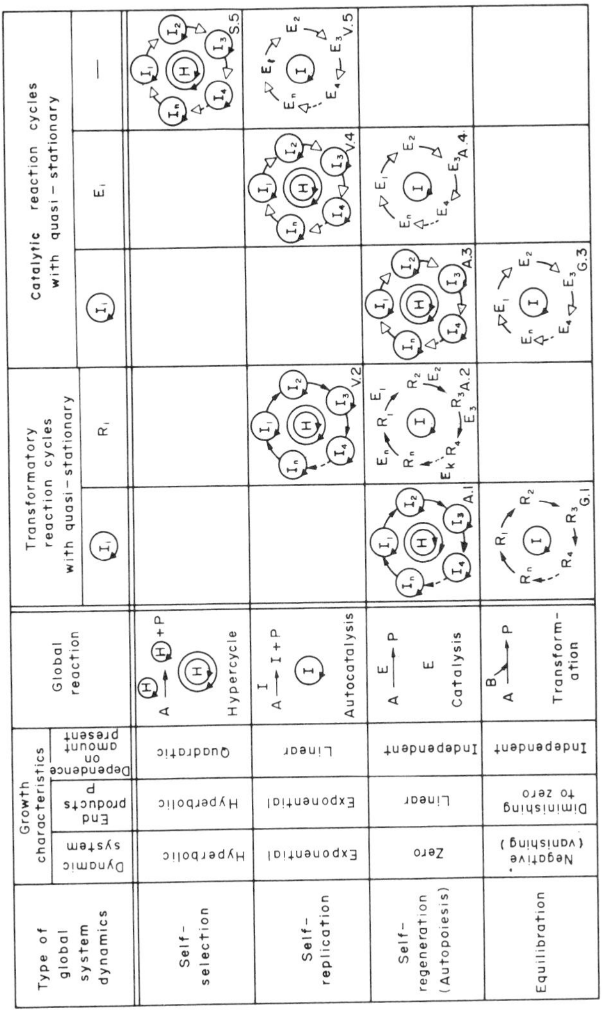

Fig. 33. A generalized scheme of reaction cycles with various degeneration characteristics.
    ⟶ transformatory reaction
    ⟼ catalytic action

entropy production connected with self-organization. Also, there is always a metabolism; in other words, autopoiesis is always accompanied by allopoiesis.

"Pure" examples of cyclically organized systems are not always easy to find. But this does not mean that cyclical organization is unimportant in nature. It is often hidden in the linking of several partial cycles. Its micro-scopic representation may surpass by far the complexity of the simple hypercycle. It depends which degree of resolution of details is required. With corresponding sacrifices in detail, all self-organizing systems may be repre-sented by the types of cycles appearing in Fig. 33 and reaching up to the hypercycle.

The transformatory and catalytic reaction cycles G.1 and G.3, which appear in Fig. 33 at the lowest hierarchical level, devolve toward their equilibrium. In the equilibrium state, the reactions become reversible and the system oscillates around its equilibrium state. In a macroscopic perspective, its dynamics vanishes. For the process cycle to turn always in the same direction (in Fig. 33 clock-wise), dissipation is required.

Autopoietic, self-regenerative systems

The next level up characterizes autopoiesis at zero net growth, i.e. pure self-regeneration. Viewed as a whole, an autopoietic system acts as catalyst for an open metabolic reaction chain in which "food" is transformed into "waste".

The subtype A.1, corresponding to a *hypercycle of transformatory reactions,* in which the *autocatalytic units I_i are quasi-stationary,* includes the *dissipative structures* such as the Belousov-Zhabotinsky reaction whose organization scheme is depicted in Fig. 3 (see p. 33). It is the space-time structure which regulates the system in such a way that generation and degeneration of the autocatalytic substances are globally balanced. Such a structure may evolve even while the organization scheme remains the same.

The same hypercyclical organization with global dynamic stabilization is exhibited by a mature *ecosystem* in which practically all participating matter—including the metabolic end products—is recycled (Fig. 34). Herbivores eat plants, carnivores eat herbivores (plus plants, possibly), predator carnivores eat prey carnivores, and so forth, until the last carnivore in the chain which is nobody's prey dies in a natural way. Its matter is broken down to its original elements and molecules by all kinds of animals and micro-organisms and fungi and, like the waste products of the metabolism, is recycled in the system by way of the plants. Viewed as a whole, this cycle catalyzes a metabolism in terms of solar energy which transforms energy-rich photons into energy-deficient ones. Autocatalysis is in this cycle represented by self-reproduction of organisms.

A particularly interesting hypercycle of the same basic type may be

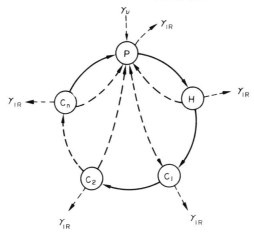

Fig. 34. A mature ecosystem is organized as a hypercycle of transformatory reactions in which all matter is recycled. The dotted arrows within the cycle indicate that also the metabolic end products, as well as the decay products after death, are recycled by the plants. Viewed as a whole, the cycle catalyzes the transformation of energy-rich photons in the region of visible light (γ_v) into energy-deficient photons in the infrared or heat radiation region (γ_{IR}). P, plants; H, herbivores; C, carnivores.

recognized in the *Gaia system* (see Chapter 6) in which several process cycles, especially one which links oxidizing and reducing reactions, are managed by the prokaryotes and their descendants, the organelles in eukaryotic cells.

The subtype A.2 refers to a *transformatory reaction cycle in which the reaction participants R_i are quasi-stationary.* Such a cycle has been mentioned in Chapter 5 when the *Bethe-Weizsäcker cycle* was discussed which plays an important role in the transformation of hydrogen into helium in stellar evolution (see Fig. 23 on p. 90).

Many of the basic biochemical reactions which participate in the energy processes of the cell, exhibit this type of organization. However, practically each reaction step requires catalytic support from the outside. Such catalytic support takes the place of an autocatalytic "motor" which may be shown to be a prerequisite for the pure self-organization of a dissipative structure of the subtype A.1. Eigen and Schuster (1977/78) cite as an example the so-called citric acid cycle for the oxidation of energy-rich molecules. This cycle, also named the *Krebs cycle*, for its principal investigator, appeared in evolution already 2800 million years ago. Another example is the *glycolytic cycle* which extracts energy from glucose and stores it in ATP; it is linked to the Krebs cycle.

The subtype A.3 represents *catalytic hypercycles with quasi-stationary autocatalytic units I_i.* A particularly interesting example is the long-range *evolution of the influenza virus A* which has been clarified by the work of Stephen Fazekas de St. Groth (Staehelin, 1976). This virus evolves for several years within a subtype by replacing an amino acid by a bigger one. This "forward" evolution with which the virus successfully evades immunization by co-evolving antibodies, sooner or later leads into a *cul-de-sac*. When this happens, the basic form of the subtype (which has survived in a small number) mutates in another amino acid position and thus forms a new subtype. With the appearance of the first population of the new subtype world-wide pandemics occur which are characterized by the most severe cases of illness. The appearance of the subtype A_5 in the year 1918 caused the death of 20 million people within a few months only and has become known as the "Spanish epidemic". Only in those areas in which an intermediary step, necessary for the evolution of the new subtype, has already caused local epidemics, has the human immune system partly caught up with this process and is able to protect the organism to some extent. Because of the limited number of amino acid positions available for mutations, the evolution of subtypes does not proceed indefinitely but forms a closed cycle (Fig. 35). It is hardly accidental that the period of this cycle is approximately 70 years and thus matches the average life-span of the human "prey". If it were shorter, the antibodies which form with each infection and remain in a life-long personal library of anti-bodies would not leave the viruses much chance to become effective. The autocatalytic units in this cycle, the individual subtypes, are, of course, not strictly stationary but reach population maxima in the times of epidemics. Viewed as a whole, however, the cycle is over long time spans autopoietic and self-regenerative.

The subtype A.4 refers to *catalytic reaction cycles with quasi-stationary catalytic units* E_i. There are many cycles of this type in cell metabolism, regulating the production and activity of enzymes, or biological catalysts. The production of an enzyme is catalyzed with the help of nucleic acids, but the substance which the enzyme catalyzes in turn may enter into a feedback coupling with the enzyme and inhibit its further activity. Inhibition and activation may be viewed as equivalents of catalytic action. Each positive or negative feedback coupling with a predecessor in the process cycle contributes to the autopoietic character of the cycle and to the creation of a dissipative structure which generally becomes visible through oscillations.

An interesting cycle of this type underlies some forms of chemotaxis and thus includes intercellular communication. This is the case in the synthesis of cyclical AMP (cAMP) which causes the aggregations of slime molds already discussed in Chapter 4 (Fig. 36). ATP-pyrophosphohydrolase (E_1) is activated by cAMP and catalyzes the production of 5' AMP from ATP. Adenyl cyclase (E_2) is activated by 5' AMP and catalyzes the production of cAMP from ATP. A third enzyme, phosphodiesterase (E_3), regulates the transformation of cAMP into 5'AMP and a fourth one, 5' nucleotidase, regulates the decomposition of 5'AMP and its exit from the cycle. This cycle connects through the secretion of

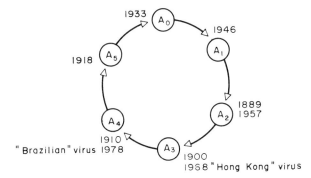

Fig. 35. The cyclical evolution of the subtypes of the influenza virus A forms a catalytic hypercycle extending over 68 years and thus corresponding to the average human life span. In this way, most of the subtypes evade the immunization by antibodies stored in the human organism from former attacks. Modified from T. Staehelin (1976).

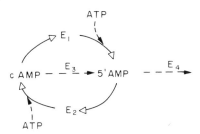

Fig. 36. The catalytic cycle mediating intercellular communication and chemotaxis between the amoebae aggregating to form the slime mold. E_1 pyrophosphohydrolase, E_2 adenyl cyclase. The scheme is highly simplified.

acrasine, whose active element is cAMP, the individual unicellular amoebae. If food is getting scarce, the secretion of acrasine occurs in rhythmic pulses which give the signal for aggregation. Viewed as a whole, this catalytic cycle acts as a basis for the self-organization of the individual catalysts—the amoebae—each of which attracts all others and is attracted at the same time. We recognize here again the effects of biological, pragmatic information in which, as has been discussed in Chapter 3, each receiver is at the same time a sender.

Systems with growth dynamics

If generation and degeneration of the participants in the cycle are not in balance, exponential or hyperbolic growth may result in typical cases. In *exponential* growth, the increase is proportional to the amount present. A fixed capital investment which draws a specific interest grows exponentially. The doubling time remains constant; e.g. the capital doubles in the first 10 years and quadruples in 20 years. *Hyperbolic* growth, in contrast, exceeds exponential growth. It increases with the square of the amount present and the doubling time is halved at every doubling.

Exponential growth is characteristic of that hierarchical level in Fig. 33 at which the system is capable of *self-reproduction* in the sense of a net increase (V). Hyperbolic growth, as discussed in Chapter 6, also includes the capability of *self-selection* (S), as Eigen (1971) has shown.

The subtype V.2 refers to a *hypercycle of transformatory reactions in which the individual reaction participants R_i are quasi-stationary.* Such a cycle is, for example, characteristic for material recycling in economic systems which participate in the general growth. Material recycling is a principle which is only partially realized in industrial economies. However, there are partial process cycles in which valuable metals are recycled or in which the products of one step are required for the maintenance and extension of the products at another step (for example, coal and steel).

If the recycling of matter is an ecologically sound principle, the recycling of energy, in contrast, is a principle which is ecologically sound only in an open cycle, as in ecosystems. In a closed cycle it leads to an equilibrium system in which the dynamics eventually vanishes. In modern economic systems, there is an increasing number of cycles in which energy production is made possible only by the reinvestment of a large share of the produced energy in the mining and processing of fuel as well as in the construction and operation of power plants. A cycle is maintained with a huge effort which is merely self-serving to an extent which is usually under-estimated. If it is supposed to go on turning, energy and material have to be continuously invested from the outside. Entropy is produced without doing much good. In some of these cycles, especially in electricity generation, the systemic net efficiency is as small as 10 per cent.

The subtype V.4 represents a *catalytic hypercycle with quasi-stationary catalytic units E_i.* An example for this type is a young ecosystem which evolves in the direction of higher complexity. Also, certain economic growth systems belong to this type. A growing agricultural system, for example, may be represented by such a cycle. A part of the harvest and the newly born animals is consumed by humans, but a constant share is reinvested in seeding and animal breeding in such a way that the overall system grows always by the same percentage—including the human population depending on it, which is also part of the system. In the service sector of a modern economy, too, there are catalytic hypercycles. Examples are the stimulation of tourism by the development of transport systems which, in turn, results in further stimulation for the transport systems to grow.

The subtype V.5 refers to *catalytic reaction cycles without degeneration at all.* Eigen and Schuster (1977/78) cite as an example the enzyme-free reproduction of single-strand RNA which is based on the mutual instruction for the formation of positives and negatives in template action. Metabolism is built in insofar as energy-rich nucleotide-triphosphate molecules are used as starting products and energy-deficient pyrophosphate is ejected. Viewed as a whole, the cycle acts as an autocatalyst of positive and negative forms.

Another example may be recognized in the cyclical growth of activity and knowledge in new scientific and technological fields. A certain degree of activity leads to publications which in turn generate more activity and thus a further increase in publications. The nearly exponential growth of scientific and technological literature in the 1950s and 1960s may be ascribed to the cumulative effect of many such hypercycles. In more "mature" fields redundancy and other "degeneration effects" become felt and they reduce the growth of activity as well as of publications.

Finally, at the highest hierarchical level in Fig. 33, the level of self-selection, hyperbolic growth rules—at least in intermittent pulses. The subtype S.5 is characterized by *catalytic hypercycles without degeneration effect*. It is exemplified by the types of hypercycles on which, according to Eigen, precellular evolution is based and which have been discussed in Chapter 6.

At the present stage of evolution, such a hypercycle—also pointed out by Eigen and Schuster (1977/78)—acts in the *RNA phage (virus) infection of bacterial cells*. The virus lacks both the metabolic and the translation systems for its self-reproduction. Therefore, it invades a host cell and uses its translation mechanism to first synthesize a protein which, together with other proteins of the host cell, forms an RNA replicase which, in turn, becomes the basis for the production of positive and negative forms of the virus. For this task, the "tricked" host cell even lends its metabolic system. The result of this simple hypercycle, constituted by the autocatalytic virus and the catalytic replicase, is hyperbolic growth of the virus until the metabolic support becomes exhausted. The host cell bursts and approximately a hundred new viruses enter the world.

The most conspicuous phenomenon of hyperbolic growth, however, occurred in the growth of human world population over the past 300 years, and perhaps already in some earlier periods. In the scientific-technological age, this type of growth may be explained by improved food and health technology which removed "natural" limits of population growth. New land was used for food production and the yield increased markedly. If, at the same time, infant mortality decreases, the result is a "population pyramid", which is no longer a pyramid but widens in a trumpet shape in the direction of younger people. If the precarious autopoietic balance in the reproduction rate is exceeded only by 1 or 2 per cent, the population doubles in ever shorter periods. Only in our days is hyperbolic growth gradually being broken by the cumulative effect of several factors. These factors include not only limits in available resources, but also limits which form in consciousness. They are based, for example, on the possibility of increased participation in cultural (or pseudo-cultural) activities in an increasingly urbanized environment so that children are considered a disturbance. On the other hand, however, the social development which points in the direction of the welfare state removes some of the worries connected with old age. In developing countries, a large number of children is still often the only possibility of establishing a kind of private "social security" and also provides cheap labour. At present, the growth of the world's population is a little less fast than exponential growth (1.7 per cent annual growth rate in 1977, compared with 1.9 per cent in 1970).

Co-evolution of cyclical system organization

Process cycles may evolve in two respects, by mutation of the reaction participants and by the development and integration of new processes. The first type is usually a prerequisite for the second. In prebiotic evolution as well as in later genetic reproduction, errors in the information transfer may result in mutants which are favoured by natural selection and—growing hyperbolically—condemn other mutants to extinction. What is selected in this process, however, is not a single individual or a sharply determined molecular species, but always—as Eigen and Schuster (1977/78) emphasize—a statistical distribution of individuals, a "quasi-species". The horizontal connection established by the type of information transfer which is characteristic of bacteria (see Chapter 6) contributes obviously to such a statistical distribution.

Eigen and Schuster have also shown that the type of cycle and the functions which it includes determine the limits of complexity in information which may be transferred by template action. For enzyme-free RNA reproduction as it corresponds to subtype V.5 in Fig. 33, the number of digital units (of individual nucleotides) is limited to about 100. The reproduction of single-strand RNA by means of a specific replicase, as it corresponds to the simplest case of subtype S.5 and acts in the described example of an RNA phage infection of a host cell, extends the complexity—at the basis of the same error probability—to approximately 10,000 nucleotides. The reproduction of double-strand DNA via polymerase and including "proof-reading" by exonuclease—a hypercycle of third degree—is limited to about 5 million nucleotides which comes very close to the maximum number of nucleotides found in large bacteria. Sexual recombination of DNA in eukaryotic cells, finally, which represents the coupling of hypercycles, extends this limit to approximately 5000 million nucleotides, a number which is only twice the number of nucleotides found in human DNA. The conclusion may be drawn that genetic development, at least on a sexual basis, has almost reached its limits in the human organism. Perhaps this may serve to explain that the physiological evolution of man has stopped—with the important exception of the brain, and especially the neocortex. Epigenetic evolution, based on the flexible utilization of genetic complexity, goes far beyond genetic evolution, but is ultimately limited by the latter. Beyond, however, lie the genealogical and epigenealogical processes of social and sociocultural evolution which are still far from reaching their inherent limits of complexity.

Epigenetic development as well as any other epigenealogical development leads to the co-evolution of two or more systems as it is schematically presented in Fig. 37. As a rule, co-evolution proceeds in an open way and not in cycles (as in the case of the influenza virus A, discussed above). Generally speaking, co-evolution transforms the cyclical organization of self-organizing systems into a helix which becomes visible over longer time spans. This corresponds to the idea already introduced of the *ultracycle* proposed by Ballmer and Weizsäcker (1974). According to this idea, each autocatalytic unit in a hypercycle represents a niche within an ecosystem. Here, the notion of the niche is to be understood in such a way that each niche, in turn, represents a smaller ecosystem. Organisms in ecosystems usually participate in more than one niche. Each mutation occurring in a niche—whether it is a genetic mutation or the addition of a new species or the establishment of new dynamic relations—stimulates (catalyzes) changes in other niches, too. In particular, an increase in the complexity of one niche stimulates corresponding increases in the complexity of neighbouring niches. In this process, organisms which participate in more than one niche play a significant role. We may also say that the co-evolution of niches is linked by positive feedback. The result of such a co-evolution at the level of the subsystems is the evolution of the overall system.

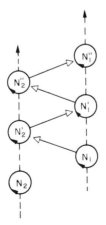

Fig. 37. Co-evolution of two niches in an ecosystem according to the ultracycle concept of T. Ballmer and E. von Weizsäcker (1974).

Complex economic systems and supersystems evolve through similar ultracycles. Certain sectors and groups of sectors may co-evolve in the framework of the national economic system. The national system, in turn, may co-evolve with other national systems in the framework of economic blocks or regional systems.

The ultracycle is a model for the learning process in general. Learning is not the importation of strange knowledge into a system, but the mobilization of processes which are inherent to the learning system itself and belong to its proper cognitive domain. Demonstration may act as a catalyst. Imitation, however, is not the acquisition of strange properties but the activation of potential properties of the dynamics of the system itself. Learning may generally be described as the co-evolution of systems which accumulate experience—a capability already characteristic of simple chemical dissipative structures. In the ultracycle information is not only transferred but also produced. With this, we come full circle and we may now repeat with enhanced conviction Ernst von Weizsäcker's formulation which has already been introduced in Chapter 3: Information is that which generates information potential.

Evolution is never total adaptation. It always requires destabilization, the reaching out, the self-presentation which offers new symbiotic relations, the risk accompanying all innovation. Evolution at all levels includes freedom of action as well as the recognition of a ubiquitous systemic interdependence. This interdependence, however, cannot be adequately grasped by a structure-oriented system theory. In such an older system theory, Ross Ashby's (1956) "law of requisite variety" claimed that for the control of its environment a system required a greater variety of relations, i.e. higher complexity, than the environment. But in life, the issue is not control, but dynamic connectedness. As Christine von Weizsäcker (1975) put it, "co-evolving systems . . . , play between adaptation and non-adaptation. Total adaptation and total non-adaptation are both lethal. In ecology, a niche fits the species sufficiently, without defining it; the species, in turn, fits the niche sufficiently, without defining it. What else is fitting, but not defining each other, than an emancipated relation." The basic non-equilibrium principle of dissipative self-organization, which holds for the dynamics of single dissipative structures, now reappears as a systemic prerequisite for the co-evolution of several systems.

From such considerations, Thomas Ballmer and Ernst von Weizsäcker (1974) have deduced their "general statement of evolution" which claims that the increase in complexity is maximized in any given moment on the surface of the earth. Communication, however, is the key to complexity. In the following chapter, we shall take a closer look at the connection between various types of communication on the one hand and the differentiation of form in the major phases of evolution on the other.

11. Communication and Morphogenesis

> The effect of a cause is inevitable, invariable and
> predictable. But the initiative that is taken by one or
> other of the live parties to an encounter, is not a cause;
> it is a challenge. Its consequence is not an effect; it is a
> response. Challenge-and-response resembles cause-
> and-effect only in standing for a sequence of events.
> The character of the sequence is not the same. Unlike
> the effect of a cause, the response to a challenge is not
> predetermined, is not necessarily uniform in all cases,
> and is therefore intrinsically unpredictable.
>
> Arnold J. Toynbee, *A Study of History*

A generalized scheme of types of communication

The world is full of vibrations originating in the manifold dynamics of dissipative self-organization. These vibrations are the communication media of a dynamic world. In Chapter 9, a distinction has been made between genetic, metabolic, neural and biomolecular communication. This scheme may now be generalized.

In the cosmic phase of evolution, there is already a kind of communication which results from the interplay of the four physical forces. From the macroscopic end, gravity brings about such an increase in the local energy density that nuclear exchange forces are brought into play which, in turn, acts back on the ontogeny of the macrosystem (the star). In the same phase, however, an exchange of energy and matter also starts between the system and its environment on the one hand and between different systems on the other.

Genetic communicaton is much more tangible because it transmits information which is stored in conservative structures. We may now speak in a general way of *genealogical communication* in which the conservative structures may be DNA as well as neural memory, books or works of art, buildings, roads and other artifacts. These artifacts need not be of human origin. Already the stromatolites, the reef-like, laminated deposits of unicellular prokaryotes, which have been mentioned in Chapter 6, transfer some conservatively stored information across thousands of millions of years. Other durable examples from the animal domain are coral reefs, or from the domain of the plants, fossil fuels. Conservative structures are equilibrium systems which essentially do not participate in exchange processes and change only slowly and partially. On the basis of these properties, conservatively stored information may be transmitted over very long spans of time. However, this is not always the case. A large part of the conservative information in ecosystems, such as paths, burrows, birds' nests and the structures of insect colonies, is not very durable and may be quickly destroyed by new processes. Therefore, there is only a rudimentary form of phylogeny in ecosystems.

Metabolic communication is the basic mechanism of systems capable of ontogeny and ranges from chemical dissipative structures all the way to ecosystems. It acts within the organism as well as between organisms, sociobiological and ecological systems. In many cases, chemical processes play the main role. Within the organism metabolic communication is based on information transfer by means of hormones, and between organisms, enzymes, pheromones and other active substances play the same role. In human society we may count the economic processes as a form of metabolic communication, including money flow. Metabolic communication usually acts relatively slowly because it depends on material transport. The travelling time of a hormone with the circulating blood along the full length of a human body is about 1 minute.

The notion of *neural communication*, too, may be generalized in such a way that it applies to communication within organisms as well as between organisms. Electrical and chemoelectrical processes play the major role here. They act in fractions of a second and are therefore in typical cases about a thousand times faster than metabolic communication. Within an organism, these processes use a nervous system or its predecessors. I call neural communication *between* organisms the direct, face-to-face visual/acoustical communication, which is greatly enhanced by language, on the one hand and the electronic transmission of information in the modern human world on the other. The result of the former both in the human and in the animal world is a group dynamics which may lead to a self-organization dynamics of the group as may never be attained by fragmented communication. This holds, for example, for collective decision-making processes from insects to man.

The visionary Italian architect Paolo Soleri builds Arcosanti, his compact city of the future, in the Arizona desert, approximately 100 kilometres from Phoenix. Since 1976, the first inhabitants have moved in. With 25 floors it is built for a "vertical" life style and for the future Soleri even envisages 300 floors. This compactness not only facilitates the optimal, multistage exploitation of solar energy, resembling the trophic levels in an ecosystem, but it is also meant to set neural processes in motion. If, in sociobiological evolution, compactness was the consequence of spatially limited metabolic group processes, it here furthers reversely the self-organization of a sociocultural dynamics.

The result of electronic communication over great distances, however, is the spatial extension of such a group dynamics and thus a kind of space-binding. In America there is considerable experimentation with so-called computerized conferencing, a group discussion over long distances which uses a common information-storage facility (Johansen, *et al.*, 1978; Hiltz and Turoff, 1978). As I write these lines, the power of space-binding has been proven in a new way. In the critical phase before the visit of the Egyptian president to Jerusalem, American television juxtaposed two separately recorded interviews of the two heads of government which were conducted from Washington by telephone. Thus, invitation and acceptance were linked in front of the public. Media diplomacy had a much more dramatic effect than the usual secret diplomacy, even if its success may vary.

Neural communication leads to a kind of self-organization of information which may result in emancipation from physical and metabolic reality and generate a new neural reality. This emancipation is achieved in the two decisive steps of apperception and anticipation, in other words, across spatial and temporal symmetry breaks. Apperception is fully developed only in reflexive mentation and anticipation only in self-reflexive mentation. But the metabolic mind, too, seems to be capable of some forms of apperception and even anticipation. The capability of apperception is clearly demonstrated by unicellular organisms exhibiting learning behaviour. Of course, there are no neurons yet. Learning always requires a decision among several possible views of reality.

For the human world, the system philosopher Ervin Laszlo (1974) has found typical attitudes toward the future, depending on the nature and the scope of the system. The difference in these attitudes may be explained by the differences in the dominant types of communication. With neural communication, the individual is capable of the clearest and farther-reaching anticipation. As Laszlo puts it, the individual is pro-active, which is to say he is capable of acting in a long-range anticipatory way. With the increasing scope of the social system, pro-active attitudes turn increasingly to reactive ones. This may find an explanation in the slowness of the metabolic mind which results in unclear images of the future. Mankind as a whole system only reacts *post festum*. Each of us is wiser and surer in action than the society in which we live.

However, significant changes may be in the offing, as is already becoming evident in the initial phases of introducing a societal electronic (neural) communication system. The days already seem far behind, in which wars and other major events affected only a part of the world. Today, the awareness of world-wide interdependence increases at a fast pace—which is partly due to the daily reporting of the electronic media. Fluctuations such as student unrest and abductions of airplanes, but also protests against the slaughter of seal pups and dolphins and other ecological crimes spread quickly around the whole world. The same is true of scientific concepts and discoveries. Perhaps the self-organization paradigm will, in such a way, result in a new existential attitude once it has broken out from the equilibrium structures of the academic world.

The development of memory

If metabolic communication acts alone, it merely serves ontogeny. Dynamic systems, such as chemical dissipative structures or ecosystems which practically do not possess effective mechanisms of conservative information storage, are limited to ontogeny, to the evolution of their own individuality. In this ontogeny, however, the ephemeral dissipative structures already form the basis for an ontogenetic memory which may be observed even in chemical reaction systems (see Chapter 3). Bacteria already show the formation of an ontogenetic metabolic memory based on conservative information storage by means of protein formation as it also plays a role in the neurons which appear much later in evolution. Bacteria are the first organisms with a conservative short-time memory.

The development of higher complexity along the axis of phylogeny requires conservative information storage which permits each step to build on the achievements of the former steps. Genetic communication works with such a conservative storage as does generally all genealogical communication over many generations. Sociocultural genealogy uses the outer storage facilities of buildings, scripts and works of art. They contribute to the phylogeny of the neural mind. Besides the genetic storage and the "tool storage" of technology (Schurig, 1976), the "image storage" of our symbolic world view with its archetypal figures now starts to play an equal or even dominant role (Schurian, 1978).

For the ontogeny of the organism, neural communication does not require outer information storage although it, too, plays a certain role in form of notebooks and diaries or other artifacts. But neural communication is capable of generating its own conservative long-time storage in the brain. It seems to do this by combining metabolic and genetic communication in concerted action. According to recent results obtained by the leading Swedish brain scientists Holger Hydén (1976), the beginning of the neural learning process is characterized by changes in electromagnetic fields which stimulate the short-lasting synthesis of two special brain proteins (so-called membrane differentiation). The calcium content of the involved neurons increases and acts as a mediator between electrical phenomena on the one hand and molecular changes on the other. The formation of a short-term memory takes seconds to minutes in animal experiments; the consolidation of a long-time memory, up to many hours. The type of new information and the level of emotional excitation play a role, too. Eight to 20 hours later, a soluble macromolecular protein is formed in various parts of the brain, only to disappear again after a few hours. It probably contributes to the consolidation of information in a long-term memory. Neural learning or the formation of a conservative neural information storage, according to this empirically founded model, is thus coupled with an epigenetic process. The unfolding of a multilevel process reality becomes clear in this coupling. Neural communication co-ordinates the metabolic and genetic types of communication which originated in earlier phases of evolution. The protein differentiation in brain cells is triggered in the genetic mechanism by metabolic communication. The genome (the totality of genetic information) available in the brain cells disposes over a sufficiently complex repertoire of protein production programmes to code the incredible wealth of our experiential space! However, and this may be of crucial importance, the coding uses not only protein and membrane forms, but also the macrostructures consisting of them.

The information storage produced in this way is not a punctual data storage in the usual sense, but a storage of structures which acts in associative ways. Hydén assumes that during information retrieval, the same kinds of stimulants which had triggered the differentiation of the brain cells may now again activate the same brain cells. A whole macrostructure of cells is activated which "exhibit the same quantitative and qualitative values in spatial-temporal co-ordination". This identity of stimulation in the coding and retrieval of information recalls us again the role of coherent light in holography and in Pribram's (1971) already mentioned holographic model of brain functions. The formation of a neural memory and remembering may thus be understood as a multilevel epigenealogical process which, at one level, also includes the epigenetic process.

Epigenealogical process—dissipative and conservative principles in interaction

Conservatively stored information is a dead equilibrium structure. It has been brought into a particular form and is in-formation in a true and direct sense. Only in a semantic context, in connection with the meaning of the formalized information in a specific situation, may information serve life. This context may be generated by metabolic as well as neural communication (see Fig. 38). We may call this linkage generally the epigenealogical principle.

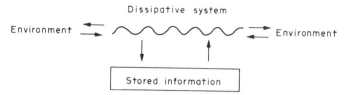

Fig. 38. Schematic representation of the epigenealogical process, in which conservatively stored information is selectively retrieved and used in dependence of a live, ever changing semantic context.

It characterizes all cognitive or knowledge-using systems, if the latter are defined as systems which, through their processes, "discover" genealogically transmitted dynamic rules (e.g. knowledge derived from experience), impose these rules upon themselves and thereby generate a variety of ordered morphological and behavioural patterns (Goodwin, 1978).

This way of using old information in a new semantic context plays a decisive role in biological ontogeny as well as phylogeny. In Chapter 8, examples for the epigentic process, the interplay of biological metabolic and genetic information, have been given. But the same type of information storage and retrieval also plays a very important role at the macroscopic branch of sociocultural evolution. It has already been mentioned that Lévi-Strauss traces "steam-engine" societies back to their use of writing. But information contained in books and scripts means nothing as long as it is not selectively retrieved and applied by the live processes which originate in a vision or a holistic idea. The same "reliving" of experience under new live circumstances takes place in the experience of art. Performing artists and the public become involved in the life of the creative artist by placing the frozen form of a poem or a drama, a painting or a sculpture, or the even more abstract form of a musical score into a new live context.

It already becomes evident here that evolution is basically a gigantic, multifaceted learning process, in its overall effect not a heuristic process in which the goal (as in school) is given, but an open learning process. Genealogically determined chreods which become particularly visible in genetic canalization, but also in social and cultural traditions, impose certain restrictions. The dynamically acting chreod may, of course, be interpreted as a sequence of intermediary goals—but such a view is beside the point. The difference between a given goal and heuristic alignment of processes on the one hand, and open, but canalized processes on the other is subtle but essential. It corresponds to the difference between structure- and process-oriented thinking.

Symbiosis

In the evolution of life the processes through which sytems enter mutual exchange are of great importance. Exchange with the environment is essential for all self-organizing dissipative systems. If there are other self-organizing systems in the environment, special types of exchange result. These types may be briefly characterized as follows. They also play a decisive role in the transition between levels of evolutionary processes as will be discussed in the following chapter.

We may distinguish here different notions which characterize the basic types of this exchange (see Fig. 39). Communication marks just one end of a broad spectrum of possible relationships.

If the exchange takes place between a system and an environment which is not structured at that level—for example, between a dissipative structure and an environment of non-co-operative molecules—we may use the neutral term *interaction*. A finite environment will be changed by this interaction, for example by the transformation of certain energy and matter reserves, whereas an autopoietic system maintains the same dynamic régime as long as the exchange does not break down due to the exhaustion of the environmental support. Finally, the system degenerates when it is no longer supported by the environment. With Maturana (1970) we may call the set of interactive processes (or, in other words, the mind) a niche or *cognitive domain*. An autopoietic structure "knows" which interactions it has to maintain with the environment in order to maintain itself. But cognition here is one-sided. The unstructured environment does not have a comparable cognitive domain and is incapable of maintaining itself.

In the exchange between two autopoietic structures, relations may be identified which, in their general form, correspond to symbiosis, and in their extremal forms to communication and fusion. If autonomy is maintained on both sides, we may speak of *communication* in the sense which Maturana (1970) has given this notion. In his model, communication does not include any transfer of products or knowledge from one system to another, but is based on the reorientation of the indigenous processes—in other words, the cognitive domain, or the mind—of a system by the self-presentation of another system and the processes which are indigenous to it. The verbal description of a colourful sunset transmits nothing of the real experience, if not by way of remembering a comparable experience of one's own. In other words, cognition falls here together with re-cognition, presentation becomes re-presentation.

Communication is possible only where the cognitive domains of autopoietic systems overlap sufficiently. In intellectual discussions, too, a "dialogue of the deaf" only too often results. The other system has to have the possibility, in energetic and functional respects, of partially realizing the same dynamics. Communication is not giving but the presentation of oneself, of one's own life, which evokes corresponding life processes in the other. This is the way in which whole living systems communicate with each other. It becomes evident, for example, in metabolic communication with the aid of hormones which stimulate and catalyze local processes. In a physical analogy, communication may best be compared with the phenomenon of resonance in which oscillators, practically without any energy transfer, stimulate other oscillators to do their own thing. Energy has to be spent only for the transfer medium (acoustical or light waves). But the communication of dissipative

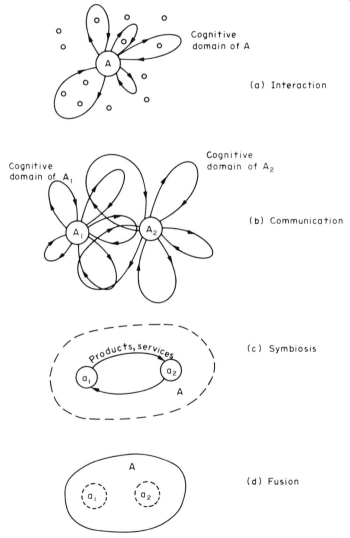

Fig. 39. Schematic representation of various types of relationships between two systems.

structures goes beyond mere resonance. Not only a simple frequency is induced in the other system, but a whole self-organization dynamics is stimulated and brought to its evolution. Each system has to make its experiences by itself, has to cope by itself with its structural problems and has to itself secure the energy flow to unfold its life. Since, in Chapter 9, self-organization dynamics has been equated with mind, we may characterize communication generally as interaction between mind and mind—not only of a neural, but also of a metabolic kind.

It has been reported that termites, even without their antennae with which they usually touch each other frequently, may become full-scale group animals if they are touched frequently enough by other termites (Watson, *et al.,* 1972). Separated from their colony, they become aggressive and begin to drink senselessly.

The birth process is an example of a process which is not produced but may only be furthered in its self-organization. The French gynaecologist Frederick Leboyer (1975) has developed a birth technique which represents an outstanding example of sensitive communication with a living system in the most difficult transition of its life. The most important thing, says Leboyer, is to give that living system the feeling of being welcome and to give him the certainty of continued relationship with the former matrix (the uterus). Low lights, maximum silence, putting the newly-born on the mother's belly are some of the basic rules of this birth technique. In particular, however, soft massage, in which one hand follows the other and gives the impression of continuous movement, evokes the memory of the wave-like motions of the uterus. Leboyer sees in such a massage generally a correspondence to the act of love which is to "rediscover the primordial slowness, the blind, all-powerful rhythm of the internal world, of the great ocean. Making love is the great regression."

A researcher at the University of California at San Francisco (Kenneth Pelletier, personal communication) has carried out experiments with the brain-wave patterns (EEGs) of a "healer" and his patient. A healer tries to restore the disturbed psychosomatic balance of a patient by laying on hands or other holistic techniques and to leave the healing of sickness or other disturbances to the endogenous processes of the patient. During the treatment, the EEGs of both persons showed very different patterns. Before the treatment, however, when the healer concentrated on his patient in order to "figure him out" intuitively, the healer's EEG for a few seconds assumed the precise pattern of the patient's. Cognition became recognition out of himself.

The same holds for communication in the sociocultural domain. True learning is never rote learning, but always stimulated experience by oneself. We may perhaps say generally that pragmatic information always requires life processes to become effective. Openness toward novelty, which characterizes effective pragmatic information, finds its correspondence in the openness of life.

Communication between autopoietic systems includes the possibility of the self-organization of knowledge by mutual stimulation of the exploration and extension of the cognitive domains. A true dialogue is never the exchange of readily available knowledge, but also active organization of knowledge which was not in the world before.

If the exchange between two autopoietic structures includes the essential mutual utilization of transformation products or services (such as mobility, or the elimination of parasites), we may speak of *symbiosis* in a narrow sense. In symbiosis, each system sacrifices part of its individual autonomy and gains in exchange the participation in a superordinate system and a new level of autonomy, with which the superordinate system establishes itself in the environment. Autopoiesis becomes modified in such a way that it plays now simultaneously at two semantic levels, the level of the individual subsystems and the level of the overall system. Thus, symbiosis leads to the formation of hierarchical organization in which, however, the lower levels partially maintain their autonomy. Societies and ecosystems are special forms of symbiotic systems. Just as the symbiosis of organelles secures the metabolism of the cell and the symbiosis of cells the metabolism of the organism, symbiotic systems of organisms secure the metabolism at the sociobiological or ecological level. Therefore, it comes as no surprise that a few modern architects—in particular, the Japanese Kisho Kurokawa (1977) and the German Wolf Hilbertz (1975) who lives in Texas—develop an architecture of metabolic symbiosis between man and his habitat which is based on a live correspondence between structure and function. Hilbertz even experiments with self-organizing building materials (for example, at the basis of ion exchange under water).

Symbiosis is usually defined in structural terms, that is to say by the relations between two or more entities, such as organisms. But it is perhaps more highly justified to speak of a *symbiosis of processes*, for example in a predator-prey relationship which destroys the entities of the prey species, but not its evolutionary process. On the contrary, as we have seen, both species benefit in a dynamic view. Thus, the notion of a process symbiosis leads straight to the notion of co-evolution.

When the autonomy of the subsystems is totally abandoned, symbiosis may lead to the extreme of *fusion*. The superordinate system remains as the only autopoietic unit. This happens in sexual fertilization, but also in fusion experiments between two fertilized egg cells—for example, of a grey and a white mouse, with zebra-striped young mice resulting—and even in the fusion between a human and a plant cell (Dudits *et al.,* 1976). One might assume that such a fusion of two different cell systems, each of which maintains at least a thousand chemical reactions, is extremely difficult. This is not so, however. A new complex microscopic order becomes established almost at once. It becomes possible, as in sexual recombination, also by virtue of the flexibility of the eukaryotic chromosome and its continuous regeneration and degeneration in the chromosome field. Nothing perhaps demonstrates more clearly the reality of order through fluctuation.

The biggest gain in complexity or knowledge occurs in the symbiosis of partially autonomous systems with the resulting two-level autopoiesis. This case corresponds to a balance between novelty and confirmation in pragmatic information exchange. Communication and fusion mark the extreme cases of symbiosis in which one level of autopoiesis becomes dominant in an absolute way—the level of the participating subsystems in communication, and the level of the superordinate system in fusion. These extremes rarely occur in a "pure" form which is already demonstrated by the endosymbiosis of the prokaryotes in the eukaryotic cell. Superficially observed, this is a clear case of fusion. And yet, the organelles to some extent maintain their autonomy in the framework of the larger cell, they keep their own genetic material and even the management of the world-wide Gaia system. "Pure" communication is equally rare. Metabolic as well as neural communication is usually accompanied by the exchange of matter, energy and information. Between fusion and communication, novelty and (far-reaching) confirmation, symbiosis represents a widely varying domain in the balance of novelty and confirmation which characterizes all life.

All living systems are characterized by symbiosis of some kind. In the hypercycles of the Eigen type we have already recognized the symbiosis between two kinds of molecules (see Chapter 6). Symbiosis may result in total mutual adaptation which represents a significant loss of autonomy without generating anything new in fusion. Such a loss in autonomy may be equated with a loss in consciousness, whether two animals become overly dependent on each other in their biological functions, or two humans in their psychological functions. In such a case, confirmation is maximized at the expense of novelty. The system approaches an equilibrium which sooner or later ends in biological or psychical death.

Communication in the major phases of the co-evolution of macro- and microcosmos

In the chapters of Part II, I told the history of evolution as viewed from the angle of view of the co-evolution of macro- and microcosmos. Summarizing the different phases of this co-evolution, an interesting picture emerges of a process which starts three times anew to generate complexity (Fig. 40). Not only do the communication processes between macro- and microcosmos, which are responsible for the spatial connectedness of the systems, change in the course of evolution, but also the processes which guarantee the coherence and continuity of evolution over time along each of its branches. Whereas the former processes primarily determine ontogeny, the latter make phylogeny possible. Only from the interaction of both does the co-evolution of macro- and microworld emerge.

In *cosmic evolution,* ontogeny is regulated by the interplay of physical forces and the results, both in the macro- and the microbranch of co-evolution, are

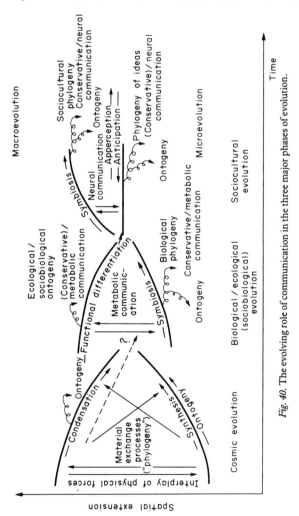

Fig. 40. The evolving role of communication in the three major phases of evolution.

passed on in the form of matter. Whereas macroevolution, in this phase, is at first characterized primarily by the condensation of matter, and therefore by conservative self-organization, different processes play a role in the synthesis of matter in microevolution; they also result in equilibrium structures (stable nuclei and atoms)—or it seems so at least from a macroscopic or intermediate angle of view. The increasingly active co-evolution of both branches seems to bring dissipative self-organization of macrostructures—cores of galaxies and stars—into play. Ontogeny dominates in cosmic evolution. However, there is a kind of unordered "phylogeny" in which matter is transferred criss-cross to new evolutionary sequences. As in the later biological phylogeny, complexity is thereby furthered, here in the form of planetary systems. Also, a controlled long-time burning of smaller stars, such as our sun, is ensured by the carbon cycle which depends on the "phylogeny" of some carbon. This controlled burning, in turn, makes the development of biological complexity on our planet possible. The units of this early phylogeny are highly normalized. The particularities of the history of matter transferred in such a way may be reconstructed only in vague contours, primarily by isotopic ratios which, for example, permit the exact dating of a nearby supernova explosion. In this way, we have only recently discovered something about the supernova which acted as a midwife for our solar system.

With the *origin of life* on earth, other types of processes are called into play. The connection between macro- and microworld is primarily ensured by the evolutionary ultracycles which have been discussed in the preceding chapter. At first, they become effective in coarse-grain, one-sided action, such as the transformation of the earth's surface and its atmosphere by the prokaryotes. This, however, facilitates the more finely tuned, continuously acting epigenetic processes in the eukaryotic organization of life's microevolution. Metabolic communication plays the major role in this second major phase of evolution. The results of this co-evolution, however, are no longer transferred directly in the form of matter, but enter in the form of information a process of true phylogeny. What evolves in phylogenetic development to higher complexity, is *organization*—an organization which, in principle, may be realized independently of time and space, as long as the environment is favourable.

This information transfer along the time axis does not happen in the same way on the macro- and microbranches of evolution. The development of the macrosystems is based on the development of macrostructures of metabolic processes which corresponds to a long-drawn ontogeny of dissipative structures. Ecosystems do not conserve the structure of their circular processes as Margalef (1968) has pointed out. Extinct species are replaced by new ones which establish different relations. The traces which the extinct species leaves on earth become quickly erased and hardly influence the evolution of the dynamics of the system. Viewed as a whole, however, an ecosystem which is still alive represents a web of dynamic relations, a space-time structure in which history is expressed in the same ways in which it is expressed in a dissipative structure (see Chapter 3).

Experience is transferred in an ecosystem in terms of dynamic rules or functions. What emerges is always shaped by systemic conditions. An ant hill, for example, is built in such a way that it can fulfil certain functions in relation to environmental processes, in particular temperature control and ventilation. Normally, and under constant local conditions, each generation uses the same architecture. This observation has led to the erroneous assumption that the genetically determined behaviour (the instinct) of the involved organisms blindly generates the same morphological structure which, in Darwinian selection, becomes confirmed or is changed over longer time-spans. The Austrian Nobel laureate Karl von Frisch (1974) reports in his wonderful book *Animal Architecture* the case of a termite nest which has been covered by a plastic tent, resulting in diminished ventilation. Within 48 hours, the animals had developed additional structures of a novel design which restored the old ventilation rate under the changed conditions. Obviously, information is transferred at the macrobranch of the evolution of life in terms of functions which may evolve in dependence of the exchange relations with the environment.

In the phylogeny of biological microevolution, in contrast, information is stored in conservative structures and transferred through them. However, it is not rigidly transferred but becomes effective in the interplay of genetic and metabolic communication in the epigenetic process. From this interplay emerges an intimate, creative relationship between ontogeny and phylogeny, a non-equilibrium between evolutionary processes which underlies the development of higher complexity.

Sociocultural microevolution, the evolution of the individual in the last of the three major phases, is based on the continuation of an outer differentiation in the differentiation of a symbolic inner world. The ontogeny of structures of the neural mind includes the organization of information which may be conservatively stored both in the outer world and the inner world (memory). However, it may also be generated in direct contact with other individuals and their mental structures. The phylogeny of mental concepts may take place in one or more individuals, in short or very long time-spans. The gain in flexibility, as compared with biological information, is evident.

The wide opening toward novelty, first in the present and the past (apperception) and subsequently in the future (anticipation), is always of prime importance. This holds for short-range plans as well as for the anticipations of religions which span whole eons. Cosmic and biological/ecological evolution have, in their phylogeny, transferred information which consisted primarily of confirmation, whereas novelty found an entry in the ontogenetic processes of the present (in the interplay of the physical forces and in metabolic communication). In sociocultural evolution, novelty increases in the experience of the present and the past, but may, above all, also break in from the future into the present.

Thereby, it is again emphasized that in the sociocultural phase of evolution the individual becomes co-responsible for macroevolution. The process of mentation originates in the individual, but the autopoietic structures of the neural mind form their own systems of relations which become translated into sociocultural macrosystems such as communities, societies and civilizations. In the same way, the neural mind shapes a world of equilibrium structures, such as buildings, machines and roads and interferes creatively with ecosystems, for example, by introducing agriculture.

If chemotaxis was a rigid control system for sociobiological behaviour, emotions now form a more flexible regulatory system for sociocultural dynamics. It becomes clear how a new level of evolutionary processes opens up a new level of indeterminacy and freedom. Although emotions are connected with certain physical correlates, such as biochemical reactions and changes in blood pressure, they clearly transcend the framework of physical exchange. The same holds for the attractive or repulsive effect of ideas, plans and visions. The symbolic re-creation of the world out of structures of "pure", materially unbound information determines the dynamics of self-organizing sociocultural systems in the first line. Thus, the transition has been achieved from the evolution of matter to the evolution of an immaterial, symbolic mind. The evolution of matter is followed by the evolution of the organization of matter, and the latter, in turn, is followed by the evolution of mental structures and webs of relations which have become emancipated from the material world. Mind and matter are complementary aspects in the same self-organization dynamics, mind as dissipative and matter as conservative principle. But mind transcends its own matter systems and is capable of the symbolic re-creation of the entire outer world in the matter system of the brain. Mind underlies self-transcendence and may evolve itself, as will be shown in the following chapter. But first, a few further dimensions of horizontal communication will be discussed briefly, which seem to play an important role in ontogeny and which enrich the simple picture of metabolic and neural communication as it has been developed so far. They place life on earth in a cosmic context.

The cosmic connection

The scheme represented in Fig. 40, showing successive phases of the co-evolution of macro- and microcosmos, obviously oversimplifies things. A more correct, but also more complicated representation would have to make visible the simultaneous action of all three phases. The evolution of life on earth, after all, is closely linked to the evolution of the sun and the solar system. All primary energy on earth (with the exception of nuclear energy and a small cooling effect of the earth's core) originated in the sun, be it in the form of direct radiation or stored in fossil fuels. If the matter aspect of the organization of living systems on earth corresponds to a terrestrial angle of view, the energy aspect cannot be separated from the sun. Thus, life is a cosmic phenomenon at the same time as it is a terrestrial one.

In the same way, sociocultural evolution is closely linked to biological and sociobiological evolution. In the description of the functioning of the neural memory earlier in this chapter, it already becomes clear that neural communication works together with genetic and metabolic communication. At present, however, it is also becoming clear that sociocultural evolution is linked with cosmic evolution. Space travel, the search for extraterrestrial intelligence and plans for space colonization are only the most conspicuous aspects of such a connectedness of which we are becoming increasingly aware. The self-image of man-in-the-universe which is designed by the self-reflexive mind, stands for a more subtle and spiritual connectedness out of which grows a sense of responsibility as well as meaning. We shall come back to this aspect in the last part of the book.

In recent years, science has become interested in a growing domain of phenomena which originate in cosmic influences on the systems of life on earth. Part of these influences act by way of the weather. As I write these lines, San Francisco is hosting a big conference of the American Geophysical Union in which the discussion of such phenomena, traditionally shunned by science, plays a major role (Petit, 1977). The focus of interest is on the influence of the sun's magnetic field and particle radiation—the "solar wind"—on terrestrial weather. It has long been known that geomagnetic storms with effects such as *aurora borealis* and disturbances in telecommunication are closely linked to events on the sun's surface. Recently, however, connections have also been found between the 11-year sun-spot cycle and storms, frequency of thunderstorms—and even the California drought. The important point is that the effects stimulated on earth involve 10,000 and more times the energy invested in the triggers. This makes one think, on the one hand, of the regulation of huge energy flows by means of very small energies in cybernetic systems, and on the other of the role of fluctuations in the self-organization and evolution of the autopoietic systems.

Of even greater interest—but also still taken less seriously by science—are direct cosmic influences on terrestrial life. There is no doubt any longer that certain cosmic and biological rhythms are coupled. It would be mistaken, however, to speak in every case of the "control" of biological and cosmic rhythms, as is usually done. Oscillation is a basic phenomenon of each self-organizing system and therefore also of a life. The relatively new branch of science which is called *chronobiology* investigates these oscillations; more of that in Chapter 14.

The coupling of the endogenous rhythms of biological systems (i.e. the rhythms due to their own dynamics) with cosmic rhythms is an essential aspect of the co-evolution of life and its environment. It expresses itself in macroscopic "biological tides" (Lieber, 1978). In many cases, such a coupling serves the better adaptation of life. All eukaryotic uni- and multicellular organisms, for example—and probably also the prokaryotes—exhibit circadian rhythms and rhythms of higher frequency. Circadian rhythms (meaning in literal translation "approximately of the period of one day") have, of course, to do with the alternation of day and night, light and darkness, which is of decisive importance for most life forms. Many basic activities of life are organized in circadian rhythms, ranging from biochemical processes within the cell and in the communication between cells to co-ordinated process systems in the organism as a whole. The hatching of the fruit fly *(drosophila)*, for example, is synchronized by light for large groups of eggs. On the other hand, the delicate one-day blossoms of my spider plant close and die precisely with sunset, whether the evening is bright and glorious or the incoming fog has brought early darkness. It is interesting that the circadian rhythms of the human organism seem to be synchronized more strongly with the cycles of social life than with the bright/dark cycle (Scheving, 1977). Here, too, sociocultural emancipation has resulted in making the mental reality independent of the physical one.

Besides the mechanical rotation of the earth around its own axis and around the sun there are apparently electromagnetic phenomena of planetary dimensions also which play a role in terrestrial life. An electromagnetic wave of the full length of the earth's circumference, i.e. 40,000 kilometres, has a frequency of approximately 7 Hertz (Hz) or cycles per second—frequency multiplied with wavelength equals the speed of light of 300,000 kilometres per second. Electromagnetic waves around the whole earth with frequencies of this order of magnitude do indeed occur. This reminds one of the fact that the "idling frequency" of the electrical brain waves, the alpha rhythm, has a similar frequency of about 8 to 14 Hz and that the brain wave rhythm gets even slower in sleep and dream states. Stroboscopic light of such a frequency induces alpha rhythm in the brain waves and sensitizes humans for resonances with other humans over a distance. This resonance effect, however, is not experienced consciously in the sense that one would become "rationally" aware of it, but expresses itself in changes in a number of vital rhythms, such as heart beat or breathing, or also in the degree of skin moisture, all factors which are connected with certain "sub-threshold" emotions. At the Stanford Research Institute in California it has been found that under such circumstances (stroboscopic light) resonances of these types occur when another person in a remote room lives a strong experience, such as an electro-shock. This seems to be the first experimental arrangement which has succeeded in making so-called parapsychological phenomena reproducible.

During the night, in particular in connection with Northern lights, electromagnetic waves occur with a frequency of around $\frac{1}{8}$ Hz and even longer waves with a frequency of $\frac{1}{40}$ Hz which corresponds to a wavelength of 12 million kilometres. They seem to originate in the depths of cosmic space. The longer the wavelength, the lower the energy loss in wave propagation. When electromagnetic storms occur on earth, the atmosphere acts as a huge resonator with the effect that certain resonances—7.8 Hz, 14.1 Hz and 20.3 Hz—may oscillate during long periods of time practically without absorption loss (Taylor, 1975).

Acoustical oscillations (in other words, mechanical oscillations of the air) which have the same low frequencies may be dangerous and even lethal to the human body, obviously by generating resonances in certain endogeneous frequencies of the body. We do not know with which dynamic level these endogenous oscillations are to be identified. It is known from electromedicine, however, that muscle fibres may best be stimulated with frequencies lower than 10 Hz and that blood circulation and pain relief respond best to a frequency between 90 and 100 Hz. The normal rhythm of the human heart beat is around 1 Hz.

Extraterrestrial influences upon terrestrial life may originate in the effects of radiation or gravitation. Science has studied some of these effects only with great hesitation and only to the extent that they are thought to originate within the solar system. A good, if not always sufficiently critical, summary has been given by Lyall Watson (1973). There are oysters which had been transplanted from the Long Island Sound to Evanston, Illinois, and which at both places opened precisely when the moon was highest in the sky. The tidal rhythms were thus not responsible for the opening of the osyters, but it was caused by the gravitational effect of the moon. The life of the crustaceans—their sexual cycle, the renewal of the shells and other regeneration processes—also seems to be connected with the effects of gravity. In the cases of other marine animals it seems less certain whether their sexual cycle, which is correlated with the moon phases, responds to gravity or to light. And the menstruation cycle of the human female? In peoples which live close to nature, it is not only coupled in its 28-day period with the cosmic rhythm of the moon, but also in phase with it (onset of menstruation one day before the full moon). In modern urban life this phase correlation may get lost to a large extent—again, sociocultural emancipation becomes effective. However, a strong coupling of the rhythm has been observed with the inhabitants of big dormitories in American universities, a kind of mutual "tuning in" of human oscillators which may be of biological or sociocultural origin.

The influence of the sun is not limited to seasonal oscillations (which are due to the orbiting of the earth). Plant growth as it is recorded in its history in tree rings, for example—a conservative storage of metabolic information—also shows a clear correlation with the 11-year sun-spot cycle and an even longer cycle of 80 or 90 years period. Solar eruptions may influence terrestrial life by way of increased ultraviolet radiation and the resulting increase in the ozone content of the atmosphere, or also by way of the magnetic field of the earth which reacts to solar events. It is known that many organisms respond in very subtle ways not only to changes in gravity but also to changes in the strength and direction of the magnetic field. Disturbances in the orientation sense of migratory birds, for example, can be ultimately explained by solar eruptions. In this context there also seems to be a certain effect due to the revolution of the planets around the sun and their changing configurations. These effects have not been well researched yet. They may not be due to mechanical or gravitational effects, but to the interaction between clouds of plasma (ionized gas) around the sun and the planets on the one hand, and between solar and planetary magnetic fields on the other. There seems to be a certain correlation between planetary configuration and sun-spot activity which, in turn, influences the occurrence of geomagnetic storms. What formerly has been rejected as unscientific astrology is now about to find at least to some extent plausible scientific explanations.

The most dramatic cosmic influence upon terrestrial life, however, may result at the molecular level by changes in the structure of water. All life forms consist of a high percentage of water and thus may be touched at this level simultaneously—perhaps not only in a disturbing sense, but also in the sense of a global regulation of system dynamics which would have a scope comparable to the self-regulation of the Gaia system.

In experiments carried out in the Biometeorological Research Centre in Leiden, The Netherlands (Tromp, 1972), the sedimentation rate of blood colloids—which represent non-equilibrium systems in full evolution—was measured as well as the production of albumin, gamma-globulin and antibody substances in the immune system. The observations showed cycles which corresponded on the one hand to daily and seasonal oscillations, and on the other to periods of 3, 6 and 11 years. By comparison with measurements carried out in other countries, climatic effects were eliminated and extra-terrestrial influences gained in probability for the explanation of the cycles of longer periods.

In the cosmic phase of evolution it is not difficult to see how new levels of evolutionary processes are called into play. The unfurling of the physical forces determines their importance in certain constellations and phases of evolution. Things are not so simple with the evolution of life. The self-transcendence of evolutionary processes, the reaching out beyond boundaries, is largely still a mystery. But it is possible to see the contours of systemic interdependencies which might bring some light into the dark. The key notion seems to be symbiosis which elevates the microevolution of life to higher levels. In the following chapter, I attempt to sketch the emergence of complex biological and mental life in a ladder of symbiotic steps. Each opening up of a new level of evolutionary processes corresponds to a particular break of a spatial or temporal symmetry. With each level, new basic characteristics of life come into play which result from new criteria for the self-determination of optimal stability.

12. The Evolution of Evolutionary Processes

> The individual is going to be universalized, the universal
> is going to be indiviualized, and thus from both
> directions the whole is going to be enriched.
>
> Jan Smuts, *Holism and Evolution*

System dynamics in macroscopic and microscopic perspective

The dynamics of a particular system may always be viewed under two perspectives: a microscopic one—processes which play in a system—and a macroscopic one—the behaviour of the system viewed as a whole. In this way, the collisions between molecules in an isolated system—a microscopic perspective—find their corresponding macroscopic view in the devolution of a system in the direction of its equilibrium. In the same way, the chemical reactions in a system far from equilibrium and including an autocatalytic step correspond to the macroscopic order of a dissipative structure. In this case, however, the maintenance of an exchange between system and environment is essential, so that a further, superordinate macroscopic perspective may be introduced which includes the system together with its environment. In the figures for this chapter, the corresponding perspective is graphically indicated with the help of dots. One dot signifies the holistic, macroscopic view taken of a system together with its environment; two dots signify exchange between a system and its environment, or between two systems; and three dots, finally, signify a microscopic view focusing on the processes in the system.

Dissipative structures mark the lowest level at which phenomena of spontaneous structuration occur in true dissipative self-organization. In a macroscopic view, such structures may be described as being autopoietic and evolving. Viewed from this angle, only the self-regeneration and the coherent evolution of the cyclically organized system and its structures are of interest. The transformations of energy and matter in the metabolic processes of the system appear as secondary. If, however, the products of the transformation of different autopoietic systems belonging to the same level, join to form new autopoietic systems, the self-organization dynamics may continue at a higher level. What is happening in such a case is nothing but endosymbiosis or symbiosis which is coming close to fusion. As has been discussed in the preceding chapter, symbiosis generates a new semantic level. In this way, evolutionary processes which belong to consecutive autopoietic levels may be linked in an evolutionary chain—a metaevolution. This is indeed the approach with which the origin and the evolution of life may be described in the self-organization paradigm. Symbiosis of molecular species in the hyper-cycles of the Eigen type leads to prokaryotic cells, symbiosis of prokaryotes leads to eukaryotic cells, and symbiosis of eukaryotes leads to multicellular organisms.

Figure 41 shows schematically how the interweaving of processes in this meta-evolution may be understood. Three dimensions of evolutionary processes may be distinguished which are characterized by coherence and, globally viewed, continuity. Firstly, each autopoietic system—for example, an organism—evolves in its ontogeny through a sequence of space-time structures. Secondly, the systems evolve at a particular autopoietic level of existence along the complex webs of phylogeny, for example, in the branch-ings which characterize the evolution of richly differentiated animal and plant domains. And thirdly, the evolutionary processes evolve themselves and bring new autopoietic levels into play. It is this third dimension which interests us in this chapter.

In metaevolution, the vertical transition to an adjacent level may be under-stood as symmetry break which implies an increase in complexity. Each evolving system is in all process dimensions also capable of *re-ligio*, the linking backward to the origin. Thus, *re-ligio* restores the broken symmetry and unity.

The emergence of complexity

The scheme depicting in Fig. 41 the emergence of an evolving hierarchy of structures, may also be applied to the evolution of knowledge by "symbiosis of information" (see Fig. 42). Knowledge, as organized information, may also be interpreted as complexity.

The result of each experience is expressed in information which has been brought into a specific form—it is in-formation in the true sense. This in-formation corresponds to a specific dynamic régime of a self-organizing struc-ture. It may consist of a particular bioenergetic process pathway, such as photosynthesis, or at the macrobranch of evolution of the relations between different niches in an ecosystem. For the original autopoietic system, such information is mostly confirmation. At a new level, however, information stemming from different autopoietic systems may enter into self-organization.

This self-organization of information may be achieved by means of biomolecular, metabolic or neutral communication. For the new autopoietic structure resulting from the symbiosis of information, this information is at first primarily novelty. But as soon as the new structure has become established, it seeks to confirm itself and increasingly retrieves and uses only that information which serves its exchanges with the environment. Nevertheless, this metaevolution of information results basically in a net transformation of confirmation into novelty. Symbiosis may also be represented in terms of such a principle which reverses the usual transformation of novelty into confirmation in life which stays at the same autopoietic level of existence.

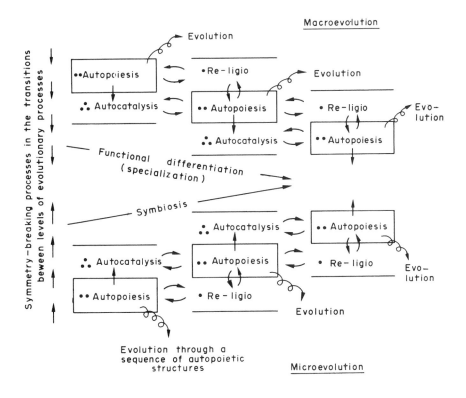

Fig. 41. Co-evolution of evolutionary process levels, expressed in process terms.

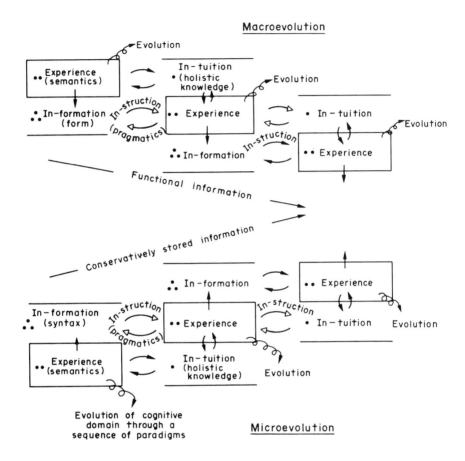

Fig. 42. Co-evolution of evolutionary process levels, expressed in terms of information.

The holistic knowledge of the system's own evolution which corresponds to *re-ligio* and which may already be observed in chemical dissipative structures, may be called *in-tuition*, which is literally learning from within. Intuition is not structural knowledge, but knowledge of one's own historical process. In this way, intuition becomes the only factor to guide direction when in processes of fast change, the orientation by means of stored information and by means of interpreting the exchange with the environment all fail.

Intuition is nothing but that holistic system memory which, in process systems, becomes visible in the form of hysteresis, or in other words, in the form of those small differences in the pathways which distinguish evolution and (forced) retreat of a system. Intuition may refer to metabolic processes as well as neural processes. Research in these phenomena has advanced so rapidly in the past few years that the German Bunsengesellschaft devoted the theme of its annual meeting in 1976 to the storage and retrieval of information by physical-chemical mechanisms (Fuchs, 1977). Besides the storage on conservative structures (such as in DNA or in computers) process systems of a chemical and biochemical nature were discussed which are capable of intuition in the sense here applied.

Whereas a neural memory is familiar to us, it is not so easy in Western thought patterns to imagine a metabolic memory. However, even unicellular microorganisms exhibit learning behaviour, that is to say, their biochemical processes assume besides other functions that of a memory as well. In plants which have no nervous system, memory may be connected to the formation of lignin (Falkehag, 1975). Lignin is an extremely stable substance which also protects the plant from stress imposed from the outside. Henri Bergson (1896) opposed the opinion that human memory has its seat in the brain exclusively; memory, according to him, is also a function of the organism as a whole. The distinction between metabolic and neural memory goes back to the communication processes involved. The existence of a human metabolic memory became confirmed when Wilhelm Reich showed in the first half of our century that traumatic experiences are recorded not only in the human psyche, but also in the form of muscle contractions. A wide spectrum of therapeutic techniques under the name of bioenergetics deal with this metabolic memory. These techniques include deep massage, rolfing (a technique developed by Ida Rolf and based on the breaking up of fascia around the muscles), structural patterning (the structural realignment of muscle groups) and a kind of induced self-release of tensions in the body.

In the functions of the brain, a state-specific neural memory, besides a punctual memory, is of great importance. We remember certain events sometimes only in association with whole situations, moods, or special states of consciousness. In Chapter 18, I shall return to the state-specificity of the neural memory.

All knowledge is based on experience. In a holistic way, this experience is accessible by intuition, if only partially and often in vague contours. It is much clearer when it is stored in conservative structures and transferred with them. Along the phylogenetic chain, however, this experience is not simply imposed on the newly emerging system. In epigenealogical processes, it becomes involved in ever new semantic contexts and in this way serves new life which develops its proper dynamics. In this way, knowledge itself becomes an evolving system which manifests itself in a sequence of different structures. In other words, the emergence of complexity includes differentiation as well as the open evolution of the differentiating process systems.

Metaevolution in symmetry breaks

The history of universal evolution may be described in terms of symmetry breaks as I have discussed in Part II. Each symmetry break unfurls a new space-time continuum for the self-organization of structures. Each new kind of symmetry break, however, also implies the transition to a new level of evolutionary processes and thus self-transcendence of evolution in meta-evolution. The symmetry breaks may be interpreted as alternating breaks of temporal and spatial symmetries.

In cosmic evolution, this is not so clearly evident and I cannot give a tight scheme. Nevertheless, some of these symmetry breaks are not difficult to recognize at the macrobranch of cosmic evolution. The big bang established an expansion of the universe which, for our purposes, may be regarded as irreversible. Thus, a temporal symmetry between the past and the future is broken. The origin of matter from an excess over antimatter breaks a temporal as well as spatial symmetry. The condensation of macrostructures in a multilevel hierarchy breaks the macroscopic spatial symmetry of the originally homogeneous universe. And with the appearance of stars, a further time symmetry is broken, perhaps already with galaxies. Individual evolution commences. The energy generation in the transformation processes of matter has a beginning and an end and runs through a specific sequence of qualitatively different phases.

In a process paradigm, even the microevolution of subatomic particles may be understood as some sort of endosymbiosis between sets of systemic properties which Bastin and Noyes (1979) call *Schnurs*, suggesting con-catenating strings. A meta-evolutionary hierarchy emerges in a similar way as it does in the microevolution of life through endosymbiosis of prokaryotes to form eukaryotes and of eukaryotes to form multicellular organisms. Bastin and Noyes obtain a combinatorial hierarchy in which the number of possible states at each level increases according to the sequence $3, 7, 127, 2^{127}-1 \approx 10^{38}$ which looks like a striking illustration to the creative hierarchy evoked in Lao Tzu's *Tao Teh Ching*:

> Tao gave birth to One,
> One gave birth to Two,
> Two gave birth to Three,
> Three gave birth to all the myriad things.

Bastin and Noyes (1979) interpret the first level as expressing the three absolute conservative laws (baryon number, lepton number and charge in the universe), the second level as pertaining to the quantum states of baryons and the third as referring to baryon-lepton interactions, whereas the fourth level would indicate a wealth of possible unstable configurations. But they also recognize in the numerical sequence the inverse of the superstrong, strong, electromagnetic and gravitational coupling constants, so that the symmetry breaks in the transitions from one level of the hierarchy to another would refer to the unfurling of the physical forces, setting the space-time stage for cosmic evolution.

This kind of consideration, however, becomes much more interesting when it is applied to the history of life on earth. Figure 43 attempts to present a rough sketch of some of the basic aspects of evolution from the biochemical/biospheric to the sociocultural phase. The macrobranch of co-evolution is autopoietically structured at least since the emergence of the Gaia system. It is interesting that the first differentiation of the solidifying earth surface may be meaningfully described with the four elements of Greek natural philosophy, earth, water, air and fire (the lightning which was so important for starting chemical evolution as predecessor of biochemical evolution). Ecosystems, in the true sense of a web of vital relations, appear only with heterotrophy. Social systems soon start to structure themselves in hierarchical ways. This becomes an important feature of insect societies. The recently discovered colony of red forest ants in the Swiss canton of Vaud, the largest known in the world, has a population of 200 to 300 million insects, but is subdivided into 1200 ant hills which extend almost over one square kilometre. And colonies, too, seem to be connected in space and time. With mammals, the hierarchy ranges from species and regional populations to kinship groups and core families. African elephants, for example, appear primarily in matriarchally organized kinship groups (Douglas-Hamilton, 1975) which, in turn, form groups of hundreds and thousands of animals.

The transfer of functional ecological information along the macrobranch of evolution assigns a special role to the holistic system's intuition in the transition to new autopoietic levels. Since information, generally, is not con-servatively stored at the macrobranch, the systemic memories of the macro-systems are the principal links in an evolutionary learning process. A sociobiological system, or a niche, which differentiates out of a more compre-hensive ecosystem, orients itself according to the total structure of dynamic relations within which it sets out to establish its autonomy. Self-consistency, the principle ruling subatomic particles (see Chapter 2), also reigns here. The macroevolution of life is determined in essential aspects by the develop-ment of higher complexity along the microevolutionary branch. Its role is primarily a balancing one whereas microevolution plays a more innovative and outreaching role on the basis of conservatively transmitted complexity.

Along the microbranch it is not so evident which are the true autopoietic units forming distinct levels of existence. Organs, for example, are not autopoietic but are, with others, specialists in the service of the organism as a whole. The frequently cited hierarchy cell-organ-organism would be mis-leading here. However, as will be discussed in Chapter 14, there are many intermediary steps of autopoietic systems which are co-ordinated in the

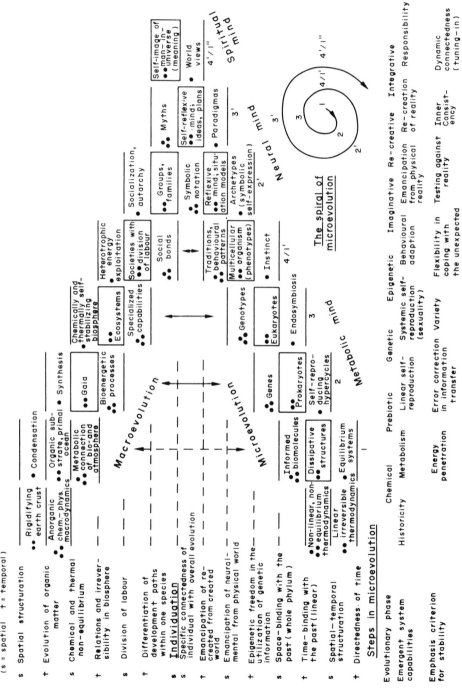

overall framework of the organism; they do not appear independently and have not become integrated by symbiosis, but originated in the hierarchical differentiation of complex organisms. The hypercycles of the Eigen type did appear independently but they still belong to the level of dissipative structures. A sharp transition between these levels cannot be identified so easily and may not have occurred at all.

The evolution of the essential dynamic *system capabilities* from historicity to the responsibility for the macrosystems shows how the basic characteristics of life developed in a specific sequence. This progression is also called *anagenesis*. It appears almost arbitrary from which step of this intricately interwoven evolution of morphogenetic processes the beginning of life is counted. The usual minimum requirement for the three-fold characteristic of metabolism, self-reproduction and transfer of mutations is already fulfilled by molecular systems evolving in hypercycles of the Eigen type. With sexuality making its appearance with the eukaryotes, variety is generated in a systematic way. If linear self-reproduction implies an indefinitely extended autopoiesis which in cell division does not know natural death at all, sexuality opens up the possibility of true phylogeny over many generations. Waddington (1975) speaks in this connection of "metagenetics", the evolution of genetic systems. It ranges from the organization of genes in chromosomes through sexual reproduction to "sociogenetic" mechanisms. The latter evolve in turn from the simple learning of traditions to the "learning of learning". This sequence which includes the neural mind may perhaps be better called *metagenealogical development* through many levels in order not to confuse the biological forms of genealogical communication (genetics) with sociocultural forms.

At each level of autopoietic existence (emphasized by a frame in Fig. 43) there is a holistic criterion—or perhaps more than one—for the self-determination of the system with regard to its stability in the presence of fluctuations and thus to the space-time structure which it chooses. Some of these criteria are tentatively indicated in the scheme. A particularly important criterion is increased flexibility to cope with the unexpected. This increase is revealed not only in genetic over-determination, which in humans is about ten- or fifteen-fold the required information, but also in the emergence of an immune system which improves during the whole life-span of the organism. The human organism, for example, possesses a practically complete library of all antibodies which it has produced during its lifetime for defence against bacterial and viral infections. An invader (antigen) is tested by these antibodies as to its

Fig. 43 (on opposite page). Schematic representation of the co-evolution of evolutionary process levels in the sociobiological and sociocultural phases of evolution. Each transition between two levels of autopoietic existence is marked by a specific break of spatial or temporal symmetry. For the explanation of the dynamics see Figs. 41 and 42. Framed fields indicate autopoietic system levels.

molecular form and is recognized by the antibody "in charge" (a protein molecule) whereupon a process is set in motion by which the normally acting inhibitions against the production of the specific antibody are cancelled and large numbers of the required antibodies are produced in the so-called immune reaction. In such a way the organism cannot protect itself against the novelty of a bad experience, but against its confirmation in repetition. The sequence of newly arising capabilities and optimization criteria expresses an increase in autonomy *vis-à-vis* the environment, and thus in *consciousness*. An aspect of this increasing autonomy is the already mentioned step-wise emancipation of the individual from the collective. Gregory Bateson (1972) has pointed out another aspect: evolution proceeds from the "adjusters" (for example, poikilothermic animals which adjust their body temperature to the temperature of the environment) through "regulators" (for example, homoio-thermic animals which maintain constant body temperature) to "extra-regulators" (for example, humans with their capability of creating an artificial environment in the form of heated or cooled shelters). The earliest life forms were by far the best adapted. If the meaning of evolution was in adaptation and in increasing the chances for survival, as is so often claimed, the development of more complex organisms would have been meaningless or even a mistake.

Each transition to the next higher level of microevolution involves a *symmetry break*. In addition to the temporal symmetry break in equilibrium thermodynamics—the past becomes distinguished from the future—the further transition to nonlinear non-equilibrium thermodynamics (dissipative structures) marks a break of spatial symmetry becoming visible in spon-taneous structuration and polarization. In the evolution of a dissipative structure each instability threshold with a transition to a new structure marks the break of a further spatial symmetry. In the transition to higher levels of microevolution temporal and spatial symmetry breaks alternate. In the first pair to follow, the temporal and spatial distribution of past experience is bound in such a way that it may become effective in the present. In the next pair, the autonomy of the evolving system from its environment becomes enhanced, at first by the increasing importance of the epigenetic process and subsequently by the establishment of an autonomous inner world. In the final pair it is at first the symmetry between the processes creating the outer and the inner world which becomes broken, and then the connectedness of man with the evolving universe becomes structured in a specific way. These four pairs of symmetry-breaking processes may also be correlated with four phases of the microevolution of life: thermodynamic/chemical, biological/genetic, epigenetic and neural (sociocultural) phases.

An even more comprehensive dynamic order in the evolution of life emerges when the steps, in Fig. 43 numbered 1 to 4, are combined into biological/metabolic evolution, steps 1' to 4' into neural/mental evolution and steps 1", . . . into spiritual evolution. The fourth steps always falls together with the first step of the following group. Four is the "powerful retrograde connection to the primal one", as Marie-Louise von Franz (1974), in developing C. G. Jung's thoughts, has found confirmed in many mythologies. Evolution is basically not the linear progression as is suggested by Fig. 43 for the sake of simplicity of the graphical representation. Considered from whatever angle of view, evolution is always a spiral as is indicated in the side sketch to Fig. 43. The connectedness over time and space, the unity of evolution as a total phenomenon, thereby becomes even more sharply accentuated.

Hierarchical securing of openness

The opening up of new semantic levels in microevolution may be described to great advantage in the Weizsäcker information terms of novelty and confirmation. In learning a language, for example, sequences of letters, words, idioms, short phrases, longer modules and so forth are repeated, i.e. increasingly confirmed. The lowest levels (letters, words) are first confirmed to a high degree whereas with each opening up of a new semantic level there is at first much novelty, for example when entire phrases are formed. In order to learn an elegant, highly cultured style in a strange language, years are required if this goal is attained at all.

The inclusion of higher semantic levels reduces novelty at the other levels and transplants it to a higher level (E. v. Weizsäcker, 1974). This becomes manifest in the high degree of normalization, i.e. confirmation, at the lower levels of microevolution. There are less than a hundred chemical elements. Life requires twenty amino acids for building proteins and four nucleotides in triplet arrangement, thus $4^3 = 64$ codons, for building DNA. All life on earth is based on the same molecular structures for the transfer of genetic information, plants as well as fungi and animals, unicellular as well as multicellular organisms. Even such a complex mammal as man consists of cells which belong to no more than 200 different types and which man shares with his closest relatives, e.g. the chimpanzees. The behavioural patterns of animals originating in early phases of evolution are limited. In contrast, sociocultural evolution entered by the more highly evolved animals and man in particular, seems to open endless frontiers of novelty.

These considerations suggest a bold hypothesis. If it is the task of higher semantic levels to reduce novelty at the lower levels, the described marked normalization results perhaps only when the newly opened level of evolutionary processes has become included in a hierarchy. There were many millions of different "species" of prokaryotes, more than the 2 million animal species today, but the few organelles they formed in eukaryotic cells are highly normalized. Out of the large number of eukaryotic protozoa emerged the limited number of cell types on which multicellular metazoa are based. Perhaps there have also been several approaches to self-reproduction which became reduced to the universally valid DNA only in fully developed unicellular organisms—not as the result of a Darwinian fight for survival, but of their usefulness in the framework of a complex hierarchy.

With the inclusion of the reflexive and self-reflexive mind in the sociocultural phase, this rule changes. It is true that the lower levels of social dynamics become even more strongly ritualized and therefore normalized, a process which is further enhanced by technology. The individuals themselves, however, do not become normalized in order to become integrated into society. The sociocultural level of dominant novelty acts *within* us. We are biologically normalized, except for a few racial differences such as skin colour. Mentally, however, we are, above all, the carriers of culture, the endless frontier of creative transformation of novelty into confirmation.

It is an open question whether the same principle of novelty reduction also holds for macroevolution, here, of course, in an upside down sense for the higher levels. The massive transformation of the earth's surface and atmosphere in the earliest phases of life on earth and the subsequent astonishing stability of the Gaia system seem to point in this direction. If the higher levels of macroevolution are indeed initially wide open toward novelty, the development of life on a planetary surface does not require sharply defined conditions but may originate in a broad spectrum of conditions. As we have seen, life itself generates to a large extent the conditions for its continuation and evolution. However, this point will only become clarified when we succeed in finding extraterrestrial life.

The transformation of novelty into confirmation may be observed at all levels of the micro- and macroevolution of life on earth. Ecosystems develop from an emphasis on novelty or fast change to an emphasis on confirmation of the same dynamic régime in maturity. Margalef (1968) expresses this, as already mentioned, in such a way that each community seeks information from the environment, but only to use this information to prevent the assimilation of more new information. The digital circuits of information processing which usually establish themselves in the left half of the human brain may also be impressed upon the right half up to an age of 2 years, but not later. Plants which encounter an obstacle to their growth may remain crooked for their whole life even when the obstacle (for example, my shower curtain) was in their way for only half an hour. And we also know from human physiology that highly specialized cells, characterized by a high degree of confirmation, cannot regenerate themselves.

In the fourth of the books in which Carlos Castaneda (1975) transmits the world view of the shaman Don Juan of the Mexican Yaqui Indians, there is a striking parallel and generalization of this principle. According to Don Juan, reality is divided into two aspects, one of which (the tonal) comprises the regularities of a world ordered by our concepts, whereas the other (the nagual) represents the unexpected. The latter aspect may be mastered by creative thought and action and by spontaneous decisions (i.e. by free will). Thus, the task of life is the never-ending transformation of the nagual into the tonal, of novelty into confirmation. The British Nobel laureate in physics, Brian Josephson (1975), has pointed out that this implies a new expression for the directedness of time, for the irreversibility of life processes.

The gradual increase of confirmation with the opening up of new semantic levels—"higher" levels from a point of view of microevolution and "lower" levels from a macroscopic point of view—result in differentiation, in increasing complexity. Just as in process thinking evolution is viewed not as an evolution of entities but as an evolution of organization (Eigen and Winkler, 1975), it may also be understood as an evolution of knowledge, or of the organization of information. Rupert Riedl (1976) has proposed such an approach. Like the quantity of energy, the quantity of information in the universe is also assumed as constant—roughly estimated, around 10^{91} bits. (A bit is the number of binary steps required for the complete description of the distribution in macroscopic states. The universe would be accordingly totally knowable after 10^{91} questions answered by "yes" or "no".) But just as the organization of energy, the organization of information may also be upgraded, at least locally. Riedl defines order as complexity times number of occurrences. The result of an upgrading by evolution is then the appearance of ever more complex structures in ever smaller numbers. In an ecosystem, for example, the number of participating organisms seems to follow roughly an inverse square law with respect to their body length (May, 1978).

It may be impossible to define reality in terms of solid building blocks (see Chapter 2). But certainly, the globally stable states of process structures form distinct hierarchies founded on confirmation—even if it is novelty which calls them into being in the first place.

The appearance of a high degree of novelty in the opening up of a new semantic level in microevolution is tantamount to the appearance of new indeterminacy, new degrees of freedom. It has already been mentioned that this indeterminacy plays a much more important role in our everyday world than the microscopic indeterminacy at the quantum-mechanical level. Up to now, all attempts to find valid formulations for morphogenesis at each level are based at best on a view which considers the interaction of stochastic and deterministic factors from an angle of view pertaining to a single level only (see Chapter 3). All processes which impinge on this level from another level are considered as random. But what is the meaning of "randomness" in the context of a multilevel evolution in which each level brings new ordering principles into play? How random is the fluctuation which is introduced into a system by one of its members or by an outsider if this individual is itself the product of a long evolutionary chain and of its own ontogeny? It seems that we frequently confuse indeterminacy and chance. Indeterminacy is the freedom available at each level which, however, cannot jump over the shadow of its own history. Evolution is the history of an unfolding complexity, not the history of random processes. Out of this fog emerge the contours of a world in which nothing is random but much is indetermined and free within limits.

Life on earth is connected with all phases of evolution. Matter which becomes organized through life originated in far-away cosmic times and places. The dynamics of life is tuned to the dynamics of the cosmos—perhaps in a mutual interdependence which we do not yet fathom. The alchemists believed in the possibility of influencing physical reality in a psychic way—a principle which seems to anticipate the modern cybernetic principle of steering macroscopic processes with a minimum of effort.

What we may observe directly, however, is that aspect of evolution which may be called time- and space-binding. Events which are widely separated in time and space, are presented on the stage of life in such a way that the Aristotelian dramatic unity of space, time, and action is established to a high degree. This unity makes it possible for evolution to act in a pragmatic, effective way. Life is intensity. Evolution acts in the direction of enhancing this intensity. This will be briefly elaborated in the following chapter.

13. Time- and Space-binding

I want the future today!

Vladimir Majakowsky

Mutual correspondence of space and time in communication

The role of communication in evolution has been evoked in many aspects in this book. The most important aspect, however, is perhaps the mutual substitution of space and time dimensions. Information at the level of syntax, the arrangement of units, may be transferred by means of templates in a spatial-holistic way as is done in the copying of genetic information. But it may also be dissolved into a spatial-temporal process structure as in epigenetic development and generally in each epigenealogical process in which conservatively stored information is retrieved and synchronized by dissipative processes. At the level of semantics the context of meaning may be transmitted holistically by resonance-like communication. But it may also be transmitted in a temporal sequence of information which refers to a sequence of situations (as, for example, in the individual frames of a film). At the level of an evolving system, finally, the totality of the evolutionary process may be experienced in a four-dimensional way or it may be represented by a sequence of changing sets of dynamic rules.

Life always chooses the holistic transfer of information when the same level is addressed, and the dissolution into sequences when this level is transcended. Transfer by means of templates acts in genetic communication, resonance-like communication in metabolic communication. In neural communication, both transmission modes play a certain role—but, ultimately, what is approached here is the third holistic transmission mode by direct four-dimensional experience. Evolution seems to facilitate such a four-dimensional experience to an ever higher degree. Not only the universe as a spatial structure is becoming increasingly self-reflexive, its evolutionary process itself is becoming self-referential.

231

The fine-structure of time

Klaus Müller (1974) and Georg Picht have proposed that time be viewed not only in its flow from the past into the future but also in a kind of fine-structure which it has in every moment. This fine-structure may again be represented by assigning to each moment aspects of the past (P), the present (N—for "now"), and the future (F). The present of a dynamic system not only has a present (NN) which consists of the immediate experiences of the moment, of the horizontal processes, but also a past (PN) which includes the vertical evolution process which has led to the present structure of the system, and a future (FN) which corresponds to the options in further evolution. On the other hand, there is also a present of the past (NP), the effectiveness of past and conservatively stored experience in the present, exemplified by genetic and generally genealogical communication. And there is also a present of the future (NF) which corresponds to the causative effect in the present which a plan or a vision of the future or in principle any anticipation may have for action in the present.

It is interesting that time-binding already appeared in the 1920s as a central notion in the theory of General Semantics by Alfred Korzybski (1949). General Semantics was a predecessor of a cybernetic theory of living systems and considered man as a holistically acting organism in an environment. However, Korzybski as well as Picht and Müller reserve time-binding for the human sphere.

I am using here the general scheme and the notation proposed by Picht and Müller without adopting their interpretation, too. In particular, I do not regard time-binding as limited to the human domain, but as a universal characteristic of evolution. Furthermore, I supplement time- by space-binding which I indicate by the suffix s. Linear, purely vertical self-reproduction in cell division is an example for pure time-binding (NP). The horizontal gene transfer among bacteria adds an element of space-binding in the present (N_sN_s) where the first suffix is meant to indicate that it is the evolution of a holistically viewed gene pool, rather than the evolution of well-defined species, which is happening here. With prokaryotes, both forms occur in an *ad-hoc* mixture. In the sexual reproduction of eukaryotes and more complex creatures, both forms appear in a systematic connection via an ancestral tree (NP_s). Evolution is characterized by time-binding as well as space-binding.

In applying these notions, a few basic characteristics of evolution should be kept in mind which are graphically depicted in Fig. 44. Most important is the basic openness of evolution toward the future. If we speak of the future of the present (FN), there is practically always a multiplicity of future structures and of processes leading to them. The same is the case if we speak of the future of the past (FP); we obtain a multiplicity of structures in the present which

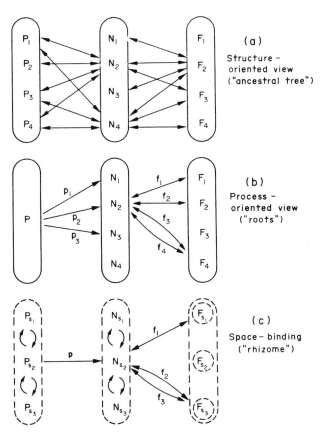

Fig. 44. Basic possibilities of time- and space-binding. (a) In a structure-oriented view, various configurations of the past (P) may have led to one and the same configuration in the present (N). The resulting image is that of the ancestral tree. (b) In a process-oriented view, a specific process p has led to the present N, whereas different processes f_i lead into a future F which, in principle, is open; also, several different processes may lead to one and the same future. This view corresponds to the image of the root. (c) In space-binding, indicated by the index s, the experience of many self-organizing systems may be bound together in one system. Here, the image of the rhizome holds.

would have been possible in principle, viewed from the perspective of the past. If, however, we speak of the present of the past (*NP*), there is only one actually realized present. If the openness of evolution is emphasized under the aspect of a future acting upon the present (*NF*), the future of the present (*FN*) refers to a specific future which will be realized.

Time- and space-binding may in principle be understood in terms of two familiar images. The "ancestral tree" branches toward the past as it corresponds to genetic communication in sexual reproduction. The "root", in contrast, branches toward the future as it corresponds to a common origin in cell division or also in the history of Adam and Eve and their descendants (or, according to the Bible, the common origin in Adam alone). Both images cannot be thought through to their ultimate consequences. The ancestral tree gets lost in an infinite variety at the beginning, the root in a singularity. Both views penetrate each other, are complementary. They need a third image, however, which brings into play the live relations of the present.

This third image is the rhizome, proposed by Gilles Deleuze and Félix Guattari (1976). A rhizome is an underground sprout such as a bulb, not a root, but a stem. Its oldest parts die off in the same measure as it rejuvenates itself at the tip. Therefore, it does not grow indefinitely as other plant stems, but renews itself autopoietically. The rhizome continuously generates new relations—not copies, as the ancestral tree, but a map, as Deleuze and Guattari call it: "The rhizome cannot be reduced, neither to the one nor to the many. It is not the one which becomes two, nor the one which becomes directly three, four, five, etc. It is neither the many deduced from the one, nor the many to which the one is added ($n + 1$). It does not consist of units, but of dimensions." In other words, it is a self-organizing, globally autopoietic process sytem—like a dissipative structure, the gene pool of bacteria, an ecosystem or the Gaia system.

All three images together—ancestral tree, root, and connecting rhizome— suggest how time- and space-binding acts in evolution. It is not introduced all at once, but acts step-wise. The completion of time- and space-binding seems to be a basic dynamic principle, a *post hoc* observable "purpose" of evolution as a total phenomenon.

The steps of time- and space-binding in evolution

In cosmic evolution, the hot initial phase of the universe in which radiation, matter and antimatter were continuously transformed into each other, may be characterized as NN_s; in that phase, the universe had only one macroscopic state, only a present which, however, expressed a multifaceted dynamics of the present which varied over space. The "freezing out" of baryons in micro-evolution also is the present of a space-bound present (NN_s). With the synthesis of helium and heavier nuclei in the course of further cooling of the expanding universe, the present of a space-bound past (NP_s) comes into play. The notation NP_s generally characterizes structuration in the framework of a condensation model, or conservative self-organization. Therefore, it characterizes the formation of the mesogranularity of the universe (superclusters, galaxy clusters, galaxies, stellar clusters and protostellar clouds), as well as of planetary systems. The evolution of stars is more complex and depends both on the present of nuclear processes and the past of transferred initial matter, always in space-binding. Stellar evolution itself occurs in the universe in many structures simultaneously and is even structured within the star in different dynamic régimes (onion-skin model). Therefore, the correct notation may perhaps be $N_s(N_s,P_s)$. The resulting heavier atomic nuclei, atoms and molecules, in contrast, have only one present $N(N_s,P_s)$.

The stabilized structures which emerge from the synthesis of matter along the microscopic branch of cosmic evolution have little historicity. Subatomic particles and atoms of the same isotope are not distinguished from each other. Only their relative abundance provides some faint hints as to their history. Imperfect crystals, as they normally occur in nature, already bear clearer testimony to some events in their history. The evolving macrosystems, however, exhibit a much higher degree of "funicity" as the Austro-American physicist Viktor Weisskopf called it jokingly in a lecture in Berkeley. He coined this term after the unhappy figure in a story by Jorge Luis Borges; Funes is incapable of forgetting.

Interaction enhances funicity. It becomes more marked along the micro-evolution of life on earth. Horizontal processes, however, partly erase funicity as, for example, in the bacterial gene pool. More complex organisms, on the other hand, not only represent their own history, but also the history of their species and their whole phylum back to the origins of their evolution. Ecosystems, societies and civilizations, too, become the more expressive with respect to their history the more complex and mature they become. We may also say that complexity and exchange with the environment enhance the individuality and therefore also the consciousness of systems.

Table 5 presents an overview over time- and space-binding along the micro-evolution of life, including sociocultural evolution. This table may be compared to Fig. 43 in the preceding chapter. I should like to leave it to the reader to re-enact the individual thought steps which have resulted in this table—or to correct them. They mirror the evolutionary processes described in Part II of the book and are characteristic of the autopoietic levels of existence. With human development, the matrix of possible simple steps of time- and space-binding exhausts itself. The four-dimensional "total experience of evolution", as it is attained in the last step of integration with the dynamics of the universe—the step of meaning—is a theme which in itself is inexhaustible.

Time- and space-binding acts in the inner as well as in the outer world. In the introduction to this book, I have already pointed to the tremendous extension of direct observation in modern science; today, it extends over 41 orders of magnitude in the time dimension and 43 orders of magnitude in length. Thus, science is a mental system which acts by means of time- and space-binding and gives rise to the self-organization of holistic paradigms.

It is also becoming clear now how evolution, starting from present-oriented physical and chemical structures, gradually integrates the past in biological development and the future in neural development into the life processes of

Table 5

Time- and space-binding along the chemical/biological branch of microevolution and its continuation in sociocultural evolution. Each new level includes the time- and space-binding of all preceding levels and opens a new dimension

	Re-ligio	Autopoiesis	Self-transcendence
Autopoietic level			
	Relative position in Fig. 43 (p. 224)		
Equilibrium systems	PN	NN	FN
Dissipative structures	P_sN	N_sN	F_sN
Prokaryotes — Horizontal genetics	P_sN_s	N_sN_s	F_sN_s
Prokaryotes — Vertical genetics (cell division)	PP	NP	FP
Eukaryotes — Sexuality (horizontal/vertical genetics)	PP_s	NP_s	FP_s
Phenotype — Epigenetics	P_sP_s	N_sP_s	F_sP_s
Phenotype — Organismic mind (self-representation)	PP_s	NP_s	FP_s
Reflexive mind (apperception)	PN_s	NN_s	FN_s
Self-reflexive mind (anticipation)	PF_s	NF_s	FF_s
Integration (man-in-universe)	$P_s(P_s,N_s,F_s)$	$N_s(P_s,N_s,F_s)$	$F_s(P_s,N_s,F_s)$

the individual systems. It is interesting that neural (sociocultural) evolution re-enacts the integration of the past into the present at a new level—and not in the same sequence as biochemical and biological evolution (see Table 5). It is also remarkable that two steps meet at the decisive "seams" between chemical and biological evolution (in the prokaryotes) on the one hand, and between biological and neural evolution on the other. I do not know what the meaning of this is.

Similar steps of time- and space-binding also occur in the genealogical and epigenealogical processes on the macrobranch of sociocultural evolution.

Interpretation: the evolutionary "purpose"

I believe that the evolutionary trend toward ever more complete time- and space-binding may be explained in terms of a three-fold, self-organizing and *post hoc* recognizable "purpose". *Primo,* the result is an extraordinary intensification of life. Not only the experience of past evolution, but also the experience of anticipated future evolution vibrates in the present. This is the true significance of the frequently misused and misunderstood slogan of living in the "here and now". While watching the sunset during his

Easter Sunday walk, Goethe's Faust expresses this intensity in images of time-and space-binding:

> Ah, that no wing can lift me from the soil,
> Upon its track to follow, follow soaring!
> Then would I see eternal Evening gild
> The silent world beneath me glowing,
> On fire each mountain-peak, with peace each valley filled,
> The silver brook to golden rivers flowing.
> The mountain-chain, with all its gorges deep,
> Would then no more impede my godlike motion,
> And now before mine eyes expands the ocean
> With all its bays, in shining sleep!

Is it to be called a mere coincidence that the great scientist and inventor Nicola Tesla recited the same verses when he watched a sunset in a Budapest park in 1882—and, like a sudden vision, received a flash of insight which covered all the principles of alternating current polyphase power which he was to patent in 1888 and which today form the basis of our electricity system (Spurgeon, 1977)?

Ivan Illich (1976) has addressed the same intensity when he defined health not as a specific state, but as the intensity with which an aware organism copes with its environment. Such a process definition of health reminds one of the old Chinese tradition to pay doctors as long as they keep one healthy and not when one has become ill.

Secundo, the variety of the phylogenetic past and the openness of the future impart to life in the present a profundity which results from the concentration of a practically infinitely rich potential of possible life structures. In the life of each complex individual the unfolding of that initial, undifferentiated core is re-enacted which is called *shunyata* in Buddhism and which consists of the purest of qualities (which will be transformed into quantities only during the unfolding). This unfolding, however, is not a linear process, but a "dancing out" of the potential in a big loop between the origin and the present. This most profound of all cybernetic life processes is called *tantra* in Buddhism.

If the step-wide symmetry breaks in evolution may be understood as the unfurling of time and space for the self-organization of structures, time- and space-binding acts in the direction of restoring these symmetries. In this way only *re-ligio,* the linking backward to the origin, becomes possible. In *re-ligio,* each system becomes its own origin and the centre of evolution—or, in reverse, evolution places its origin and centre into each self-organizing system. In the processes of structuration, evolution is open as a matter of principle. In the combination of symmetry breaking and time- and space-binding, however, evolution itself becomes a circular process in a four-dimensional space-time continuum.

Tertio, finally, it becomes obvious now that the universe with its complex life forms is becoming increasingly self-reflexive and self-knowing not only with respect to its morphology, but also with respect to its morphogenetic dynamics. Morphological structures may be studied in the present, dynamics only in a temporal extension. With the inclusion of the entire time dimension of evolution, from a distant past into a (perhaps less distant) future, and with the partial inclusion of events in distant space, the total process of evolution becomes increasingly subject to direct experience. In a certain sense, we may let the process of evolution act within us in a holistic way, especially in meditative states of consciousness and in the highest intensity of life, in love.

We may summarize all this perhaps in such a way that we state that time- and space-binding is the evolutionary way to the direct experience of a four-dimensional reality, the space-time continuum which is, on the one hand, created by evolution and in which it unfolds on the other. In this direct experience, new dimensions seem to open up for the future of mankind. More of this in the last two chapters of the book.

Time- and space-binding come into effect step by step. In the result of evolution, however, in complex space-time structures at the higher levels of life, it acts simultaneously with all the partial steps. Autopoiesis and evolution of multilevel structures require the synchronization of many levels of self-organization dynamics. The ubiquitous fact of such a synchronization may be deduced from the systemic connectedness not only of structures, but, above all, of their homologous dynamics. It is another manifestation of the basic principle of self-consistency. The little which we know about this will be summarized in the following chapter.

14. Dynamics of a Multilevel Reality

> We make our destinies by our choice of gods.
>
> Virgil

Multilevel autopoiesis

The differentiation of the macro- and microworld in the course of their co-evolution is not a "jumping" from stage to stage. Each stage remains and evolves further so that the levels of evolutionary processes increase in number and complexity increases not only at each level but also in the hierarchical stratification of these levels. None of the levels vanish although they may become restructured. Today, there are still archebacteria and prokaryotes as well as the Gaia system, forms of organization which originated thousands of millions of years ago. The domain of insects has become practically stabilized for 100 million years. Particular species may become extinct, but the great lines of evolution, such as animals, plants and fungi, remain intact with their essential subdivisions, such as mammals, or birds.

A fragmented way of viewing things concludes that evolution is primarily about the preservation of stationary states. In reaction to the Darwinian image of a steady morphological development by ever renewed adaptation, the equally misleading image of a "punctuated equilibrium" has been proposed, a basic equilibrium state in which chance developments occur here and there. Both extremal views result from a one-sidedly microscopic view. In the framework of a co-evolution of macro- and microsystems there is never equilibrium, but autopoiesis in non-equilibrium in which fluctuations may break through at any time and any place. Complexity does not just increase in each of the individual microsystems but, above all, in the ways in which a world of dynamic relations, stratified in many levels, evolves.

What, then, is the difference between a more highly developed creature, such as man, and a less highly developed? The image that man has climbed to a hierarchically higher level would be wrong. The correct image recognizes that man includes many autopoietic levels of microevolution and, by way of the neural mind, also levels of macroevolution. If the microevolutionary steps

239

laid out in Fig. 43 (see p. 224) are folded together so that a single column results, we obtain the following hierarchy which forms the basic structure of man:

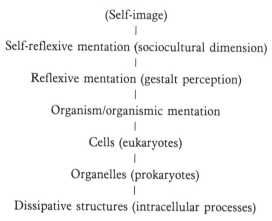

(Self-image)
|
Self-reflexive mentation (sociocultural dimension)
|
Reflexive mentation (gestalt perception)
|
Organism/organismic mentation
|
Cells (eukaryotes)
|
Organelles (prokaryotes)
|
Dissipative structures (intracellular processes)

Each of these levels has its own self-organizing dynamics, its own mind. In the total person this dynamics becomes co-ordinated in the same way in which the triune brain is co-ordinated in its domain. The result may be called multi-level autopoiesis. Mentation appears here as an integral aspect of a dynamic image of man. It is not directed against evolution, but is its high expression.

The idea of such a multilevel autopoiesis and evolution of man is basically not new. The Hindu system of the seven chakras stipulates a correspondence between the levels at which man enters into exchange relations with his environment on the one hand, and certain body structures which are arranged along the spine, on the other. In these chakras, the so-called Kundalini energy is supposed to manifest itself in a focused way. The structures in the body are viewed as opening toward the front of the body like funnels and penetrating the body. The seven chakras may be roughly characterized as follows: (1) Crotch; physical survival. (2) Pelvis; sexuality. (3) Solar plexus; power over environment and other people. (4) Heart; transpersonal love, connectedness with whole mankind. (5) Throat; seeking god. (6) "Third eye" on the forehead; wisdom. (7) "Crown" above the head; unity with the divine principle. Most people live according to one of the three lowest chakras. The ideal, however, is not to simply climb higher steps, but to activate and harmonize as many chakras as possible at the same time.

The concept of multilevel life presents considerable difficulties to Western thinking. This becomes visible in the ways in which Abraham Maslow's concept of a value hierarchy is applied in Anglo-American psychology and social psychology. Maslow's hierarchy corresponds chiefly to three main levels which introduce the following aspects from the bottom up: (1) physical values (physiological needs, physical security); (2) social values (belonging, esteem); and (3) spiritual vales (self-fulfilment). This scale may roughly correspond to a phylogenetic and ontogenetic sequence. But it is now misused to construct a priority scale in all sorts of contexts and to impose it upon the mature person. The result has already been anticipated in a Brecht song: "First comes food and then comes morality!"

The story of the young Beethoven would then perhaps read as follows: He comes from Bonn to Vienna and his first concern is getting a well-paid job which permits him to live comfortably in a cosy apartment or even to buy his own house. Next he looks for a bride, marries a woman of a good family and, using the newly acquired relations, gains entrance to noble circles whose darling he becomes. Once his financial and social status is secured, he still feels a slight unrest as if there was something left to do. He becomes pensive. Finally, revelation hits him, he sits down and composes his First Symphony (or even the Ninth immediately).

Reductionists cannot be convinced by such stories; artists are crazy in many ways, they say, shrugging their shoulders. But the Viennese psychologist and founder of logotherapy, Viktor Frankl (1978), observed and learned himself in Nazi concentration camps that keeping alive the sense of a higher meaning (for example, in the form of strong religious beliefs) is the most important element in surviving such extreme conditions. Aleksandr Solzhenitsyn (1971), in his first novel *One Day in the Life of Ivan Denissovich,* has powerfully described how improvised self-fulfilment leads not only to physical survival, but also to inner freedom. By accepting the badly organized compulsive labour for the hated régime as a challenge to his *own* creativity, imagination and personal ambition, the prisoner in a Siberian camp becomes the master of his own fate.

Even with so-called primitive peoples, such as the Asmat in West Irian who live in the mud and have been blessed only a decade ago with the first challenges of civilization in the form of a few Catholic missionaries, I have always found that their simple life is determined to a much higher extent by the structures of their myths and rituals than is the life of peoples enjoying abundance, such as Western industrial civilization. Not poverty and need, but excess and greed reduce life to a merely material level. Natural life is co-ordinated and vibrates in many levels and thereby gains dignity and beauty.

The idea of materialism, growing out of a reductionist Western science, has made a powerful impact on the re-creation of the world by the self-reflexive mind. But it is being led *ad absurdum* in our days by that same science and even its "hard" core, physics. "The use of the expression 'scientific materialism' should nowadays be tolerated only with reference to a set of methods or to an attitude of mind", writes Bernard d'Espagnat (1976), the noted philosopher of quantum mechanics. "With reference to the general conception of the world, it has become a meaningless association of words."

In a multilevel dynamic reality, each new level brings new evolutionary processes into play which co-ordinate and accentuate the processes at lower
hierarchical levels in particular ways. Therefore, reduction to one level of
description is never possible. In order to understand self-organization and
especially the phenomenon of life it is not only necessary to recognize different
levels, but also to understand the relations between them. In other words, the task
is now to understand autopoiesis and evolution of systems which include many
levels of existence and co-ordination.

The dynamic system capabilities which have developed step-wise (see
Fig. 43 on p. 224) now become co-ordinated from a higher level. While, for
example, dissipative structures already exhibit metabolism, the bioenergetic
processes of organisms are basically co-ordinated at the level of the cell. At the
level of the organism as a whole, however, there is further accentuation by the
selective acquisition or production of food. While biomolecules marked the
beginnings of self-reproduction and thus also of conservative information
storage, the latter capability becomes significantly enhanced by sexuality at
the level of eukaryotic cells and becomes further accentuated by the selectivity
which plays in sexuality at the level of the total organism which, in turn, is
accentuated by sociobiological and sociocultural factors. If a human has
culture he will not simply pair off sexually. A high view of love will attempt to
realize by way of the partner a mental and spiritual connectedness with all mankind and even creation as a whole. But these high ideals do not function when
there is a basic biological incompatability of the partners which apparently is
determined at subconscious (neural?) levels of the multilayered person.

There are perhaps electromagnetic and other frequencies and field effects
which have hardly been researched, but which may play a role here. Such
exchange effects may at least be easily demonstrated by so-called Kirlian
photography which (probably) records electron emission in high-frequency
fields (Krippner and Rubin, eds., 1974). I underwent such experiments in the
laboratory of Thelma Moss at the University of California in Los Angeles. An
academic colleague put his finger next to mine and the emission from my
finger tip radiated freely to all sides—it seemed that I liked him. He, in
contrast, cut off my emissions a few millimetres from my finger in a perfectly
straight line, apparently by means of a "defensive" field. Perhaps he did not
trust me fully. This effect increased markedly when I pretended to be cross
with him, although this had been previously agreed upon in this way. It
diminished when I put my arm around his shoulder and called him my dear
friend. On the wall there was a similar photograph taken with a student
couple in love. Both finger tips were surrounded by a single corona of
radiation in a true symbiosis of the fields. . . .

At the same time as the system capabilities become modified and partly
reorganized in multilevel autopoiesis, a "fine-structure" of many autopoietic

subsystems appears which, fitting the hierarchy, co-ordinate and are co-ordinated. Their basis seems to be the self-organization dynamics of cell groups which offer an always available basic activity which is normally suppressed. The motor activity is of particular interest in an evolutionary context. It is not, as has been previously assumed, generated by sensory impact which only exercises a control function. The motor activity of invertebrates is either regulated by the oscillatory activity of single nerve cells (pacemakers), or by the oscillations of a whole group of nerve cells none of which oscillates in isolation. The walking systems of animals usually seem to be based on regulation by whole groups of neurons. Experiments with the cockroach have shown that the extensor motor neurons are always excited, but normally suppressed by periodic bursts (limit-cycle behaviour?) of a neuron system which is known as flexor burst generator (Pearson, 1976). In vertebrates, the motor activity is always co-ordinated by such complex neuron systems.

The immune systems—of which, according to recent results, humans have at least two, an inner and an outer—also represent examples of normally suppressed basic activities which, if needed, are brought into action and may increase very quickly in a non-linear way. Although they act at a molecular basis, their self-organization pertains not to the molecular level, but to the level of populations of lymphozytes (cells which produce antibodies). The interplay of inhibitor and activator substances in the head formation of freshwater polyps has already been mentioned in an earlier chapter (see p. 60f.). In similar ways the human skin regenerates itself. The cells of the epidermis are always ready to divide, but are normally prevented from doing so by inhibitors, the so-called chalons. In the case of an injury the inhibitor runs off so that there is no longer an obstacle to cell division. A stimulating substance seems to play a role, too. After the closing of the wound, the local concentration of the inhibitor rises again and suppresses further cell division. Finally, as has already been mentioned, neuron populations in the brain always show some basic activity comparable to "idling" and which becomes manifest in the alpha rhythm of brain waves (8–14 Hz). Other self-regulatory intermediary systems include, for example, the endocrinal system, the sexual reproduction system (with the orgasm organizing itself once the "plateau" régime has been reached), digestion and blood circulation. If I understand Chinese acupuncture correctly, it is about the improvement of the self-regulation of such intermediary, autopoietic systems of the organism by means of non-linear reinforcement at suitable places.

Hierarchically co-ordinated dynamics

The new field of chronobiology (Bünning, 1977; Scheving, 1977) deals with the biological rhythms which formerly have also been called biological or physiological clocks. Such rhythms result from the dynamics of self-organizing systems in organisms as well as in their interaction with the environment. In the former case, they are called *free-running* or *endogenous rhythms.* The observed periods range from fractions of seconds to the order of 10 years. Thus, they span at least nine or ten orders of magnitude, perhaps more, because the fastest rhythms are not well researched yet. An upper limit for the frequency of brain waves, for example, has not been found yet; the usually cited figures of 30 to 40 Hz are determined by limitations in the instruments used, not in the observed phenomena. Human blood cells are used up and replaced in a rhythm which has a period of about an hour. Certain body cells divide about once a year, others never. All cells continuously degenerate and regenerate whereby the metabolic system (the self-regulation system of the enzymes) oscillates in rhythms at periods starting with about 1 minute. At the subcellular level, there are also the epigenetic oscillations which result from the rhythm of the enzyme synthesis which, in turn, depends on the genetic control mechanism; the corresponding period is approximately 1 hour. New sprouts on my house ferns permit me to observe such epigenetic oscillations in the form of eight-shaped "search movements" of the tips, which show a period of about 1 hour. With the help of such movements, the new sprout unfailingly finds its optimal living space with reference to light and humidity.

Whereas the regulation of the cell cycle itself is well understood for prokaryotic bacteria, it is relatively little researched for the eukaryotic cell. However, there are indications pointing to a mitotic oscillator which regulates cell division and to other oscillators which play an important role in the basic activities of the cell. One of the most important and mysterious phenomena is the capability of the eukaryotic cell to produce under conditions of radical change in the environmental substrate precisely those macromolecules which are required for the catalysis of those biochemical processes which can make use of the new substrate. It seems that the new substrate is capable of inducing the synthesis of these macromolecules (Nicolis and Prigogine, 1977). The corresponding regulatory circuit thus transcends the cell boundaries and includes environmental conditions. The already mentioned chromosome field theory provides an approach to the understanding of such epigenetic dynamics.

Non-linear oscillators always tend to synchronize themselves. In the case of disturbances, the synchronization is quickly restored. It may act in such a way that superimposed rhythms are generated. Circadian rhythms with a period of approximately 1 day may result in a simple way from the coupling of biochemical oscillators whose period is of the order of 1 minute (Nicolis and Prigogine, 1977). Circadian enzyme rhythms in human red blood cells were found even in the dissolved cell and thus may be assumed to already act at the subcellular level.

As has been discussed in Chapter 11, circadian rhythms may be largely understood as the result of adaptation to environmental rhythms. In a different environment (for example, in dark, closed rooms or in caves) many of these rhythms continue. They are then free-running (endogenous) and usually shift their period to some extent. For example, humans in caves tend to generate a somewhat longer daily cycle between 24 and 25 hours. In many cases free-running rhythms may also be influenced by light or magnetic fields. If with cold-blooded animals the temperature is the main synchronizing factor, more complex animals seems to have several regulatory systems so that originally linked rhythms, such as sleep and body temperature, may also become separated and desynchronized. It seems that part of the free-running rhythms are genetically anchored. In the fruit fly *(drosophila)*, for example, a small chromosome part has recently been identified which seems to be responsible for circadian rhythms.

In ontogeny, which even for complex organisms starts from a single cell, the zygote, the multilevel coupling of regulatory circuits becomes particularly conspicuous. Events at a molecular and subcellular level ultimately result in a multilevel dynamics of multicellular organization which also includes the environment. There is thus a chain of self-transcending processes. In eukaryotic cells, the chromosome itself exhibits a certain self-organizing dynamics. At another level of organismal development, electrical fields seem to play a role as well as morphogenic substances. Electrical fields which are generated in connection with a partial system—a little researched phenomenon—are ideally suited for positioning insofar as they may reach far beyond the physical structure of the evolving system. With respect to morphogenic substances it is not clear yet whether the phenomena involve in the first line the diffusion and the gradient in the concentration of such substances, or a phase gradient (with each cell acting as autonomous oscillator), or the fixation of substances on membranes. However, it seems clear that in the case of morphogenic substances there must be interactions between at least two substances (Nicolis and Prigogine, 1977). Usually, this interaction occurs between long-range inhibiting mechanisms and a short-range activation mechanism. The activator comes into play when it exceeds a threshold value; until then, its effects are suppressed. That time plays a role in these phenomena also seems to follow from a new model (Wolpert, 1978), according to which a cell's positional value decreases with longer residence in a "progress zone". With cell division there are always cells leaving a positionally informed zone; the first cells to leave form proximate structures, with the last ones structuration comes to an end.

The space-time structure of self-organizing, non-linear systems thus results in the self-organization of a superimposed dynamics which secures the coupling of oscillators to form superoscillators and the synchronization of morphogenetic processes at many levels. We may perhaps speak of the symbiosis and self-transcendence of the endogenous dynamics of self-organizing systems which is of even greater importance than symbiosis and self-transcendence in a morphological perspective. This phenomenon of the self-organization of hierarchically related strata makes possible the autopoiesis and evolution of a holistically acting, multilevel reality. It is not a cacophony of strata which results, but a dynamic, richly orchestrated connectedness. The brothers McKenna (1975) have elaborated on such a multilevel dynamics under the assumption that the same basic wave form (represented in the *I Ching*, the ancient Chinese Book of Changes) recurs at all levels in different frequencies, from the slowest to the fastest rhythms of the universe.

This is perhaps the most striking gestalt effect in a multilevel reality: Each autopoietic system generates its own space-time which is a fundamental parameter for many phenomena. A general system theory based on such an endogeneous space-time of systems instead of on the ticking of a mechanical clock in a space measured by a universal yardstick may become capable of unravelling deep commonalities. To speak of one day in the life of a one-day fly does not say much; to speak of its life-span, however, implies a comparison with other living systems at the basis of their endogenous dynamics. This individualistic system time does not lead to chaos, just as the joining of marked individualists in a dynamic, motivated society does not necessarily lead to chaos. Resonances and synchronizations occur in quite natural ways.

Perhaps, such dynamically coupled régimes of a higher order will facilitate the interpretation of phenomena which appear as abnormal or even miraculous. Western medicine gradually has begun to understand that its actions are geared to a single dynamic "base" régime of the organism. Healing is also possible beyond this base régime in other régimes. I have myself watched the impressive and rare "maro" trance dance in a sacred village festival in a remote Toraja village in the mountains of central Sulawesi (Celebes). After the bleeding was stopped by wiping with a red Ti leaf the profusely bleeding; deep cuts in the foreheads of the men healed within 20 minutes without leaving a scar or as much as a trace. Every year for a few weeks the Dutch meditation master Jack Schwarz, who lives in California, comes to the University of California in San Francisco to permit research on himself. He also gives demonstrations before groups of medical doctors in the course of which he pierces his upper arm with a thick knitting needle. If he is in a deep meditative state (which may be measured in terms of brain-wave frequencies), no blood appears and the wound has healed without a trace after half an hour. Once, when he became distracted during such an experiment, blood spurted from his wound and stained the meticulously white frocks of his audience.

Walking over red-hot stones on Bali or the Fiji islands, putting needles through cheeks and tongues or eating glass from India to Indonesia may all be done without danger in various degrees of trance, that is to say, alternative dynamic régimes of the organism. Walking on hot coal has even become a hobby among American academics, or perhaps a means for self-confirmation. Trance, meditation and ecstasis are so-called "altered states of consciousness" which will be brought up again in Chapter 18. With their help, a person remains open to novelty, may transform this novelty into the confirmation of his own life without having to fear destruction.

In *ecosystems*, the picture becomes more complicated. If, in the development of the organism, two types of non-linear processes play the main role, namely, genetic and metabolic processes, the number rises to at least six in ecosystems. Besides genetic processes (birth, death, mutation) and metabolic regulatory processes of a sociobiological kind (chemotaxis), there are now further non-linear metabolic processes comprising competition for niches (in particular, for food) within a species or between two or more species, predator-prey relationships and symbiosis as well as optical-acoustical communication. All these processes bring their proper rhythms into play which, for example, become noticeable in population dynamics.

In addition, ecosystems are characterized by particularly important environmental rhythms such as daily and seasonal oscillations. But here, it also becomes clear how the coupling of non-linear oscillators of shorter periods may result in oscillations of much longer periods at the level of a higher, holistic system level. A well-studied case is the 80- to 90-year period in the interaction between the spruce budworm and spruce-fir forests in Eastern Canada (Holling, 1976). The worm appears in large numbers only after a sequence of unusually dry years, though not always. In between it is reduced to small numbers by its natural enemies. If conditions are favourable, the budworm reproduces so quickly by feeding on the balsam fir that its predators cannot catch up with it. The results are stands of destroyed mature fir in the middle of less affected stands of young fir and non-susceptible white birch, besides a dense regeneration of fir and spruce. The destruction of mature fir eventually reaches such dimensions that the worm population is greatly reduced for lack of food. In the long span of time until the next burst of the budworm, the fir again takes up its competition with spruce and birch in which it has the advantage. The periodic oscillations of the entire ecosystem due to the appearance of the budworm is a prerequisite for the formation of a complex ecosystem in which spruce alone would otherwise dominate.

Not control hierarchy, but stratified autonomy

The most important aspect of this multilevel dynamic coupling in the world of the living may perhaps be seen in the maintenance of a certain autonomy at all hierarchical levels. Multilevel autopoiesis must not be confused with a control hierarchy in which information flows upward and orders flow downward. Until recently, the dynamics of life has been misunderstood so thoroughly that in system theory the notion of an "organismic system" was the synonym for a control hierarchy. In a living system, each autopoietic level may interact and communicate with the total environment in partially autonomous ways. The environment of a cell consists not only of its neighbour cells but also of the total biosphere with its chemistry and energy flows, and even of the solar system with its radiation phenomena and its gravity. Each level has its own self-organizing dynamics which gives rise to specific environmental relations.

In America it is fashionable today to attempt to influence autonomous body functions such as heart beat, circulation, electrical brain and muscle waves, or digestion, by so-called biofeedback. The enthusiasts of this technique even view it as preparation for an evolutionary jump of man. Total control of rational thinking over the body! I am absolutely horrified by such prospects even if Indian yogis have gone rather far in this domain. I am happy that my body functions by itself at many levels and that these levels understand how to reach harmony between them. I am satisfied with the multilevel dynamic reality of my organism.

Up to now in this chapter, I have only spoken of organisms and ecosystems. The same hierarchical coupling of oscillations, however, holds to an even higher degree for *sociocultural systems*. Here, the non-linear processes show an even much greater variety. Biological genetics is supplemented by other forms of genealogical communication such as tradition and laws in the social domain and books, works of art and architecture in the cultural domain. The metabolic processes become enriched by the production and distribution of energy, goods and services. *Politics* may be viewed as the complex interaction of many non-linear regulatory processes. But there are also the non-linear, self-organizing systems of emotions, mental-neural constructions (ideas, paradigms, ideologies) and value systems. The dynamic complexity of the human sphere is hardly grasped yet by these few notions. And yet, the coupling of this self-organization dynamics at many levels results in social and cultural structures of autopoietic and temporarily harmonious nature which are capable of stimulating and carrying a great deal of creativity. Sociocultural systems may be almost timeless (such as the already mentioned "clockwork" societies in Lévi-Strauss' scheme) or they may evolve quickly when the enumerated process systems become activated.

Social systems, corporations and states are natural multilevel systems and accordingly should not be organized as control hierarchies, in which decisions and orders are handed from the top down. A rigid and centralized world government is not the proper solution for the emergent world *problématique*. Elsewhere (Jantsch, 1972) I have elaborated on the principle of decentralized initiative and centralized synthesis in the management of complex human systems, as it already works in well-organized corporations. Zeleny and Pierre (1976) speak of the manager not as a decision-maker and giver of orders but as a catalyst.

The growing gap in the human world between systems which are extremely large and others which are very small (e.g. between multinational corporations and craftsmen or family enterprises) has been diagnosed correctly as unhealthy by E. F. Schumacher (1973). In the lectures on his last American tour before his death, in the spring of 1977, Schumacher told of his visits with the manufacturers of agricultural machinery. The "tractor of the future" was already on the drawing boards, a huge machine, fully automated, with fully air-conditioned cabin and stereo equipment; it is supposed to cost 90,000 dollars, in today's value of the dollar. When Schumacher inquired after machinery for agricultural family businesses, he received the classical American answer: "Forget it!" In fifteen years, the small farmer will have become extinct, anyway. A doubtful trend becomes reinforced by technological development. The recent developments in the direction of an "intermediary" or "appropriate" technology, especially for developing countries, are meant to revive the intermediary hierarchical levels which are indispensable for keeping a multilevel reality viable.

Another consideration seems to be essential for the viability of a complex human world. In Chapter 12 I have pointed to the complementarity of novelty and confirmation in the meta-evolution of a hierarchy of semantic levels of evolution. The top level was always the widest open to novelty whereas confirmation increasingly ruled the lower levels. A result of this principle is the normalization of the lower-level elements.

If this still holds in the human world, the highest level, i.e. the level of the value systems and the societal institutions incorporating them, should be the most widely open to novelty. The contrary, however, seems to hold, even in Western democracies. The task is to preserve the institutions of society in their present structure, we hear from all sides. Governments are supposed to preserve the constitution, churches the religions, universities the "objective" structures of science, and all of us the existing institutions. The values of society must not be touched, its structures not be put in question.

The result is a decrease in the frequency of innovation, of the "creative vibrations" as we climb the levels of a sociocultural hierarchy. But this is precisely the precondition for the establishment of a control hierarchy, which can only function if the controlling levels oscillate in lower frequencies than the controlled levels (Mesarović *et al.*, 1970). With the isolation of the higher levels of innovation and novelty a control hierarchy seems even inevitable. It is the urge to power as well as the lack of understanding of the dynamics of living systems which leads to the curtailment of evolutionary forces not only in the dictatorships of the East, but also in the enlightened democracies of the West. A convergence becomes visible which was not meant in this way. George Orwell, in his novel *1984*, has given it frighteningly realistic contours in the framework of a forecast which, a few years before this date. still appears possible.

In the last part of the book I should like, therefore, to pose the question of how openness is possible in the human world.

PART IV

Creativity:
Self-organization and the Human World

> The creation is . . . a blazing flame too hot to be
> handled indiscriminately. It is the other sun, the non-
> deterministic sun of becoming.
>
> Paolo Soleri, *Matter Becoming Spirit*

And how is evolution to continue in the human world? Has it, as some hold, become caught in a net of coercive factors in which it is ever more inextricably entangled with every motion? Is the evolution of man coming to its end—or is even the evolution of life on earth coming to its end with man? In this final part of the book, it is not my intention to analyze the problematics of our time or to offer solutions, but only to point to certain perspectives. I believe that the most important task today is the search for new degrees of freedom to facilitate the living out of evolutionary processes. It is of primary importance that the openness of the inner world for which no limitations are yet in sight, is matched by a similar openness of the outer world, and that it tries actively to establish the latter. I believe that sociocultural man in "co-evolution with himself" basically has the possibility of creating the conditions for his further evolution—much as life on earth, since its first appearance 4000 million years ago, has always created the conditions for its own evolution toward higher complexity.

15. Evolution–Revolution

Two dangers threaten the world: order and disorder.

Paul Valéry

Gradual change, manipulation or evolutionary fluctuation?

The theory of fluctuations as the basis of an understanding of coherent system evolution through a sequence of temporarily stabilized structures may conceivably be applied to the formulation of a new political theory which stands a good chance of giving realistic descriptions in many cases. Such a theory would transcend the framework of this book and exceeds my ambitions. Certain approaches in this direction may already be found in Marxist theory. The invalidation of the "law of large numbers", however, cannot easily be made compatible with the claim that revolutions are founded on mass movements.

The emergence of a world-wide system was already recognized by Karl Marx in the "Communist Manifesto" of 1848; his chief examples were the world market and an emergent world literature. But his thinking was, of course, caught in the equilibrium thinking of nineteenth-century physics. Marx viewed the coming world revolution as the last step in the direction of a durable, class- and tensionless society which would also mark the ultimate stage of the evolution of man and his consciousness. This ultimate stage is predetermined as it is in an equilibrium system. The initiation and acceleration of this macroprocess is a human task, but this human role in history is interpreted in very different ways by different schools of Marxist thought. If it was at first viewed as the spontaneous breakthrough of a fluctuation, it became increasingly the kind of tedious, cumulatively acting reform work which, today, we call the "long march through the institutions" and which came to the foreground with the disappointed hopes of the nineteenth century. In his foreword to Marx' *Class Fights in France* of the year 1895, Friedrich Engels rejects the idea of a revolution by surprise and wants to see revolution founded on long, continuous work out of which success is to grow "with mathematical certainty".

It was Lenin who realized again the potential of a fluctuation at the right moment—but he deeply mistrusted any spontaneity. A quotation is ascribed to him according to which "trusting is good, but control is better". In his materialistic—and therefore reductionist—view of history, Lenin believed that "history as a whole, and the history of revolution in particular, is always richer in content, more varied, more multiform, more lively and ingenious than is imagined by even the best parties, the most conscious vanguards of the most advanced classes" (quoted in Feyerabend, 1975, p. 17). But the October revolution of 1917 was ice-cold manipulation, the manifestation of a strong will to power much more than the reaching out and the creative self-organization of evolutionary processes. At least Solzhenitsyn (1976) has plausibly described it thus in his book *Lenin in Zurich.*

The difficulties met by the theory of fluctuations when it is applied to human society may have to do primarily with our inadequate understanding of interpersonal communication. A phenomenon such as the emergence of a new civilization—which usually breaks through very quickly, in a matter of decades only—seems to be well described with the self-reinforcement of internal fluctuations. It is startling, however, that such structures also emerged in historic situations characterized by the slowness of communication over long distances. The founder of Christianity, who died at the age of 33, was obviously not in a position to spread his message face to face. Rather, a resonance in consciousness may be assumed which was possibly enhanced by the social situation created by Roman imperialism in the Mediterranean region and in the Near East. In many cases the human core out of which a new civilization is born seems to be organized in relatively small tribes of several hundred or thousand members which share their life in a communal spirit. The Essene scrolls which were found in the caves near the Dead Sea give us a picture of this type of organization. Similar approaches are visible in the structuration of the counter-culture in the 1960s. As in dissipative structures there seems to exist a critical size for the onset of self-organization dynamics. On the other hand, the correspondence of structure and function here also prevents the group from growing too big. The growth process of such a civilization in its infancy probably occurs through multiplication of such units rather than by their growth.

In such groups, whose size still makes direct contact possible, social processes may unfold in great intensity. One of these processes is imitation. People, write Doris Lessing (1975), do not grow "by an acquisition of unconnected habits, of isolated bits of knowledge, like choosing things off a counter: 'Yes, I'll have that one,' or, 'No, I don't want that one!' But in fact people develop for good or bad by swallowing whole other people, atmospheres, events, places—develop by admiration. Often enough unconsciously, of course. We are the company we keep."

Metastability of institutions

Just as an individual may act as a fluctuation in a small group, larger autopoietic units may in turn act as fluctuations in bigger, spatially extended societies. The phenomenon of *metastability,* however, seems to play a significant role here. It already occurs in simple chemical reaction systems, as has been discussed in Chapter 3. With a high degree of flexible coupling of the subsystems, as it is increasingly introduced by the processes of human communication in the technological age, the old structure remains stable far beyond the point at which, from a macroscopic point of view, it is supposed to become unstable.

The question of the limits to complexity is again posed here. In principle, these limits have to be looked for where stability ends. Stability, in turn, is limited by the degree of coupling with the environment. In a static view, higher complexity implies a loss of stability as also most mathematical models (May, 1973) would predict. Self-organizing non-equilibrium systems, however, may be unstable and yet exist—by evolving. It is sufficient that the processes between subsystems are fast enough to damp smaller and medium fluctuations and to maintain the system in a state of metastability. Thereby, the shift to a new structure is delayed during a finite period which is sufficient for the unfolding of life processes. Metastability is delayed evolution. Through metastability, a dissipative system itself structures the space-time continuum for the unfolding of its endogenous dynamics. No complex system is ever truly stable; it is always, as long as it maintains its structure, metastable. This kind of dynamic existence, therefore, makes a tremendous increase in complexity possible. With the abandonment of permanent structural stability, evolution becomes open and unlimited. No end is in sight, no permanency, no telos. "We can no longer speak of the end of history," write Ilya Prigogine and Isabelle Stengers (1975), "only of the end of stories."

The resistance of an autopoietic structure against its evolution which, at the same time, implies the destruction of the old structure, is included as an essential aspect in the comprehensive dynamics of evolution. No autopoietic structure can stabilize itself forever, but it has, nevertheless, to defend itself to its utmost and to damp the fluctuations. If it would not do this, nothing much would come of evolution. The higher the resistance against structural change, the more powerful the fluctuations which ultimately break through—the richer and more varied also the unfolding of self-organization dynamics at the platform of a resilient structure. The more splendid the unfolding of mind, as we may also put it.

To live in an evolutionary spirit means to engage with full ambition and without any reserve in the structure of the present, and yet to let go and flow into a new structure when the right time has come. Such an attitude is meant by the supreme Buddhist virtue of "non-attachment" which so frequently is

confused with non-engagement. The political *baisse* into which the metafluctuation of the 1960s has dropped may be traced to a large extent to such an
attitude of non-engagement in personal and societal relationships.

But, we may now ask, does the complementary development of higher
stability and ever-increasing fluctuations not lead into dangerous regions where
destruction looms? Do the fluctuations which are potentially ready in the
nuclear arsenals of the big powers not already threaten life on the whole
earth? This is true, alas. If development were to continue in the same ways,
each quantum jump of evolution toward new social and cultural structures
would potentially unleash such destructive forces that the substrate—the
systems of the biosphere as well as the biological and sociocultural experience
stored in gene pools and libraries—would inevitably suffer. Are the ultimate
limits to complexity reached here? I do not believe so.

From quantum jumps to "gliding" evolution?

In self-reflexive systems, fluctuations may be anticipated and act in the
mental constructs of the present even if only in a primitive form of fear (of the
atomic bomb, of environmental pollution, or the "limits to growth"). We may
learn to "defuse" the fluctuations, if not to suppress them. On the other hand,
however—and this appears to be of much greater importance—the completion of time-binding with the inclusion of the past and the future in the
present, the genealogical process of history in its strictly sequential order may
to some extent be overcome. In the last chapter, I shall return to this fascinating prospect. Here, I should like to merely indicate that this implies that it
is no longer whole structural platforms, whole civilizations, societal systems,
art and life styles which must jump to a new structure. A pluralism emerges
in which many dynamic structures penetrate each other at the same level. In
such a pluralism, there is no longer the familiar evolution in big step
functions. Change, increasing in an absolute measure, occurs not only vertically, in historical time, but also horizontally, in a multitude of simultaneous
processes, none of which necessarily has to assume destructive dimensions.
The reality of the human world becomes dissolved into many realities, its
evolution into a multitude of horizontally linked evolutions. One may think of
the evolution of a pluralistic ecosystem which does not need to make great
jumps if one of its member species becomes extinct or a new one makes its
entrance.

Up to now the evolution of mankind generally followed the mechanism of
self-reinforcement of fluctuations and spontaneous structuration of a sociocultural system acting more or less as a unity. In this respect, evolution and
revolution were indistinguishable except for arbitrary criteria and scales. At
the levels of smaller and subordinate systems, we hardly become aware of this.

At the level of more comprehensive societal structures, however, we sometimes feel the effect of big caesuras which sometimes even leave bloody traces. And yet, the restructuration is frequently less radical than the accompanying circumstances lead one to believe. The French revolution at first led back to a monarchy and at least 150 years were needed to fulfil some of its essential demands. Considering the civil rights movement in America, or the equality of women, we find ourselves still involved in processes which were initiated in this distant past. The revolution of 1848 had to wait for the end of World War I, 70 years later, to see some of its postulates fulfilled. And the Parisian students of 1968 were betrayed step by step of all for which they had climbed the barricades. They did not even get their interdisciplinary university. And yet, they have initiated important and irreversible processes.

This phase shift in the evolution of two principal levels of the human world—the level of social structures and the level of cultural guiding images— may perhaps be represented in the way attempted by O. Markley of the Stanford Research Institute in California (see Fig. 45). Each level alternately

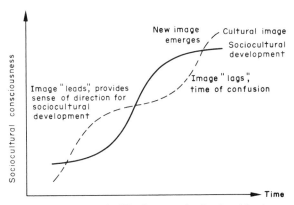

Fig. 45. Phase shift between cultural guiding images and sociocultural development. In one phase the cultural image leads, in another it lags behind. After O.W. Markley (1976).

pulls and is pulled by the other. Such guiding images run ahead of reality and seem to occur in different rhythms. The slowest rhythms with a period of one eon (more than 2000 years) correspond to the change in religions. Elsewhere (Jantsch, 1976) I have discussed how religions guide mankind through long periods of time like a light shining at the end of a long tunnel. The message of love expressed in the Christian faith has hardly been fully translated into reality but it has guided a large part of mankind. In our days, new guiding images are due at this highest level of the self-image of man-in-universe.

At the level of the most comprehensive human systems, the level of civilizations, the destructive effects of structural instability became evident already thousands of years ago. Most of the ancient civilizations have left nothing other than archaeological debris. Many myths reappear in later civilizations in new forms—in these mental structures we may perhaps recognize the most impressive coherence in the overall evolution of mankind. It is not always clear, however, whether such myths were transmitted by means of tradition or whether they go back to common archetypal roots, to common basic experiences of being human.

In the more recent cultural history of mankind, things have changed. The Christian civilization contains many live elements of the Graeco-Roman antique, in science as well as in philosophy, aesthetics, political science, theatre, sports and other domains. Here, genuine evolution took place in which many essential features entered in changed forms into new structures. A similar restructuration which, according to tradition, we count as change *within* Christian civilization—a somewhat arbitrary view—occurred with the Renaissance, the rationalists of the seventeenth century and the beginning of the scientific-technological era in which God became excluded from science. Even in our days, many features from the ancient world and the Middle Ages are still alive. The evolutionary jump toward a new cultural structure cannot be even unambiguously fixed in time. Between the discovery of America by European seafarers and the first opera, more than a hundred years elapsed; and yet they express the same spirit. Western industrialization in the nineteenth century, finally, and perhaps also cybernetization in the twentieth century, have called into play new cultural structures which grew without difficulty from the old ones and are coherent with them to a large degree.

In the twentieth century there is a marked decrease in the role of so-called "great historical figures" who—whether they were wise statesmen or bloody tyrants—triggered fluctuations which were to make a huge impact. At present, the fluctuations are more anonymous, broader and thus partly less cutting also. The same holds, for example, in comparison with the nineteenth and the early twentieth century, for leaders in technology and the economy. Their systems have become so complex that nobody has a clear overview any longer and, to use Galbraith's (1967) expression, a mid-level "techno-structure" has taken over as the effective decision-maker in technological development.

Does a "gliding" evolution in which the interweaving of many partial processes and partial jumps replaces restructuration in one, big jump, present an opportunity for the imminent social restructurations—not only in Western society, but in the whole world society with its burning inner contradictions? Perhaps we are at the beginning of a new phase of true social creativity. The urge to participate at all levels of political, economic and also partly cultural processes, the call for "parity", seems to speak for such an assumption. Up to a certain degree, social revolutions have always been carried to their end in mental dimensions first and followed up with developments over decades and centuries which were often not even well synchronized. The modern Western welfare state, whatever one is to think of it, has not been introduced by bloody revolutions.

Cultural pluralism and autonomy of the systems of human life

I believe that the future will be largely determined by two factors, besides others. One concerns the progressive weakening of control hierarchies with respect to human systems as well as the accompanying abandonment of the idea of a single, monolithic cultural guiding image. The other factor concerns the strengthening of the autonomy—that is to say, of the consciousness—of subsystems.

As has already been mentioned, in a control hierarchy the higher levels have to oscillate at lower frequencies than the lower levels which they control. Therefore, in the control hierarchies of Western democracy, the cultural ceiling is deliberately held rigid, as mobile as the social structures at medium and lower levels may have become. Social change is permitted only in the framework of unchangeable "values of society" and the latter's depositories, the institutions of society, are supposed to be preserved forever and without change. In an alive, multilevel reality, however, the issue should be to keep precisely the highest level open toward novelty.

A society which lives and evolves at many structural levels would require a reversal of the present scheme. The lower levels, which until now have been the stage of technological change leading to ever faster frequencies (in this trend, too, a reversal seems at hand), would become levels of diminished innovation and increased confirmation. In the technological environment, the achievements may become consolidated and improved in quality. The innovation frequencies in wide every-day domains of technology would slow down. At the other end, however, at the cultural level, the slowly oscillating, almost rigid structure would become dissolved into a cultural pluralism characterized by many simultaneous frequencies and in which many fluctuations would act as triggers for smaller and partial evolutions which would connect and interact without ever building a firm platform. It is significant that in a recent book for which I acted as co-editor (Jantsch and Waddington, eds., 1976), several authors independently diagnosed the beginning of such a cultural pluralism, multiple structures with high local fluctuations and a symbiotic joining of cultural guiding images and life styles. Magoroh Maruyama (1976), in this context, speaks of the "transepistemological process" in which the individual epistemologies or world views penetrate each other. In Berkeley, such a pluralism of life styles is already colourful reality, even if the colours seem to have paled somewhat in the past few years.

An alive and bubbling cultural pluralism would not only make life more interesting and exciting, but also strip the social structures of the strait-jacket of rigid values and legitimize their endogenous dynamics. In America, there is a growing discussion concerning so-called "victimless crimes" which are pursued only at the basis of moral claims corresponding to a specific world view. Sexual behaviour belongs in this category as well as the selective ban on drugs which often are more harmless than the socially sanctioned drugs (alcohol, nicotine). The decriminalization of marihuana has made much progress in America and homosexuals may openly stand to their inclination without having to fear becoming social and economic outcasts. On the contrary, the city of San Francisco even provides funds for their street festivals.

With cultural pluralism, perhaps the political structure of the Western version of representative democracy would change, too. To even touch upon this theme is considered daring or even criminal. But the contemporary form of democracy has at least two features which do not correspond to the laws of natural evolution. One feature concerns the principle of the law of large numbers (majority voting) and the negation of the role of fluctuations. The absolute majority determines the decision often before a creative fluctuation has been given a chance to stimulate resonances. The second feature concerns blindness toward longer-range developments and generally toward processes. Future structures are determined by a process of "bargaining in small steps", in which it is each side's avowed purpose to cancel out a reluctantly conceded step by taking two steps in the opposite direction. In combination with the short-range orientation toward elections—oscillations with a period of mostly not more than 4 or 5 years—this bargaining in small steps results in the stabilization and rigidification of structures, rather than in the catalytic furthering of their evolution.

The strengthening of subsystem autonomy is a theme which has recently been taken up in interesting ways. Dieter Senghaas (1977), for example, has proposed replacing the "associative" structures in the development strategies for the Third World—which are primarily meant to link up with the exchange structures of the Western world—by "dissociative" structures which would function autonomously to a large extent. An "autocentric" development of poor countries and regions would not only emphasize internal trade patterns over external trade but also seek to make culture, technology and consumer patterns largely independent of Western models, though not separate. The far-reaching alignment of production in the developing countries with Western needs as well as the inflexible application of Western technology and political models to the problems of developing countries lead to an ever more precarious situation. Many countries, including some Western industrialized countries, are incapable of producing their own food. Most of the industrialized countries depend for their energy supply on imports. All this can certainly not be changed quickly. But the joining into one big world system can no longer be the goal. The Western world is already standing too far apart with respect to the exploitation of world resources to be an equitable member of a future world family.

The above-mentioned Japanese anthropologist Magoroh Maruyama (1976), who is active in America, emphasizes non-hierarchical organization rules for human systems which he calls "heterogenistic". They determine quite a few non-Western forms of society (Camara, 1975). We may recognize here the same non-equilibrium principle which underlies all dissipative self-organization. This principle acts not only in non-hierarchical social systems, but also in the aesthetic design of the systems of human life, such as gardens and architecture. Maruyama also applies his principles to the design of the structure of human communities in future space colonies which, if they ever become realized, will be immense laboratories for the design of human systems.

Protection of all that is weak corresponds to the ethics of a self-reflexive society. But it often becomes misunderstood and misapplied in the sense of "homogenistic" equalization which slows down creative dynamics. Gregory Bateson recently stated in his position as Regent of the University of California that American universities are much more concerned with the rights of the worst students than with the 1 per cent interested in and setting out to explore the "eternal verities".

Evolution creates wholeness which interacts autonomously with other wholeness. This wholeness, however, is not only fatally violated at the level of nation states. As Ivan Illich (1978) puts it, the heteronomous ways of production (determined by others) overshadow the autonomous (self-determined) ways of production at the level of the individual. We do not learn any longer, we are taught (in particular, by the ready-made products of the media); we do not design our own environment, it is supplied by industry; we do not live in a healthy way, we are medically taken care of; we do no longer determine the values of our life, they are prescribed by experts. And so forth. Humans, who can no longer produce autonomous values, have to be supplied with them. The activities which this requires increasingly block the social system. This kind of feedback which, from a certain point, leads to diminishing efficiency of social activity, has been aptly called "specific counterproductivity" by Illich. Examples may be found everywhere. At low traffic density, a car may still perform the functions for which it had been conceived, namely, for example, to provide a means of fast transport in the city. If too many cars jam the streets of the city, walking may be faster. In Chapter 10 I have already pointed to the investment of energy in cyclically organized industrial systems which sometimes only yield a net share of one-tenth or so of the needs which originally led to the construction of such systems. The rest may be compared to friction or other forms of waste.

In his role as an integral aspect of sociocultural evolution, man has the possibility and the obligation to assume evolutionary responsibility in his proper domain—that is to say, on the planet earth thus far. In other words, he is called upon to act as a system manager at all levels. System management now appears as an activity which aids and enhances evolution. It is our task to act *with* evolution, not against it. Above all, this means the recognition and application of a multilevel ethics which transcends the level of the individual person and thus acts in a self-transcendent way. What this means in more detail, will be briefly discussed in the following chapter.

16. Ethics, Morality and System Management

> When all the world recognizes beauty as beauty,
>   this in itself is ugliness.
> When all the world recognizes good as good,
>   this in itself is evil.
>
> Lao Tzu, *Tao Teh Ching*

Multilevel ethics

In a structure-oriented view, ethics is often understood as a metaphysical category, a set of rules of behaviour which is valid in an absolute sense or at least given as a logical *a priori*. The view is further enhanced by the claim to absolute validity raised by religions and ideologies which incoporate ethics as their pragmatic aspect. In such a view, the only legitimate attitude is then adaptation.

In a process-oriented view emphasizing evolving systems, however, self-determination is in the focus of interest. Ethics refers to the dynamics of systems. But the self-organization dynamics at different levels of evolution follow different optimization criteria. At the level of chemical dissipative structures, the emphasis is on energy throughput; at the level of the first primitive self-reproducing structures on error correction; at the level of the complex sexually reproducing cell, on variety; at the level of the organism, on flexibility and the capability to cope with the unexpected; and so forth. In Fig. 43 (see p. 224) I have attempted to name such criteria. Quite generally, we can define ethical behaviour as behaviour which enhances evolution. But this implies that ethics is not given *a priori*, but emerges with evolution and follows a development which, in principle, is wide open. With each new autopoietic level a new ethics comes into play and also a new mechanism through which this ethics acts as a regulatory device. In sociobiology, this mechanism is of a primarily chemical nature. In the human world it consists chiefly of laws, rules of behaviour and taboos—but also of morality as a direct inner experience.

As an integral aspect of evolution, ethics is not subject to revelation, as is the ethics of religions with a personal god. Rather, it may be experienced directly by way of the dynamics of self-organization and creative process. I should like to call this experiential aspect of ethics *morality*. As a dynamic principle, morality is a manifestation of the mind—of metabolic mind at its characteristic levels as well as of neural mind at the evolutionary levels which include the central nervous system. The direct living experience of morality becomes expressed in the form of ethics—it becomes form in the same way in which biological experience becomes form in the genetic code. The stored ethical information is then selectively retrieved and applied in the moral process in actual life situations. The interaction between ethics and morality gives rise to an epigenealogical process comparable to the epigenetic process in which genetically stored information is retrieved and applied to life situations. In such processes, complexity is made available to life and thereby generates new complexity.

It is nonsense to claim that animals lack morality. But it belongs to another level of consciousness than does human morality. *Re-ligio*, the linking backward to the origin, not only provides a possibility of sensing the dynamics of evolution at its origin, but also makes morality effective. Quite generally, the moral experience increases with increasing time- and space-binding in evolution. Thereby, it gradually replaces the genetically determined instinct as a regulator of behaviour. In the human realm, in which possibilities for action are available in such a rich spectrum, it is primarily the mental structures—our intentions, desires and preferences—which are guided by morality and determine our overall behaviour.

At the higher levels of human consciousness which include self-reflexion, the enhancement of evolution implies more than simple self-preservation and self-presentation which are modes of behaviour already characteristic of early steps in evolution and even of chemical dissipative structures. It also implies more than simple self-reproduction and evolution of a species which already characterizes organisms that appeared early in evolution. And it certainly transcends the frequently cited slogan of the "survival of the species" which expresses a dull equilibrium principle. In the wake of a fashionable reductionist sociogenetics, which is misleadingly called sociobiology (Wilson, 1975, 1978), there also appear attempts to explain the origin of human values on a purely genetic basis. A voluminous book by George Edgin Pugh (1977), for example, discusses primary values which are genetically fixed and regulate the behaviour by means of pleasure and pain. These primary values are food and sexuality, a modern reductionism which recalls the old reductionism of Sigmund Freud and Karl Marx with emphasis on the proper functioning of humans in work and sex only. Pugh's secondary values—morality, ethics and so forth—are then supposed to be established by the primary values with a view to their usefulness in problem-solving!

Human ethics, more than any other ethics, is a multilevel ethics. Our consciousness includes transpersonal dimensions. By means of our mental constructs we design social and cultural systems. Therefore, human ethics includes the *ethics of whole systems* demanded by West Churchman (1968). But ultimately we are called upon to develop an ethics which I should like to call an *evolutionary ethics*. It would not only transcend the individual but all of mankind, and explicitly include the main principles of evolution, such as openness, non-equilibrium, the positive role of fluctuations, engagement and non-attachment. Here, at the start of a world-wide creative crisis which puts the traditional ways of our life in question at all levels, we are still far from formulating and implementing such an evolutionary ethics.

What in the Western world we call ethics is a behavioural code at the social level which is primarily geared to ensure the free unfoldment of the individual. This is the reason why there is so much talk of rights, of basic human rights as well as of rights of particular groups, minorities, privileged and underprivileged people—and almost never of responsibility. Sir Geoffrey Vickers (1973) has pointed out that rights constitute a static and defensive, structure-oriented concept, whereas the acceptance of responsibility implies creative participation in the design of the human world. The ethics which dominates in the Western world is therefore an individualistic ethics in the disguise of a socially committing behavioural code. It is not a multilevel ethics in the true sense.

Morality, in contrast, is the direct experience of an ethics inherent in the dynamics of evolution. Morality, too, uses form, a specific epistemology, a basic attitude toward the world which corresponds to a specific structure of our consciousness. The higher the number of levels and the intensity at which we live, the higher also become the number of levels and the intensity at which our morality becomes effective as an aspect of our complex consciousness. The resulting tension between ethics and morality is one of the core elements of the ancient Greek tragedy. In conflict with the law, Sophocles' Antigone buries her brother Polyneikes who had rebelled against the king and whose body is left to the wild animals outside the city walls. Her morality bids her to put the divine law of love above the law of the state. But the law of the state, too, is an expression of the divine order. Out of her own free will and with great inner strength, Antigone accepts her death. She is not crushed by a blind fate, as modern interpretations would have it. She is living out her destiny without escaping its consequences. In the tragic turn of her life she gains the ultimate freedom.

Morality is a manifestation of consciousness. A major aim in the design of the human world by humans themselves must be the increasingly improved match between ethics and the levels of a multilevel consciousness. The same holds for those mental structures with the help of which we plan our future.

At each level they have to satisfy the ethics which we ourselves have established there as the guardian of our actions.

Time- and space-binding in planning

Since the middle of the 1960s, the increasing interest in medium- and long-range planning has thrown the theory and practice of planning into a creative crisis resulting in profound change. It was not possible to simply apply the older concepts of short-range planning to a widened time horizon. The objective of short-range planning is the efficient and economic realization of a specific and relatively clearly recognized product idea within a rigid structure of planning and action. But it was impossible to aim for a long-range future along a rigid yardstick. Thus, planning became a multilevel concept. Above the level of tactical or operational planning in the short range there is now the *strategic level* at which a variety of options are imagined, tested and prepared for realization. Strategic planning creates a mental non-equilibrium structure with fluctuations fed into it deliberately to trigger further evolution in one or the other direction. And above the strategic level, there is the *policy level* at which the dynamics of the system in question (e.g. an industrial corporation) is viewed in the context of an all-embracing sociocultural dynamics. And even higher is the *level of values* which is no more subject to rational elaboration but always plays a decisive and guiding role, whether implicitly or explicitly. Open planning, therefore, is never a purely rational process.

Elsewhere (Jantsch, 1972, 1975) I have described in detail how the social dimension already enters at the strategic level. The focus is on a particular social function—e.g. transportation, communication or energy generation and distribution—and technological options are examined with a view to the consequences they would have for the systems of human life. The methodology of the systems approach has been developed for this task. In such a perspective, short-range profit maximization can no longer be the decisive criterion. Additional considerations come to the foreground which deal primarily with the continuity of the corporation, or generally the organization, in an environment of technological and social change.

Sociocultural responsibility comes into even sharper focus at the policy level. This is the stage for the evolution of institutional guiding images. The guiding image of the industrial corporation, for example, has undergone significant change over the past 20 years. From the image of a money-producing machine it has first evolved to the still dominant image of an organism, the survival of which is to be secured by diversification and other measures. But also this organismic image is about to change. It is, after all, not of decisive importance that a particular organization survives, but that the continuity of social processes such as the production and distribution of goods and services can be maintained even if that means to switch to new structures which may replace the present mixed economies of the Western world.

This multilevel planning not only links the perspectives of different time horizons, but also different basic attitudes and logic. Elsewhere (Jantsch, 1975) I have described these basic attitudes or world views with the help of a metaphor, namely, the relations which we may establish with a stream. Standing on dry land on one bank and watching the stream go by corresponds to a *rational* attitude. If we try to steer our canoe *in* the stream, in direct interaction with its forces and keeping proper distance from both banks, we are taking a *mythological* attitude—we enter into a direct relationship with the life forces around us, we deal with them at their proper level, we become involved and try to influence the overall process. But if we imagine that we *are* the stream, just as a group of water molecules is the stream and at the same time only one of its aspects, we are experiencing an *evolutionary* attitude. In the Western world, such an evolutionary world view is realized only with great difficulty, but it has given rise to the elaborate pure process philosophy of Buddhism and Taoism. Table 6 sketches the relations between levels of planning, world views, guiding images and systemtheoretical approaches. It gives an idea of how planning may be organized as an evolutionary process.

Multilevel management is resisted by the thinking and the emotional preferences of many people. Generally, we prefer to think either in "concrete" short-range terms or in "general" long-range terms. But the higher art of the manager lies precisely in his capability to think, feel and act at several levels at the same time. It is of crucial importance in evolution that the top levels are kept open for the entry and absorption of novelty. In contrast to a widely held belief, planning in an evolutionary spirit therefore does not result in the reduction of uncertainty and complexity, but in their increase. Uncertainty increases because the spectrum of options is deliberately widened; imagination comes into play. Instead of doing the obvious, the not-so-obvious is also deliberately sought out and taken into consideration. Complexity increases because the immediate domain of the organization in question, or the individual, is transcended and relations within the larger system of society, culture or the world at large move into the foreground. Reality *is* complex, and evolution manifests in the increase of this complexity. Greater complexity (which is not the same as greater complicatedness), therefore, means a more realistic attitude taken by planning.

Opening up "at the top"

With the "opening upward", the managers of the higher conceptual levels (which are not necessarily the higher power levels) become the managers of change, whereas the lower conceptual levels need administrators, the depend-

Table 6

Multi-level planning in relation to a multi-level reality. The levels of planning correspond to different time horizons and different levels of logic and system paradigms.

Level of reality	Level of planning	Time horizon	Logic	Guiding image for		Basic attitude	Levels of description
				Organization	Manager		
Values							
Policy (System dynamics)	Policy	Long-range	Evolutionary (what system dynamics in context of overall dynamics?)	Management (catalysis) of social processes	Catalyst	Evolutionary (being the stream)	Structure↔Function ↕ Fluctuation
Social functions	Strategy	Medium-range	Mutual-causal (how to structure relations in a system so as to keep it viable?)	Survival of an organism	Co-ordinator	Mythological (steering in the stream)	Structure↔Function
Tactical targets (products, services)	Tactics (operations)	Short-range	Linear-causal (how to reach a recognized target)	Profit-producing mechanisms	Receiver of information and giver of orders	Rational (outside the stream) problem-solving	Structure
Resources (people, knowledge, materials, energy, capital, etc.)							

able operators of established and only slowly changing processes. Many people prefer the static security promised by specialization and unchanging work environment. But others, in contrast, prefer a dynamic security which may be compared to riding a bicycle and which promises ever-renewed creative challenge and ever-changing tasks and rewards. It is this dynamic type of manager which belongs at the top of a system management, in the private as well the public domain.

This already works well in the most advanced-thinking enterprises in the private domain, but not so well in social management systems, such as government at all jurisdictional levels, labour unions or universities. There, it is precisely the higher levels which are closed toward novelty and geared to confirmation. This even holds where sovereignty rests in the people who have a wide opportunity to participate, as is the case in direct democracies such as Switzerland. Every Swiss citizen votes several times a year on issues at the federal, cantonal and community level. To be precise, not every citizen votes, in spite of the penalty of two Swiss francs. A study, carried out on behalf of the Swiss government and published in 1977, comes to the conclusion that only 28 per cent of the people representing a "lower socio-economic level" (defined by income as well as education) actually vote, compared with 68 per cent of the people at a "higher socio-economic level". If income is relatively higher than education, participation in voting is high; if it is relatively lower, only 25 per cent of the disappointed go to the polls. This already makes it pretty clear that management is a matter of consciousness.

But in the type of questions put to the voters, the direct democracy does not show a very high consciousness. A few years ago, the city of Zurich voted on a new urban transportation system. The strategic decision (for a subterranean streetcar system) was made by the City Council and also the tactical decision (a specific track pattern) was made ahead of the vote and elaborated in the form of a single, expensive, detailed plan. The question was then put to the voters: Do you want a subterranean streetcar system with this particular track pattern and at this specific cost? The answer was simply "No". It could not have been anything else but "Yes" or "No". But what precisely did not please the voters remained obscure. The City Council again started off out of the blue by ordering the next arbitrary plan to be elaborated.

The dialogue with the voters can only be held at the highest level of values. I am aware of the lack of any tested, effective mechanism for such a dialogue. TV appearances of politicians are monologues, and the New England style "town meetings" in which President Carter likes to appear occasionally are at this stage no more than symbolic gestures. It is important that the citizen affected by decisions is given the opportunity to participate in a discussion on the impact these decisions will have on the values shaping his future living environment. Strategic decisions depend on complex technical evaluations and should be left to the Parliament or the City Council; they require full-time professional managers. The tactical and operational decisions are best made by specialists, experts, or administrators. But at the top, openness toward innovation is of crucial importance. Heads of governments, city mayors, labour union leaders and university presidents should not simply be administrators, but the major agents and managers of openness and change. It is clear that reality is far from such a desirable image. But the discussion around nuclear power plants has at least made it abundantly clear that technical experts have nothing more to contribute at the level of values than any other citizen. The State Senate of California has stated this explicitly after hearing a dozen Nobel Laureates, half of which were as furiously against nuclear power as the other half was for it.

Evolution questions the principle of democracy in a very profound way. A democracy can only be creative if it admits and even furthers fluctuations. But this requires a new attitude toward the majority principle which basically is on the side of confirmation and meets novelty at best with distrust or even open hostility. Evolutionary creativity always renders invalid the "law of large numbers" and acts in an elitist way. Wherever democracy has functioned well, the role of individual imagination has been tacitly tolerated or even supported. But it is time now to realize in explicit terms that the unqualified belief in the majority principle—or, in other words, in the rule of the average —can only serve to transfer the motor function of sociocultural dynamics from the creative individual to an impersonal system. "Thou'rt shoved thyself, imagining to shove", Mephistopheles exclaims mockingly. Perhaps the most profound political paradox of our time lies in the need for "elitist" fluctuations to turn self-determination into evolutionary, creative self-transcendence. The only alternative is equilibrium—the equilibrium of spiritual, social and cultural death.

Process planning instead of structural planning

Planning in an evolutionary spirit—which, at the same time, implies an ethical spirit—also meets with further difficulties. Anticipation, that important faculty of the self-reflexive mind, permits the selection of a specific goal. But this goal is often taken as a fixed structure and alternatives become excluded from the further planning process. The extrapolative type of planning in which econometricians indulge follows a given process into the future under the tacit assumption that the underlying structure will not change over longer periods of time, which is *a priori* unrealistic. Normative planning stipulates "good" scenarios for a particular time horizon and attempts to align action in the present to the realization of a scenario structure envisioned for the future. In our days, a naïve *teleology*—aiming straight at a recognized goal—is sometimes replaced by a more sophisticated *teleonomy* which realizes that some goals may only be reached by taking detours and by following possible process networks. But in either case, the values of the future are stipulated in the present, although they are inherently defying rational prediction. Each structure implies a particular system of values which often is hardly visible in the present. Therefore, Edgar S. Dunn (1977) is right in speaking in a somewhat derogatory way of "utopian engineering". These structural approaches did, indeed, originate in the area of technological planning and engineering.

In an evolutionary spirit, creative processes ought to be permitted to interact freely and to find their own order of evolving structures. Dunn calls this approach "evolutionary experimentation". Its original image is the movement of bacteria. Running phases in one direction alternate with tumbling phases. This is due to the flagella which rotate like turbines and alternately join together in one direction and fling loose in all directions. The direction taken in the running phase is arbitrary. And yet the bacterium shows a net movement in the direction of optimal food concentration. While in a running phase, it measures the concentration in the environment of its body. If food concentration remains the same or decreases, the running phase is of normal duration, about 1 second. However, if the concentration increases, the running phase lasts a little longer. Eventually, such a random biased walk, as it is called in technical terms, leads unerringly to the highest food concentration (Boos, 1978).

In process-oriented management, the role of the manager would resemble that of a catalyst. His task would be primarily the prolongation of those processes which seem to run in a creative direction, to stop those which appear unpromising and to eliminate those which he deems uncreative. At the same time, it would be his task to stimulate and further the interaction between creative processes.

I remember a little story from the time I worked as a physicist and engineer in the headquarters of a multinational electrotechnical company. Three engineers had worked for a number of years on the technical realization of a new principle, but had no results to show. The department head ordered their work stopped. But the three engineers, greatly alarmed, asked for and were granted an interview with the president. They talked to him in great excitement for half an hour, after which the president interrupted them. He had not really understood their technical arguments for an imminent breakthrough in their work, he told them, but he had understood something else: that there were three people full of enthusiasm and confidence in the ultimate success of their work. That was all he needed to know in order to let them continue.

Modern corporate long-range planning, as I have sketched it above, helped indeed to introduce the flexibility which is also a characteristic of open evolution. The formulation of punctual and structural goals frequently is not taken literally and absolutely, but as an indication only for the direction in which to start. As experience accumulates in pursuing such goals, they become modified, they change almost beyond recognition, and sometimes they are totally abandoned. "Aborting the corporate plan" was the title of Stafford Beer's contribution to a planning symposium which I organized in Bellagio 10 years ago (Jantsch, 1969).

In this kind of open-ended planning, as in all open evolution, the purpose is not waiting at the end of our path into the future, but is immanent in the process itself. Knowing is ultimately possible only by means of doing. The oscillations in the evolution of the sociocultural dynamics enhanced by planning are then comparable to the endogenous, free-running rhythms which the organism develops under environmental constraints. Planning, in this view, imposes new rules upon sociocultural dynamics and thus gives rise to the self-generation of new patterns. Policy and sociocultural dynamics constitute a complementarity in the same sense as genotype and phenotype, the former in both cases being rate-independent and the latter rate-dependent (Pattee, 1978). And the systems approach to planning which continuously tries to match the direction taken with the environmental conditions would be the sociocultural equivalent of the epigenetic process—the process which selectively and in a synchronised way retrieves and utilizes conservatively stored information in relation to the needs of the system in coping with an ever-changing environment which itself is in full evolution.

Process planning in an evolutionary spirit ends the dualism between the planner and the planned, organization and environment, corporation and society, culture and nature. We have not fallen from grace by developing the capability for rational thought—if we use the latter in its proper context. On the contrary, we are called upon to enhance the evolutionary process within and around ourselves by assuming responsibility for it at our appropriate level of evolution, the sociocultural level. Planning becomes an instrument of evolution, not just of our technological environment, but, more importantly, of our own evolution as well.

The complementarity of values

Multilevel management and multilevel ethics imply another difficulty for Western ways of thinking and feeling. We are used to distinguishing un-ambiguously between opposites, especially between "good" and "bad" or between "good" and "evil". In such a view, to act ethically means simply to further the "good" and to suppress the "evil". But in a multilevel ethics, values may become reversed from level to level. The reduction to a single semantic level is responsible for the confusion reigning in the implementation of minority rights in America. In order to increase minority participation in academic professions, admission standards are lowered for minority students. In order to restore or achieve equality in the long run, equality is violated in the short run. But instead of making this two-level policy explicit in the law, the sloppy fiction of equality as an absolute, constitutional right is upheld. No wonder that the U.S. Supreme Court decision in the case of *Alan Bakke* vs. *Regents of the University of California* met with such a confused reaction—both sides claiming victory, others becoming deeply concerned about the future of minority participation.

What is technologically good may be socially evil. What the individual desires—for example, unlimited gasoline consumption for his car—may appear undesirable at the level of the national economy. Microscopically viewed, war is always evil, and yet I feel that the last World War brought something very good to my generation—an alternative to living in a "Thousand Year's Empire". The bubonic plague diminished the world population by one-third and caused more personal suffering than we can imagine today; but evolution was not thrown back by it. The life processes, including the population growth of mankind, continued the way they would have taken without interruption. In evolution, *homeorhesis* is generally a very important and significant phenomenon. This is the term which Conrad Waddington applied to the tendency of processes to continue in their original patterns, even if temporarily disturbed. Homeorhesis is probably more important in evolution than homeostasis, the tendency of spatial structures to remain the same.

Perhaps, evolution will not be too deeply disturbed by a turbulent development possible in the near future, with the loss (or, to be precise, the premature loss) of many lives. An organism which has reproduced itself, may be spared by evolution. Perhaps the major catastrophes on the planet earth mean not much more to evolution than the weeding out of a garden which permits more beautiful flowers to grow. If, as Bertrand de Jouvenel (1968) holds, we are gardeners of the planet earth, we ought to understand this ambition of evolution.

Problem-solving presupposes the existence of an unambiguous answer in the quest for the good and the right. But such an answer is possible only at one specific level of a multilevel reality. A technological problem may be solved; the solution, for example a new material or a new construction, satisfying specific requirements, remains valid and applicable. It can be shelved for later use. In the dynamics of dissipative biological, sociobiological and sociocultural processes, however, there are no problems which may be solved once and for all. There is only a dynamic, evolving *problématique,* as Ozbekhan (1976) has called it, and which appears at many levels in different and changing aspects. There is no "solution" for poverty or injustice. A dynamics, however, which gives rise to such a *problématique*, is open to some effort of regulation. A valid approach to understand this dynamics is to pose questions not only at one level, but at as many levels as possible. An answer is not an end, it does not terminate anything. To pose questions at ever new levels of discourse corresponds to an opening up of consciousness toward a multilevel reality.

A dynamic ethics of becoming, in contrast to a static ethics of being, does not know possession. The inhabitants of the Indonesian island of Bali do not view themselves as possessors of the island, but as guests on the "island of the gods". Guests which have to work hard every day for their living. And yet, this spirited feeling of being the guests of the gods never fades. It is the never depleted source of a happiness which manifests in continuous gratitude, in sacrifices to all gifts of life—water, fire, food, everything. In the Western world, we do not live much longer than the Balinese do; our possessions vanish no less. But the Western proprietary ethics suggests that the meaning of life is not in the process of living to the fullest extent, but in the accumulation of money and goods, of knowledge and influence.

In a multilevel, evolving reality, opposites vanish ultimately. There is no "good" and "evil". It is utterly childish to feel ashamed of evolution like the Nobel Laureate whom I recently heard profess to it with much empathy. With the partial exception of photosynthesizing plants, all life lives off other life—and we still do, even if we do not devour our prey in the open fields, but employ slaughterhouses and specialists for the preparation of our food.

Process thinking does not know any sharp separation between opposite aspects of reality. It also transcends a dialectic synthesis of opposites, that clumsy Western attempt at making a rigid structure of notions move and overcome its dualism. In process thinking, there is only *complementarity* in which the opposites include each other. Friedrich Hölderlin, in his Sophocles distich, has perhaps given this thought the most profound expression:

> Many seek vainly, joyously to express joy.
> Finally I apprehend it, here in my sorrow.

17. Energy, Economy and Technology

> We are not tuned to the larger system that contains us.
> Nor is the ecological frenzy of the present, an ecological
> wafer for the temporal and spatial mass of ecology, of
> great significance. The ecological role is historical like
> anything else. In the light of history (evolution), its
> meanings and its lessons are quite different from the
> instant takes we make of it.
>
> Paolo Soleri, *Matter Becoming Spirit*

Time-binding in the exploitation of energy sources

Like any dissipative phenomenon, life unfolds in an energy stream, transforming free energy into entropy. In the case of our planet, the sun-earth-space system provides the temperature gradient in which the solar flux from the sun penetrates the atmosphere and the biosphere (the Gaia system) and becomes degraded in the process. By this inclusion of the cosmos as an energy source as well as a sink, the environment takes care of the energy exchange and becomes practically inexhaustible. In this way only does dissipative self-organization on earth become possible on a large scale.

The total solar radiation impinging on earth at any moment is roughly 173,000 terawatt (TW or 10^{12} watt or a thousand million kilowatt). Out of this, almost 52,000 TW are returned immediately to space by direct reflexion, without having done anything useful on earth. 81,000 TW are transformed into heat, 40,000 TW become invested in the evaporation of water and 370 TW are used for weather mechanics such as wind, waves, convection and currents (Hubbert, 1971). A small share only—95 TW according to most recent estimates (Hall, 1978)—maintain via photosynthesis the various life processes, with the reject heat in photosynthesis (about 100 times as much) already deducted. This life energy, however, streams through complex ecosystems with many trophic levels and is used in many steps.

Mankind participates in the 95 TW energy flow of life at present with about one half per cent or 0.5 TW (4000 million humans times an average of

125 watt heat generation). The biological energy flow, however, is by far exceeded by the technological energy flow which runs through the human world. At present, this technological energy flow amounts to almost 10 TW, equivalent to one-tenth of the total energy flow of terrestrial life. In this way, technology has increased the biological share of the human world by a factor of 20. The United States exceeds the world-wide per capita average of more than 2 kilowatt six-fold, whereas the Indian per capita consumption is less than one-fifth of the world-wide average.

Since the harnessing of fire about 450,000 years ago, with which the first phase of human technology started in a real way, man has made use of that "short-term storage" of solar energy given by the biomass. Wood and organic waste, however, can yield only a maximum of 5 to 10 TW when this storage facility is continuously used; less than half of that seems economically practic:able (Starr, 1971). The other continuously acting "short-term storage" of solar energy which comes into play with the evaporation of water may yield a maximum of 3 TW if exploited in hydro-power plants; about one-third of this is used at present. The "medium-term storage" of geothermal energy, which is not due to solar energy, contains a grand total of only 3 TWY (terawatt-years).

Of the energy streams of the biosphere which may be tapped directly, tidal energy and other phenomena could at best play a very minor role. According to Starr (1971) they can, combined with the already mentioned short- and medium-term storages, yield a total of 18 TW, out of which only 3 TW appear practicable until the year 2000.

It is no wonder, then, that since the beginning of the industrial era man has attempted, and still attempts, to make use of the various "long-term storage" facilities provided by terrestrial nature, one after the other. Since the beginning of the cambrian age 600 million years ago, when multicellular organisms appeared, a part of life has not become decomposed into its molecular parts and recycled after death, but has sunk into swamps and moors where the potential energy of the macromolecular structures was partially preserved. Out of the energy stream which pulsed through life, a small share, not much more than one-millionth, became transferred from the short-term storage of the organism to the long-term storage of fossil fuels, in which coal, oil, natural gas and oil shales accumulated. This storage is estimated at about 15,000 TWY out of which at best one-half seems recoverable at present; 89 per cent of this storage is kept in the form of coal. Out of the minable reserves at best one-fifth, or about 1500 TWY, may be exploited at costs which do not exceed the present ones by more than double (Hubbert, 1971; Starr, 1971). The fossil storage still continues to regenerate itself at about the same scale as always, that is to say, at a maximum of 100 megawatt (million watts) which is only one hundred-thousandth of present use.

More energy is contained in the long-term storages stemming from processes of cosmic evolution preceding the origin of the solar system. These are the storage facilities present in atomic nuclei which are akin to being used either for fission or for fusion. Based on fission reactors with direct one-way burning of uranium, the fission energy storage is of the same order as the storage of fossil fuel energy. With breeder reactor technology, however, these reserves are up to 100-fold as valuable in terms of energy and thus correspond to many years of the total solar radiation influx. The size of the reserves on which these calculations are based depends in a very sensitive way on the price which can be paid for 1 pound of uranium oxide. Expressed in pre-1973 prices (which have risen four- to five-fold until 1978), the picture is the following: At 10 dollars per pound the reserve is about 1000 TWY, at 100 dollars per pound 3000 to 30,000 TWY and at prices up to 500 dollars per pound about 1,500,000 TWY. Uranium may also be extracted from the oceans; there, the reserves amount to about 100,000 TWY (Trueb, 1974).

Much more powerful is the other nuclear long-term energy storage from which fusion energy may be liberated. As long as the simpler deuterium-tritium reaction is envisaged, the total potential is limited by the available reserves of lithium which, in the reaction system, becomes transformed into tritium. Lithium may be extracted from sea water. With an extraction of 2 per cent only, the resulting fusion energy potential amounts to about a million terawatt-years and is thus comparable to the reserves in fission energy. If, however, the higher temperature deuterium-deuterium reaction can be mastered, the picture changes dramatically. The oceans of the world contain one deuterium atom per 6700 simple hydrogen atoms; each cubic metre of sea water contains 34.4 gram deuterium the energy potential of which is equivalent to the combustion heat of 300 tons of coal. Since the volume of the world's oceans is 1500 million cubic kilometres, or 1.5×10^{18} cubic metres, even an extraction of only 2 per cent of the deuterium amounts to an energy storage which is of the order of 10^{10} TWY and thus about a million times larger than the fossil-fuel storage (Hubbert, 1971).

This regression to energy storages which stem from ever-earlier phases of evolution and whose potential increases with the time distance constitutes another aspect of time- and space-binding. In the liberation of materially stored chemical and nuclear energy an ever-farther removed past becomes active in the present. With the production of environmental pollution and waste, however, the present also becomes active in the future. This becomes dramatically clear with the example of radioactive waste, where the better utilization of the fission-energy storage by means of the plutonium breeder technology also greatly enhances the undesirable effects in the future. Is this a "punishment" for having intervened in the storage activity of evolution, or is our mental horizon too limited to think in such evolutionary dimensions? Is nuclear energy "sinful" or is its use in line with evolution?

When I listen to the heated discussion about plutonium, I cannot help—even though I become concerned—but remember the prokaryotes. If they had their own science council and their own environmental protection agencies, they will certainly have warned: "There is an extremely poisonous stuff, called oxygen, which burns organic matter and would mean the end of all of us if it ever occurred freely. Some among us speak of better energy efficiency with the help of this substance. The irresponsibility of such arguments is self-evident." And they were right, of course. Some of the descendants of the prokaryotes which did introduce oxygen to the atmosphere were capable of adapting. Others, however, have to live underground and be protected by plants from the poisonous oxygen. They are no longer the masters on earth and many have lost their independence. But with their self-transcendence they have created the prerequisites for all more complex life—ourselves included.

Over the past hundred years, the "energy-genealogical" way of time-binding in the exploitation of energy resources has been the only one taken seriously. It is opposed to the "energy-autopoietic" way which emphasizes the exchange with the environment and thus the direct utilization of solar energy. Only that part of solar radiation impinging on earth which is normally transformed into heat may be used, and of that part only the share which hits the solid surface of the earth. This gives a maximum of 28,000 TW. Counting a 10 per cent conversion efficiency, today's energy consumption would already require that one-third of a per cent of all land areas in the world be covered by solar collectors. Unless, if it becomes possible to realize plans for extra-terrestrial solar collectors which would transmit the transformed energy by means of microwave beams. But just as in the liberation of stored energy on earth, the energy household of the earth would be burdened by this amount, in contrast to the tapping of the natural solar flux. It is not yet clear which limitations will come into play first, limitations in energy generation or limitations in energy consumption.

Energy-intensive economy

In Chapter 8 it was mentioned that with the maturing of ecosystems the specific energy required per unit of biomass decreases. The energy flowing through the ecosystem is increasingly well utilized, especially by the formation of a hierarchy of trophic levels in which the same energy flow is recycled from step to step. Accordingly, the primary production of plants per unit of biomass also decreases in mature ecosystems. In an overall way, however, biomass increases, which is to say that the ecosystem becomes increasingly more complex and dense. In this way, the energy stored in the system between its entry and its exit also increases with more highly organized systems. This is a general law of life (Morowitz, 1968). The result is typical curves such as those shown in Fig. 46. The maximum yield, as may be seen from this figure, is reached for ecosystems of relatively low maturity.

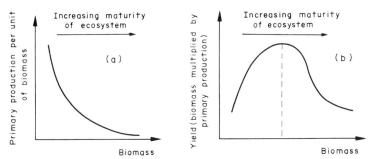

Fig. 46. Primary production and yield of an ecosystem of given spatial extension. (a) With the maturing of the ecosystem, the biomass increases in an absolute sense, but the primary production per unit of biomass decreases; this means that the ecosystem learns to improve its economy. (b) Therefore, the yield reaches a maximum in an ecosystem of relatively low maturity; this is the reason, why agricultural systems correspond to a low degree of maturity. After R. Margalef (1968).

Ramón Margalef (1968) has pointed out that agricultural systems are deliberately laid out for a low degree of maturity. This also has the advantage of the robustness found in younger systems. Normal agricultural systems run a much lower risk of breaking down under human exploitation than more complex systems would. Margalef believes generally that man is geared to systems of low maturity. The additional energy flow which he pumps through the ecosystems can, according to Margalef, only result in a rejuvenation of the biosphere which creates new opportunities for evolution. Even environmental pollution tends to some extent to reinforce biological activity, as may be seen in eutrophized lakes. More carbon dioxide, too, will mean more plant growth.

The energy required to take care of one unit of the human biomass (for example, one person) increases continuously. In agricultural, this is due mainly to the manufacture of chemical fertilizers and the mechanization of work, and also to a shift in the primary production (plants) in the direction òf a larger share of secondary production (animals, whereby animal feed often also includes grains and other highly nutritional primary production, fit for human consumption). Other factors are food processing and transport. If, in the United States, each unit of consumed food energy required a five-fold investment of non-food energy in 1940, this figure had risen to nine-fold by 1970.

Margalef's interpretation of the higher energy requirement in terms of a rejuvenation of the system is appealing. Industrialization brought mankind a marked acceleration of evolutionary processes. New "niches", such as new professions, have been created. Besides the agricultural sector, the industrial and the service sector developed to which is now added the information sector. As in an ecosystem, primary energy streams through many trophic levels which, the more people in the industrialized countries they comprise, the more removed they are from the primary energy transformation in agriculture. But the comparison is misleading. It would refer to a mature ecosystem in which the energy used per unit of biomass is supposed to decrease. In human systems, however, it increases with higher complexity not only per capita, but partly also (as demonstrated by the example of agriculture) per unit of production. How may this be explained?

I believe this has to do with the fact that sociocultural systems do not only act in simple autopoietic ways such as sociobiological systems or ecosystems— which is to say, they are creative in the self-realization at a specific level—but are also creative in the design of their own evolution and in the transformation of an environment which transcends by far the immediate needs of man. Probably, it is also of importance here that man is the only creature to use exosomatic tools, such as machines and vehicles, in a massive way. Their operation requires much more energy than the living parts of the system. Sociocultural systems only partly obey the laws of biological life. If self-organizing systems from chemical dissipative structures to ecosystems are self-limiting, technology represents a world of equilibrium structures whose growth is not self-limiting. Their energetic aspect refers to mixed equilibrium/non-equilibrium systems—or man/technology systems.

The American anthropologist Richard Adams (1975), in his most interesting book *Energy and Structure,* describes the urge to energy as an urge to power by domination of the environment. Even Marxist theory knew nothing better than to seek the way to the liberation of man from feudal power structures in power gains by the build-up of heavy industries. Quite a few developing countries, too, find themselves disappointed with building up heavy industry. Algeria's hope that its heavy industry, built with high priority as an "industrie industrialisante" (an industrializing industry) and initiating the self-organization of an industrial economy, has come to nothing. Technological processes are not life processes. They may at best play auxiliary roles in the unfolding of life.

The evolutionary trend toward higher flexibility becomes much more effective at the human level than in ecosystems with older and less complex life forms. There, enhanced autonomy of a mature ecosystem sometimes implies lower flexibility toward unexpected environmental fluctuations. Man is not only capable of much faster learning processes and thus of quicker adaptation to environmental changes but he is also made by evolution for a life of high uncertainty and the absorption of much novelty. Organisms which appeared earlier in evolution are normally highly specialized, for example in one or only a few types of food, in the acquisition of this food, in communication or in defence. Man, in contrast, is extremely multiverse. It is only technology which makes him forget and unlearn this.

It is interesting to remember that the massive transformation of the earth's surface and atmosphere by the oxygen production of prokaryotes, which lasted for 2000 million years, did not require complex systems. It laid the ground for the development of complex systems and the self-transcendence of life toward higher autopoietic levels. Later transformations by younger ecosystems also generated the conditions for the appearance of higher evolutionary processes. Does the temporally and spatially extremely dense transformation of the earth by sociocultural evolution announce a further step of self-transcendence? This might be the step from the planet earth into outer space.

The immediate effect, however, is a clogging of the terrestrial system. The exponential rise in energy consumption, which in recent years corresponded to a doubling time of only 13 years, is closely linked to the nature of the economic process which currently dominates in the industrialized countries. It is essentially a one-way process, also called "throw-away economy". Georgescu-Roegen (1971) has described the economic process as entropy-producing process in which entropy becomes visible not only in terms of reject heat, but also in terms of waste and environmental pollution. This is true, but is besides the point. Every other life process, too, generates entropy and waste, but this waste is practically completely recycled in biological cycles.

We are totally in the dark with respect to the impact of waste and pollution on the self-regulatory Gaia system. Will it regulate temperature somehow to stay near the optimum for life, if the already clearly measurable increase in the carbon dioxide contents of the air leads to increased heat radiation? Is there a mechanism to counter the consequences of ozone destruction in the higher layers of the atmosphere? Is Gaia generally robust or, after 1500 million years, easy to disturb by man? We do not know—because science so far has shown interest in life forms, but not in life processes.

The one-way economy has to use ever more energy to remove the waste and pollution it generates. A recycling economy according to biological principles seems possible to a great degree and is increasingly called for with every day. However, what weighs even more gravely is the fact that the technicized economic process removes people from their participation in the social process. Economic growth is prescribed to meet unemployment. A "right to work" is claimed where the basic issue really is to share the responsibility for the social and economic processes with all creative humans. The result is comparable to a helix in which unemployment, inflation, needless production and senseless consumption, waste and pollution drive each other to grow in a positive feedback system.

Originally, the flexibility of the human society was so remarkable that it was possible to match social structure and economic function to some extent. Richard Adams (1975) mentions the example of an Indian caste society of the year A.D 650. A peasant caste had the task of production, a merchant caste the task of storage and transport of goods, a warrior and noblemen caste did not deal with material things, but consumed, and controlled by wielding power. A priest caste, finally, stood apart from life processes; in the energetic scheme, according to Adams, it corresponds to entropy production. The casteless people removed the waste. In the class concepts of Western sociologies it was originally possible to match social structures with economic functions in a similar way (peasants and workers, merchants, landowners and officials). In our days, however, these correspondences have become erased to a large extent. Perhaps this is a sign for an imminent basic change in an economic process which hardly corresponds to the consciousness of our time any longer.

Economy, environment and consciousness

Economy is essentially a process system of human consciousness. It may, in principle, operate in as many dynamic régimes as we are capable of imagining. Economy is a system of subjective, dynamic relationships which essentially depends on psychological factors. This is partly becoming recognized today and new possible structures are being discussed which put economy in the perspective of human consciousness. The following five basic types seem to emerge from the current discussion:

1. *Dominance* of the environment by means of energy and matter technology. The environment is made predictable by restructuring it; this facilitates control. In other words, confirmation is maximized. An equilibrium state is sought in which the environment can no longer evolve, and with it the human world. This path leads to the clogging of the endogenous processes and to what Ivan Illich (1978) has called the already mentioned specific counter-productivity.

2. *Adaptation* of the social structure to the environment, again with the aim of attaining an equilibrium state in which confirmation dominates and static security is maximized. John Rodman (1977) has recently criticized the "popular rhetorics of survival" and pointed to the dangerous narrowing down of the images we make of nature and ourselves with the help of one-sided recipes: The conservation of resources includes a one-dimensional reduction of experience to a homocentrically defined, economic utility. The preservation of wilderness has to do with the religious search for the sacred which is mediated by a temporary, contemplative and sometimes ecologically confused aesthetics. "Nature

moralism" seeks justice for extrahuman nature but views the relations between the living species under the angle of a much too species-specific morality of rights and duties. Only an attitude, which Rodman calls *ecological resistance,* widens the image we make of ourselves and of nature. It regards differences as natural and goes beyond an attitude of precaution, reverence and morality. It corresponds more fully to the following version 3.

3. *Symbiosis* with the environment by using circular processes according to biological models, for example, tapping the direct solar energy flux, utilization of energy at many trophic levels of human life (as, for example, in the compact cities of the future envisaged by Paolo Soleri and already realized in his Arcosanti), recycling of matter and increased uses of biotechnology (not only in agriculture). The economic ideal here would perhaps be Willis Harman's (1974) idea of a "humanistic capitalism" which would be based on both an "ecological ethics" and a "self-realization ethics". It would try to base the planning and mobilization of the creative potential of as many people as possible on perhaps those principles of system management which I have evoked in the preceding chapter. This attitude emphasizes autopoiesis, but not evolution.

4. *Evolution* and exploration of new opportunities for unfolding (niches) *by extension of the environment,* or in other words, by the colonization of the solar system and other parts of outer space. In the foreground of interest here are the man-made and settled structures which, according to serious studies by the American physicist Gerard O'Neill (1977), are not only feasible, but allegedly even of economic advantage. The utilization of raw materials (metals) contained in small asteroids whose orbit is changed so that they circle the earth, plays a role in such serious speculations (O'Neill, 1978) as well as the already mentioned solar collectors in orbit around the earth. This attitude implies a "technological forward escape" which would open up great new opportunities for the currently dominant economic process (according to point 1) and also seeks to establish a broad basis for symbiotic processes. The essential design task does not only concern technological constructs, but also the design of ecosystems, human communities and cultures. In the latter domains, greater uncertainties may lurk than in the required physical technology. At stake is virtually the re-creation of new worlds in all their multilevel reality—a grandiose extension of the re-creation of the world in socio-cultural evolution.

5. *Evolution* and the opening up of new niches by an *extension of consciousness.* Here, an information technology has to be mentioned which does no longer only serve the steering of energy and production processes but,

above all, the diversification of our own personal experience. Learning would then no longer be adaptation to a specific form into which knowledge has been brought (e.g. the learning of a scientific discipline), but the formation of new and alive relationships with a multifaceted reality which may be experienced in many forms—learning would become a creative game played with reality. Instead of "utopian engineering" creative processes would be permitted to unfold and form new structures. The energy required for this would be extremely small. An information society in this sense would elevate the economic process to a totally new level, comparable to the supplementation of slow metabolic processes by fast neural processes with the evolution of the brain. The beginnings of such an information society already exist and have considerably accelerated the sociocultural processes. But the opening up of new levels and experiences of consciousness remains a wide-open frontier. To keep it in motion is the aim of this attitude. Time- and space-binding may intensify the experience in the present in practically unlimited ways. Here is an opportunity to get directly to the four-dimensional experience of the unfurling of space and time in the processes of evolution.

The first two of these five versions which, for improved clarity, have been presented here in somewhat simplified form, lead to the rigidification of life. Version 3 describes a harmonic, alive, autopoietic existence without evolution. The versions 4 and 5 refer to seemingly incompatible alternatives. The one requires more matter and energy technology, the other moves to a technology which lets information evolve at a higher level—and thus calls into being a new type of mind. Perhaps, however, it is a mixture—or, better, an ecology—of versions 3 through 5 which corresponds best to openness in sociocultural evolution, an openness which assigns a creative role to mankind in the design of its own future.

The autopoietic version 3 does away with the prejudice that only a control hierarchy may be expected to cope with the complex problems of the present and the future. On the contrary, co-operation between autonomous wholenesses is made the basic principle here. A certain division of labour may always be granted, but it must not lead to total dependence on other systems.

Evolution is never purely functional, even in the biological domain. There is always some extravagance, the pure beauty of morphological and behavioural features which seem meaningless from a utilitarian point of view. Adolf Portmann has always pointed to such functionless features and has been duly ridiculed by the Darwinians. But at the human level we may ourselves test the truth of his gestalt perception. Do we really want a predictable environment which may be rationally controlled? Who has not felt a chill in his spine when, in the Cuba crisis of 1962, the Kennedy administration made its computers spit out decisions—which, fortunately, it did not trust in the end? And, above all, do we only want to function, to make work and sexuality our total life, as Freud has prophesized? And nothing else?

The most important openness is the openness of the multilevel consciousness which establishes us as creative beings, as persons. Perhaps all three versions, 3 to 5, ultimately serve such an openness. When Gerard O'Neill, the promotor of space colonies, gave a lecture in San Francisco in the summer of 1977, thousands of young people came to listen as if there was a rock concert. Some of them were the same young people who had demanded a more human society a few years ago, and partly succeeded in this demand. The urge for cosmic connections, for a symbiosis of mankind with life and intelligence beyond our earth, cannot be simply explained away as an "escape from reality". It reflects the urge to establish a deeper connection with evolution and thus to what we ourselves represent. The extension of our consciousness is self-reference at the same time as it is cosmic reference.

Earlier civilizations have cultivated the inner path. Our Western civilization, in the search for truth as well as for creative opportunities, has primarily cultivated the outer path. At the crossroads of our time we may decide for either of the two paths—or, for the first time in the history of mankind, we may decide for a symbiosis in the complementarity of both paths. This, however, requires that we understand the creative process itself and its birth from the processes of evolution. This is the theme of the following chapter.

18. The Creative Process

> Philemon and other figures of my fantasies brought
> home to me the crucial insight that there are things in
> the psyche which I do not produce, but which produce
> themselves and have their own life.
>
> C. G. Jung, *Memories, Dreams, Reflections*

Self-organization, art and the experience of art

The usual image of the artist who creates the work of art is time and again questioned by the creative artist himself. "When the camera is running, the film makes itself totally by itself", the French film director Jean Eustache emphasizes. In a dualistic world view it used to be the muse of divine inspiration which used the artist as an instrument. In the non-dualistic world view, however, the creative process appears as an aspect of evolutionary self-organization. "What nature leaves imperfect, the art perfects", said the alchemists. And Roland Fischer (1970) writes of evolution as "the creative process in *statu nascendi*" and adds that "what we are used to call 'creative', is only a weak reflexion of the forces which have produced us".

An artistic aspect of the metafluctuation of the 1960s is the *"nouveau roman"*, which in its origins is connected with the names of Alain Robbe-Grillet and Nathalie Sarraute. It still makes its impact in France and beyond. Robbe-Grillett is not only the most important author, but also the chief theoretician of this direction. In a lecture in Berkeley in November of 1976—ironically on the same day on which the anthropologist Colin Turnbull proclaimed the self-destruction of the mind on the same campus (see p. 3f.)—Robbie-Grillet attempted to explain the dynamics of the *nouveau roman* in thermodynamic terms. The intuitively correct attempt went astray in many details since, as we know, self-organization dynamics cannot be explained in terms of the older notions of equilibrium thermodynamics. And the idea of self-organization is central for the *nouveau roman*.

Robbe-Grillet spoke of an *"ordre mouvant"*, a flowing order in the novel or film. "Mobile order" is also spoken of by Emil Staiger in quite another context, namely classical literature. Order is not established *a priori*. Right at the start, Robbe-Grillet states, non-equilibrium and polarization occur. A sentence is written down; it is still freely chosen. But it immediately entails further sentences which put it in question. If a sequence of such sentences forms a certain order—we may say, a non-equilibrium structure—this order, in turn, is put in question, and evolves to a new structure. Without himself doing anything, the author feels that he is driven across many of such instability thresholds without ever reaching permanent order.

Another feature of the *nouveau roman* is the active exchange between art and environment. The novel is not supposed to transmit any meaning, in particular, not any meaning which might be introduced by analogies from the inner world. Meaning only emerges from the interaction of the work of art with the reader or the contemplator—a *"sens tremblant,"* a vibrant meaning which can never be fixed statically. One has to let things speak for themselves. The artist is at best a catalyst.

One year after Robbe-Grillet, Jerzy Grotowski, the boldest and most essential renewer of the theatre in the 1960s, came to Berkeley where a group of his disciples want to work on his newest ideas for a "paratheatre". These are no longer the same ideas which have made his theatre laboratory of Wroclaw in Poland famous in the whole world. The incredible intensification of the performing process which resulted from the most rigorous limitation to the essentials on the one hand and from the highest ecstasy of expression on the other, drove the ultimately dualistic concept of stage representation—there are always performers and audience—into its self-dissolution. I vividly remember Grotowski's visits to Paris. The performances took place in deliberately narrow spaces, in a wooden structure erected on the stage of the Odéon with a hundred spectators (there was no space for more) looking down like into one of those corrals in which cattle is driven to be slaughtered; or in the Sainte-Chapelle, the most precious jewel of French gothic architecture. I can still hear in my mind the ecstatic chanting which grasped the whole audience with its rhythm, making the audience swing like with the movements of the ocean on a ship. But the audience still remained an audience and was permitted to sit on benches and remain silent.

Grotowski called this the theatre of poverty and he led it to its zenith. Beyond, there are the "paratheatrical phenomena" which capture his interest at present. They characterize a theatre in which there is no representation as such any longer, in which there is no distinction between performers and audience, only between "initiators" and "participants" in an encounter whose dynamics is self-organizing. But it is meant to be more than the "happening" of the 1960s which experimented with extreme novelty. The evolutionary balance between novelty and confirmation, out of which grows new order, is still to be realized in the theatre. This last consequence of intensifying the theatrical process is called "process theatre" by Grotowski. Processes are no longer played, but unfold, interact, join to form dynamic structures which undergo an open evolution. The aim is again to let things and people speak for themselves. Grotowski has realized, like Prigogine, that live order forms by itself, if processes are allowed to unfold.

But even conventional theatre is supposed to "live" and vibrate. In the dualistic split between performers and audience, this requires not only life on stage, but also in front of the stage. In an article by Fritz Thorn (1976), I found the following astonishing references to the self-organization dynamics of audience systems which enter into live exchange with the stage: "Two factors determine the degree of magical connection which is decisive for the success of the salon comedy as well as the engaged drama: the darkness in the audience space and the number of spectators which form a collective. The latter depends on such worldly things as the architecture of the theatre, the physical arrangement of persons in the front rows and their heterogeneity. Experience has shown that a homogenous audience, such as students, workers or members of the 'better classes', has to be more numerous than a heterogeneous one in order to play a role in the connection—*its* role." Compare the following statement by the pianist Vladimir Horowitz (Epstein, 1978): "An audience is a strange animal. Alone, they sometimes understand nothing at all; but when they are all together, then they understand."

Walter Schurian (1978), using the example of the contemporary Viennese painter, Ernst Fuchs, speaks of the painting as a live work of art which is characterized by autopoiesis as well as self-transcendence. For the artist, the dynamics of living systems becomes the "iridescent smile around the mouth of the sphinx". Its mystery cannot be explained; in order to solve it, the contemplator has to himself become part of the work of art. The endogenous dynamics of the painting has to be experienced directly. The contemplator even has to experience it in himself. For different viewers and at different times, this dynamic structure of the work of art appears different. It evolves with history. In the stage settings for the Hamburg Opera production of "The Magic Flute" (1977) Fuchs wanted to further the appropriation of the dynamics of a work of art by the spectator by covering time and place of the opera with a "veil of pictures and images of the fantasy", which corresponded to the everyday world of the creative artist. For Mozart and Schikaneder, Fuchs assumed, this may have been the fantastic and exotic reports from the New World which moved Europe at the time when the opera was written.

Every work of art has its formal aspect. It may be visual, as in paintings and sculptures, or half abstract and half melodious in a poem, or logical-abstract as in musical scores. Even in the formal aspect there are features of self-organizing structures. The musicologist Hildemarie Streich, for example, on the occasion of the Eranos meeting 1977 under the general theme "The

Meaning of Imperfection" (Schoch, 1977), pointed out that the magnificent development of occidental music has been made possible only by the transition from the *ars antiqua* to the *ars nova* in the Renaissance. At that time, the classical Pythagorean system was replaced by the natural-harmonic system in which the small third, until then banned, became acceptable. With this, a small rest interval, called "wolf", representing an inner non-equilibrium, was accepted. The ratios of the acoustical frequencies could no longer be transformed into each other without leaving a small rest. The result is that the different keys assumed a different character; C major has, as it is expressed, another "temperament" than, for example, E flat major. The pure tuning of the *ars antiqua* gave the latter its angelic splendour. The inner non-equilibrium of the *ars nova*, in contrast, led to the incredible wealth in expression with which music filled the human world. In this way, music assumed a new role in the self-reflexion of the occident.

In contemporary music, there is a certain parallel. Serial music which corresponds to a deterministic principle is combined with aleatoric approaches (from *alea*, dice), which represent the freedom of interpretation by random combinations of fixed modules. Thereby, the complementarity of deterministic and stochastic elements, of confirmation and novelty, is introduced which characterizes self-organization. Aleatoric techniques are not all that new, however, because there are already toccatas by Girolamo Frescobaldi from the year 1637 and minuets by Mozart, the parts of which may be freely combined.

As Schurian (1978) puts it, great art always shares in the design of the "inclusion of the world in the consciousness of the universe". This inclusion, however, is not an act of grace, but the self-transcendence of life.

The experience of art is a further example for the general epigenealogical process in which life processes put certain formal aspects into a semantic context and bring them into live exchange with the environment. Since we were just talking of music, this process may be compared with that story of Baron Münchhausen in which a postillion, riding in ice-cold winter weather, blows a joyful song on his horn. No sound can be heard. Later, however, when he has hung the horn next to the warm stove in a restaurant, the song unfreezes and the horn smathered without anybody seen playing it.

The life of a sculpture is not inherent in the form of the stone but in the processes which unfold between the work and its contemplator. The life of a piece of music has been put by the composer into a highly inadequate formal notation; and yet, the performance may be full of life. How is this transfer of dead form into live processes achieved? Nikolaus Harnoncourt (1976), the conductor which has earned great merit for his historically faithful arrangements of old music, speaks of a "generally unwritten *convention* between composer and interpreter", which, however, has to span long periods of time with their inevitable cultural changes. We have, says Harnoncourt, "read and performed during decades and in good faith all occidental music according to the conventions of, say, Brahms. But there were two essentially different, even extremal interpretations, which were based on the same prerequisites, and paradoxically even on the same attitudes: either the 'missing' indications (dynamics, espressivo, tempi, orchestration) were added, or what appeared in writing was performed in a 'work-faithful' and 'objective' spirit. Our aim, in contrast, has to be to find out *what was meant.*"

Thus, the convention has to be established at the level of meaning, or perhaps even at the pragmatic level, the level of effectiveness. "When I sit down at the piano," says Horowitz (Epstein, 1978), "I become transformed. I see the composer. I am the composer. Music gives me this feeling." There is obviously a certain correspondence between the endogenous dynamics of the artist in the creative act, the endogenous dynamics of the work of art and the endogenous dynamics of the contemplator or listener—a "tuning in" over time and space which is based on the homologous dynamics of all aspects of evolution. Which is the same as to say, it is based on the homology of the mind. A prerequisite is the possibility of resonance. "Only a good human can conduct Debussy," the conductor Ernest Ansermet told me once.

An orchestral concert is perhaps the most complex example of such a multilevel and multistage communication. There is the composer, the music, the conductor, the orchestra and the audience. And yet, in inspired moments, there is this phenomenon of an all-embracing "communion of love" of which Wilhelm Furtwängler (1955) spoke at the end of his life.

Open science

The self-organization of the work which roots in creative man but transcends him becomes evident in the structures of knowledge no less than in the structures of art. I have already mentioned Thomas Kuhn's (1962, 1977) theory of the evolution of such mental constructs or paradigms. It is essential that these structures do not accumulate knowledge in the way equilibrium structures (such as crystals) grow, but that they evolve through instability phases to new structures which at best take over subsystems of the old structure without change. The instability results not only from a falsification of the old structure. It is not the relational web of available knowledge, the static order, that is of decisive importance here, but the openness of the dynamic régime in which neural-mental processes unfold. Again, the mental structure, the paradigm, "lives" in the most intense and natural way when it is in a balance between novelty and confirmation. As Paul Feyerabend (1975) emphasizes, these mental structures change with the people who express their own life through them. Subjective and objective knowledge fall together in the framework of self-organization. The creative process wherever it unfolds, in art or science or merely in natural and effective living, falls together with the dynamics of evolution.

The openness and the inner non-equilibrium of scientific structures is mirrored in the working style of a creative scientist. Of Niels Bohr, for example, it is reported by his close collaborator Leon Rosenfeld (1967; quoted in Feyerabend, 1975, p. 24) that "he never regarded achieved results in any other light than as starting points for further exploration. In speculating about the prospects of some line of investigation, he would dismiss the usual consideration of simplicity, elegance or even consistency with the remark that such qualities can only be properly judged after the event. . . ." Since, however, science is never a completed process, Paul Feyerabend (1975) draws the conclusion that simplicity, elegance or consistency are *never* necessary conditions of scientific practice.

The self-organization dynamics of the creative work—its mind—and human self-organization dynamics—the human mind—are two sides of one and the same evolutionary process; they form a complementarity. Therefore, the relations between the scientists of a specific field resemble an evolving structure. Stephen Toulmin (1977), in an article with the significant title "From Form to Function", has compared the recent history of the philosophy of science with a square dance in which phases of marching forward separately, but in parallel, alternate with phases of "weaving" in which all participants join hands and do figures together. In Toulmin's view, this is the only way to prevent the hardening of science into a "permanent, professional scholasticism" as well as its "softening into a morass of well-meaning imprecision".

On the revolving stage of consciousness

Creativity is not a state but a process. This may perhaps be best explained with the model of the "revolving stage of consciousness" which has been proposed by Roland Fischer (1975/76) (see Fig. 47). Two ways, equivalent to each other, lead from the every-day "I" to the transpersonal "Self". Fischer calls the the way which plays the major role in the cultural history of the West, the perception-hallucination continuum of increasing ergotropic (central-sympathetic) or hyperarousal; the right half of Fig. 47 refers to it. The other way which corresponds to the spiritual practice of the Far East is the perception-meditation continuum of increasing trophotropic (central-parasympathetic) or hypoarousal; in Fig. 47, it is depicted in the left half. In simpler words, both ecstasis and meditation lead from the I to the Self. In Fischer's terminology, the Self is the seeing, recognizing, image-producing aspect of the person—we may also say, the self-organizing aspect or the mind of the total person—and the I is what is seen, recognized and imagined, namely, the world. "Thus in a mystical experience of the 'Self'-state we become conscious of what we are unaware in the 'I'-state: of being the *consciousness of the universe*" (Fischer, 1976). Schurian's above-quoted view of art as the sharing in the consciousness of the universe here becomes the sharing in our own potential.

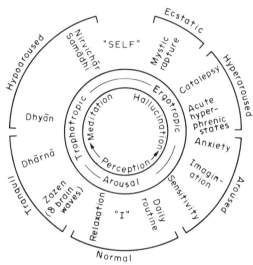

Fig. 47. The revolving stage of human consciousness. There are, in principle, two equivalent ways leading from the everyday I to the transpersonal Self, the meditative (left-hand side) and the ecstatic (right-hand side). The strange-sounding expressions in the left half of the ring signify specific states of Eastern meditation practice. In this diagram, creativity is no state which may be localized, but the feedback process of a continuous link of the everyday I with higher states of consciousness. After R. Fischer (1975/76).

The levels of subcortical arousal are often separated by complete amnesia (forgetting). Switching over "horizontally" between the corresponding ecstatic and meditative arousal states may easily occur. But the experiences made at a specific level are often state-specific. I have often observed this myself. When, for example, Alban Berg's violin concerto was performed for the first time after the war in Vienna, I did not want to miss this event in spite of considerable fever. I felt terribly sick—and through many years I felt sick when I listened to this concerto although, on the other hand, I loved it very much. Finally, I was able to overcome this involuntary association, a feat which I never achieved with respect to a similar disinclination against tea with milk. In Harold Pinter's play "No Man's Land", as well as in Charlie Chaplin's older film "City Lights", which Fischer cites as example, the sudden switches between association complexes lead to absurd situations. People who have become friends under the influence of alcohol or who have saved another person's life are either honoured or treated as tramps, depending on the similarity of the state they are in (for example, drunkenness). Roland Fischer, who has carried out the most important research of the state-specific characteristics of brain functions, once received a letter from an alcoholic: ". . . there was a lady in San Antonio . . . I could find her home when I was drunk. But I could not find it when I was sober" (Fischer, 1975). Fischer quotes in this connection the Mexican poet Octavio Paz who wrote: "Memory is not that which we remember, but that which remembers us."

Hallucinogenic drugs, with whose aid many of the states along the ecstatic and meditative paths may be stimulated, often seem to catapult consciousness on to a level which seems separated from the environmental reality. The fantastic wealth of forms and colours which is experienced in hallucinations has often led to attempts to catch some of them by drawing during the experience. The result was almost always extremely poor. It was not possible to give form to this experience so easily.

This amnesia separating levels of consciousness does not seem to apply to people which have trained themselves in mystic states throughout their lives. I heard Lama Govinda, the venerable German scholar and abbot of a Tibetan Buddhist order, tell of his experience with a ten-fold LSD dose given to him by American drug enthusiasts. Neither did he experience anything which he had never experienced before, nor did he become separated from every-day reality. However, the multilevel character of consciousness becomes clearer than ever in the drug experience of persons trained in meditation. It is possible to simultaneously hallucinate and rationally reflect on it, and to reflect on this reflexion, in turn, and so forth. The world may simultaneously undulate and remain solid. In the mirror of a water surface which is recognized as such, the whole depth of the universe may become reflected. I have also sometimes had the experience of multilevel dreams in which an act became experienced at different levels simultaneously, and even different acts were experienced simultaneously.

For some peoples the arousal induced by drugs provides an essential access to the sacred. The Huichol Indians, for example, who have lived for centuries in the almost inaccessible mountain regions north-west of Mexico City and have never been deeply touched by Western civilization, choose every year a delegation which travels more than 500 kilometres to "hunt" the sacred peyote cactus and kill it symbolically with their arrows. This pilgrimage is connected with severe exercises and the eating of the drug. The communication with the divinity takes place in rituals lasting for many days. In this communication, the fate of the whole tribe is decided for the following year. After their return, however, the Huichol pilgrims express their experiences and hallucinations in the form of "yarn paintings" (Sauter and Bertschy, 1977).

The creative process consists of giving form to a vision. The vision is holistically experienced at a level which, in Fig. 47, lies between the everyday-I and the Self and which corresponds to a certain degree of ecstatis or meditation. To give it form is a task for the rational I. Using memory, such as in the case of the Huichol Indians, is one way. In the Western practice of art, however, the creative process plays continuously between the lower and the higher levels on the "revolving stage of consciousness". Neither can the artist passively yield to the vision, nor can he focus on the mere crafts aspect. What this difficult synchronization implies may be imagined when one thinks of a sculptor who has to combine the vision of beauty with hard manual work. Only in the design by this painful process does the vision become communicable. It is reported that Beethoven rewrote the Florestan aria no less than seventeen times. He would hardly have been capable of explaining what it was that did not satisfy him. The Leonora Overture No. 2 is a masterpiece, but in listening to the third version we understand that Beethoven did not want to leave it at that.

The creative process is less an oscillating between ecstatic/meditative and rational levels than it is the simultaneous vibrating of many levels of this continuum. In this context, it is interesting to note that recent research results (*Brain/Mind Bulletin*, 1977b) have found a certain balance of electrical brain wave (EEG) patterns in states of increased openness. These patterns are chiefly composed of theta rhythms at about 4-6 Hertz (Hz) or cycles per second, which are characteristic for dream states, alpha rhythms (8-14 Hz) which characterize "pure" openness or "idling", and beta rhythms (above 16 Hz) in which the active exchange with the environment is reflected. Balanced right and left brain activity in theta and alpha rhythms characterize meditation. In a state of "lucid awareness", however, all three wave bands occur in perfect balance between both halves of the brain. This state is apparently left undisturbed by interactions with the environment such as walking, reading, thinking, problem-solving and is also not shaken by emotions. Healers, who are capable of balancing body/mind functions, for example by the laying on of hands, seem to show such a balanced EEG in many cases. It may not be erroneous to suspect in such a balanced, multilevel, autopoietic brain-wave pattern the prototype of brain functions in the creative process.

To bring the vision into a possible form is not the only task. This form has to be conceived at a level which is generally recognized, at least within the framework of a specific culture. It has to correspond to the innermost experience and memories of the performing artist as well as the audience. It has to express not only the life of the creative artist, but equally the life of those other people sharing in the act. In other words, the form has to be related to a common human basis. This relation is perhaps mediated by means of frequencies which become directly experienced in the contemplation of art and which resonate with the natural rhythms of the organism—metabolic as well as neural rhythms. This becomes evident, for example, in the "breathing" of music, or also in the dynamics of a painting which involves the contemplator.

The left half of the brain is normally the seat of the rational everyday I. The right half of the brain, however, is not the seat of the Self, but the extension of a holistically oriented total nervous system upon which the digitally acting function of the left half of the brain became imprinted by epigenealogical processes. Insofar, it is perhaps incorrect to speak of hemispherical specialization. Only the left half of the brain became specialized. Education in the Western sense focuses almost exclusively on the training of the left, analytical, half of the brain. In recent years only, approaches are developed in America, especially outside of the established school system, which aim at a balance of analytical and holistic brain functions, confirmation and novelty. In this connection, it is interesting that experiments in American primary schools have shown that occupation with drawing and painting significantly furthers cognitive learning (*Brain/Mind Bulletin,* 1977d).

The interaction of different levels of consciousness thus is basic to creativity in a human sense. At the same time, however, fluctuations may play in both directions between these levels and thereby stimulate the evolution of autopoietic structures of consciousness. Many cases of a spontaneous "conversion experience" have been reported (even in the memoirs of the avowed atheist Bertrand Russell), which changes a whole life within minutes and gives it new perspectives. Sometimes intuition seems not to be rooted in history stored in us by evolution, but in the transpersonal level of our consciousness. Knowledge in this sense is then no longer stored experience from past processes, but a kind of wisdom which sees both the beginning and the end and is capable of grasping evolution as a total phenomenon. Here, the image of the possible relations to the stream, which has been evoked in Chapter 16, may be useful again. We may stand on dry land, steer our boat consciously in the stream—or be the stream. The philosopher Jean Gebser (quoted in Fischer, 1975/76) seems to have meant the same when he spoke of the mental, mythical and magical structures of consciousness.

It is the simultaneous life of these three levels of consciousness which give depth to our life. Any reduction to one of these levels limits openness and emphasizes one-sided confirmation. Just as there is a "downward reductionism" to materialism, there is also an "upward reductionism", corresponding to a purely spiritual life which remains without consequences. The spiritual exercises of Hinduism and Buddhism have not always gone hand in hand with an open society. What is important, is to achieve multilevel autopoiesis of consciousness.

But what is "lower" and what is "higher" here, which direction is "downward" and which is "upward"? An important aspect of our urge toward the higher levels of the Self is the arousal of subcortical parts of our nervous system. The youngest and in a certain sense the highest aspect of our consciousness is rational mentation which corresponds to the lowest level of the everyday I. Art (and perhaps also scientific intuition) is, at least partially, connected with the brain structures which first appeared in evolution and in which self-expression and emotion reside.

The conclusion may be drawn that it is not individual levels which impart depth or height (both terms seem to express the same notion here!), but the multilevel vibrations of many levels of consciousness. A new level does not mean an "ascent", but an enrichment of the ensemble of possibilities of expression and the dimensions of its autonomy. At the beginning of this book, consciousness was equated with that autonomy which a system gains in the co-evolution with its environment. It may be added now that this autonomy, at the same time, means openness toward novelty. It does not suspend the relations with the environment, but accentuates them.

The intuitive presence of evolution as a total phenomenon, the knowledge of being embedded in an all-embracing unfolding, creates a kind of pre-religious experience in which mythologies and religious rituals are not yet fixed and even may co-exist side by side. The largest part of the 200 million Indonesians belong officially to one of the great religions, Islam, Christianity, Hinduism or Buddhism, but these forms and rites constitute but a super-structure built over a deep reverence for life, for the life-stream, which becomes visible in the ancestors' cult and which Western anthropologists mis-understand as "animism". The continuity of life precedes everything. Even in cock-fights, only animals are permitted which have fathered offspring. A bachelor, such as I am, will be suspended on a banyan tree until the end of the world—if he does not succeed in fathering at least spiritual children. As Emile Durkheim (1912) has shown in his sociology of knowledge, the creative processes of art and science and all culture grow out of the differentiation of such a pre-religious basis. Creativity is nothing else but the unfolding of evolution.

The question, where evolution will proceed to, occupies many people today. It may be posed in a meaningful way only where there is openness, where novelty has a chance to become woven into evolution. The last chapter will therefore be devoted to the new dimensions of openness which time- and space-binding introduces in evolution.

19. Dimensions of Openness

Such beauty, set beside
so brief a season,
suggests to our stunned reason
this bleak surmise:
the world was made to hold
no end or *telos,*
and if — as some would tell us —
there is a goal,
it's not ourselves.

Joseph Brodsky, *The Butterfly*

Intensity, autonomy and meaning—the dynamic measures of evolutionary progress

The evolution of the human world happens simultaneously at several levels. There is the biological level at which nothing has changed much for a long time with the exception of an increase and further differentiation of the brain. Connected with this is evolution at another level, namely the level of neural-mental consciousness. The third level, finally, is the environment which we design and transform, the world of social and cultural structures as well as the technological artifacts and the new environment by which they are accompanied. It is evident that the inner and outer worlds co-evolve in close, mutual interdependence, although they do not necessarily mirror each other nor agree in phase.

As has been discussed in Chapter 13, a telling measure for the state of evolution is the degree of time- and space-binding or, in other words, the effectiveness of the total phenomenon of evolution in the presence of a specific system. We may also call this measure *intensity*. Another measure is the flexibility in the establishment and organization of relations, in other words openness toward the appearance of novelty in the further evolution of the system as well as the environment. An alternative name for this measure is *autonomy,* understood here in a dynamic context. A prerequisite for autonomy is complexity, but it is not precisely the same because complex systems, too, may become highly dependent on others and see their vulnerability increased (as, for example, in the highly technicized systems of the human world). Autonomy is the active, dynamic application of complexity and thus, as has been mentioned several times already, comes close to the notion of consciousness. The notion of autonomy may also be applied to the openness of the neural consciousness, that is to say to the flexible, partial emancipation from the dictates of the limbic system (emotions) and especially of the reptilian brain (stubbornness). A third measure, finally, is the subtleness of tuning in to the multifaceted dynamics of an indivisible evolution, the degree of aligning with the evolutionary processes which aims neither at total adaptation nor total independence. We may call this measure evolutionary connectedness or *meaning.*

Intensity, autonomy and meaning generally increase in evolution with higher complexity. Whereas complexity, however, is a static notion which tells little about the unfolding of processes, intensity, autonomy and meaning are dynamic notions which circumscribe the dimensions of creative processes. If, however, as discussed in Chapter 12, complexity generates more complexity, we may now also state that intensity, autonomy and meaning, too, act autocatalytically and generate the conditions for their own increase.

In the design of the outer world, we have progressed relatively far with respect to time-binding. This becomes clear in the exploitation of energy sources which, as has been shown in Chapter 17, brings energy into play which has been stored in ever earlier phases of evolution. After the continuously renewed plant growth, fossil fuels, fissile material and soon also fusion material are brought into play. The light nuclei of matter akin to fusion originated in the first phase of a hot universe. Hand in hand with this time-binding goes an increase in the transformation of matter into energy which is expressed by the well-known Einstein formula $E = mc^2$. In chemical transformation, that is to say, combustion, only about 10^{-10} of the mass is transformed into energy; in fission, it is 10^{-3} (one-tenth of a per cent) and in fusion almost 10^{-2} (1 per cent). Thus, we have come relatively close to the energy yield from annihilation, in which matter becomes totally transformed into energy and which dominated the first seconds of cosmic evolution. The intensity of transformation becomes ever greater. But we may also put it this way that many aspects of our relations with the environment become more intense with increased energy input. We need only think of travelling or of agricultural or industrial productivity per worker. At the same time, the autonomy of many people doubtlessly increases. With more energy, we may do many things which we would not be capable of, were we still dependent on muscle power alone. On the other side, we become more dependent as is painfully experienced during power black-outs or public transit strikes. Above all, however, we build obstacles into the future in many respects. We are becoming increasingly aware of this, especially when we face the question of radioactive waste disposal involving extremely long-range effects, in some cases approaching time spans of the order of the present age of mankind.

A fascinating aspect of this time-binding of the past into the present may be seen in the fact that it does not remain limited to the past, but creates chreods, or development lines, which point far into the future. This is part of the nature of genetic as well as sociocultural evolution. Evolution continuously makes decisions which remain commitments for a very long time to come. This holds, for example, for the choice of the genetic mechanism based on DNA and RNA, and also for most other genealogical processes. At the same time, however, new degrees of freedom appear at every new level of evolutionary processes, opening up new dimensions of openness. The task is to simultaneously live novelty and confirmation in the balance characterizing life.

An autonomy gain in the present constellation is always accompanied by an autonomy loss in the choice of the path into the future. This becomes even more evident when we think of the application of the oldest energy storages to the building of a nuclear arsenal. This application is supposed to serve the enhancement of national autonomy in the present, but it limits the openness of the future in ways which now become visible. It seems to be too late for total nuclear disarmament. No side could be certain any longer that the other side does not keep a hidden reserve. Twenty years ago this would have been feasible from a technical point of view.

These examples already make it obvious that intensity is a measure of evolution which may be applied in much simpler ways than autonomy. Meaning is even much more complex and hardly accessible to simple arguments with regard to the design of the outer world.

This becomes even more obvious when considering another time-binding which has received much attention recently, namely, DNA recombination and the creation of new life forms while by-passing biological evolution. To be precise, this time-binding reaches back to that early phase in the history of life on earth in which horizontal gene transfer among bacteria was a dominant strategy of life. It continues today, too, at least among bacteria. To what extent are more complex organisms, including humans, exposed to horizontal genetic processes acting by means of viruses, is at present a wide-open question. A virus which changes the DNA in a body cell by squeezing its own DNA in does not necessarily change the DNA which is active in reproduction. Up to now it is only known that, for example, the virus which generates teck encephalitis spreads directly by means of the egg cells of the tecks, without, however, giving up its individuality.

In any case, there is some natural DNA recombination going on all the time through various mechanisms. But how does the chreod look which we are about to lay far into the future by means of our planned association of evolutionary information?

If the question concerning the increase in intensity and autonomy generally may be answered for the present, often not so easily, the question for the future prospects of autonomy can only be answered in connection with the meaning. Meaning, however, is hardly readily accessible to us through observations of the outer world. Its light shines in the evolution of the inner world, the human multilevel consciousness.

Immediacy of existence

In a civilization with a true history, which is to say with a genuine evolution of sociocultural structures, the concept of lineally progressing, irreversible time undergoes a significant modification. Not only is *re-ligio,* the linking backward to the origin, becoming the main spiritual concern, it is also becoming the core of creative action. Death and rebirth correspond to the evolutionary process of the destruction of the old structure and the sub-sequent formation of a new one. Joseph Campbell (1956) has given expression to this evolutionary process in his "monomyth" of personal development. In the myths of many cultures, the hero leaves his home, masters dangerous adventures, experiences, so to say, his rebirth and returns as an adult. From Odysseus and Siegfried to the chosen ones of the science fiction stories and films (exemplified by the films "Star Wars" and "Close Encounters of the Third Kind", both released in 1977), all heroes have gone through this cycle. Bruno Bettelheim (1976) has pointed to the same process of development in fairy-tales such as "Cinderella". In recent years, psychoanalysis, too, has discovered and applied this evolutionary process of personal dynamics (Shainberg, 1973).

The linking back to the origin not only restores strength, but also creates the possibility of recognizing and bringing into play ever new chreods, new development lines. This is the creative act as it is carried out by historical man and as it corresponds to the tantric ideal of Buddhism. The British mathematician G. Spencer Brown (1969) has pointed out that the first step toward a division of space—for example, if an observer reflects on the world and breaks the original unity—establishes the identity of organization in the inner and the outer world. Everything which follows is then no longer fully open but partially determined by this first step, or severance: "The act is itself already remembered, even if unconsciously, as our first attempt to distinguish different things in a world where, in the first place, the boundaries can be drawn anywhere we please. At this stage the universe cannot be distinguished from how we act upon it, and the world may seem like shifting sand beneath our feet. Although all forms, and thus all universes, are possible, and any particular form is mutable, it becomes evident that the laws relating such forms are the same in any universe" (Spencer Brown, 1969, p.v.).

In recent years, a consequence of quantum mechanics has attracted considerable attention which stipulates the inseparability of the dynamic behaviour of two or more photons or particles which originated in the same event—in the case of the photons, for example, in the annihilation of an electron with a positron—or which have otherwise been in interaction and now fly off in different directions. Although the applicability of the microscopic concept of quantum mechanics to macroscopic events is limited in perhaps still largely unknown ways, a principle seems to be evoked here which stipulates a basic connectedness of dynamic phenomena in a universe which, after all, originated in all of its parts in the big bang and the subsequent dense interactions (d'Espagnat, 1976). Experiments carried out in the past few years overwhelmingly confirm the validity of this principle as far as single events are concerned (Bell, 1976).

The linking backward to the origin presents the possibility of a new start. Even a simple dissipative structure dissolves its web of relations before it evolves to a new structure. It is therefore not surprising that hopes of eschatological dimensions are nourished, for example, by influencing the biological evolution of mankind by pursuing consciousness down to the levels of cells, molecules and atoms with the aim of a "descent to the first day" (Murphy, 1977). That consciousness elements of these levels are, in principle, accessible seems to be borne out by serial LSD therapy in psychotherapeutic practice (Grof, 1975).

The process of tuning in with evolution may be conceived as unification with the origin in *re-ligio*, or, alternately, by "pulling" the origin into the present. Longchenpa, the most important philosopher of the Tibetan Nyingma tradition, perhaps the purest form of Buddhism, has indicated a three-fold path to higher meditation: pure openness which lets visions and higher insights stream in; the radiation of the own heart which penetrates the whole universe; and non-dividedness, out of which grows meaning (Longchenpa, 1976). In these three paths we may again recognize the three dynamic measures of evolution: autonomy, intensity and meaning. All three lead to an immediacy of existence in which all opposites contain each other. "If there is no longer hope or fear", says Longchepa (1976), "you are free from all hindrances." It is most significant that meditation is equated here with the highest intensity, not with that "emptiness" which plays such a big role in Western misunderstandings about meditation.

In Western literature, the immediacy of existence has never been expressed more profoundly than in the work of the German poet Heinrich von Kleist. If Penthesilea is the embodied intensity, the highest radiation of the heart, Käthchen von Heilbronn is pure openness. However, writes Kleist, they are "one and the same creature, only conceived under opposite relations"; Käthchen is "a creature equally powerful by totally yielding as the other is by action". The blindfolded Prince of Homburg, however, experiences under the drums of the death march, in the highest ecstasy his connectedness with the cosmos, the meaning of his life:

> Now I possess you, Immortality!
> You penetrate the blindfold of my eyes
> With flashing splendor of a thousand suns.
> I feel great wings expand from both my shoulders
> And through ethereal stillness swings my soul.
> Now like a ship borne far by breath of winds
> When cheerful harbour towns recede from view,
> I see the whole of life go glimmering down:
> Colours and forms grow indiscernible
> And everything beneath me lies in mist.

The suspension of historical time

To escape historical time is an ancient dream of man. It grows from the urge to find meaning, to recognize oneself in a cosmic order from which a lineally progressing history seems to remove man farther and farther. Archaic cultures stuck to a concept of cyclical time, the myth of the eternal return (Eliade, 1954). Time was bound in itself and this also reflected in the spatial binding to a centre, such as a sacred mountain which is symbolically represented by temples. The order of the events was not seen in their sequence, but in their association with earlier events of the same basic type. Such a guiding image corresponding to an autopoietic dynamic structure—similar perhaps to a structure with limit-cycle behaviour—was indeed capable in such cultures of preventing sociocultural evolution over thousands and even tens of thousands of years.

In the immediacy of existence, linear and irreversible time becomes suspended. The experienced processes of the past and the visions of an anticipated open evolution are directly grasped in a four-dimensional present. Poetic reality breaks into the profane reality of everyday life.

In such time- and space-binding of consciousness the sequential order of information, corresponding to a specific sequence of events, is suspended, too. A state is generated which resembles the cyclical time concept of archaic cultures. Events are no longer connected in a sequential mode, but in an associative mode. Information retrieval no longer follows linear search procedures, but develops associative strategies such as information retrieval systems working with combinations of thematic keywords. Even before the introduction of computers, there were the "peek-a-boo" systems consisting of stacks of cards into which holes had been pierced to mark the keywords which applied to a literature reference.

The cultural pluralism which has been evoked in Chapter 15 and which is about to replace the era of uniform, committing guiding images, may be interpreted as a suspension of historical time. Individuation approaching completion leads in its ultimate consequences to a pluralism which appears to be a new version of monism. For decades, already, modern art has no longer expressed itself in specific uniform styles characteristic for a given period, but has become pluralistic—in this aspect, too, preceding the general cultural developement. More surprising to many comes the conceptual pluralism of modern science. Theory of relativity and quantum mechanics "function" in white domains of observations, but all attempts to mould them into a unified paradigm have failed so far. The pluralism of more recent concepts, especially in the physics of subnuclear particles, makes some physicists already speak of an "ecology of models" which cannot be fused to a unified model, not because we lack the necessary knowledge, but as a matter of principle.

Let us remember that the evolution of dissipative structures, too, can be described only by simultaneously employing two complementary models, a macroscopic-deterministic and a microscopic-stochastic one. And the co-evolution of macro- and microcosmos may only be grasped in the synopsis of complementary approaches. As a matter of principle, the autopoietic levels in a multilevel dynamic reality which have become separated by symmetry breaks cannot be united in a super-model, but only by way of describing the web of relations between them. "The world is far too rich to be expressed in a single language", writes Ilya Prigogine (1977). "Music does not exhaust itself in a sequence of styles. Equally, the essential aspects of our experience can never be condensed into a single description. We have to use many descriptions which are irreducible to each other, but which are connected by precise rules of translations (technically called 'transformations'). Scientific work consists of selective exploration and not of the discovery of a given reality. It consists of the choice of questions which have to be posed."

The child's world is a world of associations in which opposites are neighbours. The brain of the adult person, too, generally does not work sequentially but by way of association, perhaps in a very complex manner. Many more data seem to be stored in our brain's memory than are usually retrieved in an associative way. They surface in exalted states such as experienced by many people in dangerous situations or in the moments before death. The same associative mode rules the epigenetic process as well as all other epigenealogical processes which act in the brain but also in entire societies.

The already-mentioned technique of "computer-conferencing" attempts to condense the sequential processes of bilateral communication (of which a telephone conference call consisted so far) in an associative way, thus lifting the processes to a new level. It is no wonder that one of the pioneers of this technique, the French-Californian Jacques Vallée (1977) has recently suggested that "synchronicity" as discussed by C. G. Jung and Wolfgang Pauli (1954)—the linked appearance of seemingly unconnected events which cannot be explained by linear causality—may also be traced back to a mode of reality in which information occurs associatively and not sequentially.

Be that as it may, historically acting evolution itself with its increasing time- and space-binding has generated the prerequisites for man to suspend historical time and reconnect it in new ways. It is perhaps no longer exaggerated to speak of a process which I should like to call the *epicultural process* and which continues the line of epigenealogical processes at the highest level of evolution. The epicultural process is the learning process of mankind-at-large. Just as in the epigenetic process life may partially (and only partially!) circumvent the evolutionary sequence of genetic information, an epicultural process might partly circumvent the evolutionary sequence of events of which it makes us aware. Out of events which are remote from each other in time and space, new webs of meaning grow, a totally new four-dimensional reality. The great visions and paradigms, religions and ideologies have always made their sudden appearance in consciousness as such new webs of meaning. In our days, however, such new space-time structures may occur pluralistically and thereby deepen even further the immediate, four-dimensional existential experience.

As already mentioned in Chapter 13, we may experience evolution directly in three ways. Firstly, as an *ancestral tree* which branches in the opposite direction of historical time and makes the experience of an entire genealogical phylum accessible over time and space—be it a biological-genetic or a cultural-traditional tree; with its stereotypes, it gives us a kind of static security. Secondly, we may experience evolution in terms of *roots* branching *in* the direction of historical time and accessible by way of *re-ligio,* the linking backward to the common origin; re-ligio opens up the unformed with its wealth of open possibilities, new evolutionary development lines and therefore new realities and thereby gives us a kind of dynamic security. And thirdly, we may experience evolution as *rhizome* which cuts across the direction of historical time and in which space-time evolution becomes condensed in the present; the rhizome stands for becoming of all kinds and thereby gives us security through intensity.

In this book, I have first emphasized the genealogical aspects of sociocultural evolution, the tree of experience which binds time and space by letting the past act in the present by means of books and traditions, rote learning and imitation. With *re-ligio,* the creative aspect moved into the foreground. Only with the additional image of the rhizome, however, can time- and space-binding become effective in the present, can information become truly pragmatic. Only in this three-fold experience may historical time become suspended and destiny become subject to design.

For Deleuze and Guattari (1976), the rhizome is an image for natural language, for a new way of thinking in associations and a new way of writing books (their own psychological/linguistic/political anti-Oedipus book). It is an image for a new, pluralistic kind of living generally. A rhizome suspends historical time, not because it falls out of it, but because *it creates historical time*. Out of isolated cultural chreods, trajectories through space and time, a four-dimensional culture continuum emerges—not the monolithic crystal of a "world culture", but the alive mind of mankind in ever-renewing relationships and ever-changing expressions.

We have come full circle. What I have described as a global culture rhizome is nothing but the immediacy of existence as it is represented already by the simplest dissipative structure at its proper level. Like such a structure, a rhizome also is characterized by openness, non-equilibrium (heterogeneity of relationships) and autocatalytic self-reinforcement of fluctuations (the establishment of new relationships). Like the dissipative structure, it develops autonomy and consciousness. At the level of self-reflexion, however, there are no longer individual dynamic régimes, but an intricate, interweaving web of such régimes. What the evolution of life has ultimately created is nothing but that with what it has started.

Epilogue
Meaning

Deep in the human unconscious is a pervasive need for
a logical universe that makes sense. But the real
universe is always one step beyond logic.

Frank Herbert, *Dune*

We stand at the beginning of a great new synthesis. The correspondence·of
static structures is not its subject, but the connectedness of self-organization
dynamics—of mind—at many levels. It becomes possible to view evolution as
a complex, but holistic dynamic phenomenon of an universal unfolding of
order which becomes manifest in many ways, as matter and energy,
information and complexity, consciousness and self-reflexion. It is no longer
necessary to assume a special life force (such as Bergson's *élan vital* or the
prana of Hinduism) separate from the physical forces. Natural history,
including the history of man, may now be understood as the history of the
organization of matter and energy. But it may also be viewed as the
organization of information into complexity or knowledge. Above all,
however, it may be understood as the evolution of consciousness, or in other
words, of autonomy and emancipation—and as the evolution of the mind.
Mind appears now as self-organization dynamics at many levels, as a dynamics
which itself evolves. In this respect, all natural history is also history of mind.
Self-transcendence, the evolution of evolutionary processes, is this evolution
of the mind. It does not unfold in a vacuum, but becomes manifest in the self-
organization of material, energetic and informational processes. In this way,
the old dualism of matter and mind is overcome, which has characterized (and
sometimes plagued) Western thinking throughout the past 2000 years.

Life, and especially human life, now appears as a process of self-realization,
the outer, Darwinian aspect of which becomes visible in the experience of
resistance against emancipation, and the inner, co-ordinative aspect of which
becomes expressed in the *crescendo* of an ever more fully orchestrated

consciousness. In self-transcendence, the opening up of new levels of self-organization—of new levels of the mind—, the chord of consciousness becomes richer. In the infinite, it falls together with the divine. The divine, however, becomes manifest neither in personal nor any other form, but in the total evolutionary dynamics of a multilevel reality. Instead of the numinous, we may also speak of meaning. Each of us would then, in Aldous Huxley's (1954) terms, be Mind-at-large and share in the evolution of this all-embracing mind and thus also in the divine principle, in meaning.

In Buddhism, the most comprehensive process philosophy and religion, no dualistically conceived god comes down to earth as in the monotheistic Christian and Muslim faiths. Gautama Buddha was human. But he realized his own being to the fullest extent, and not without suffering, and in this way reached up to the divine. Mankind is not redeemed by a god but redeems itself. As C. G. Jung (1961) formulated it at the end of his life, we are no longer concerned with the dualistic opposition between god and man, but with the immanent tension in the God-image itself, expressed in the mandala of the mystic Jakob Böhme with.the two halves of the circle standing "back to back". This inner non-equilibrium, the glorious imperfection of life, is the effectiveness principle of evolution. The God-idea does not stand above and outside of evolution as an ethical norm, but in true mysticism is placed into the unfolding and self-realization of evolution. Hans Jonas (1969) has given this evolutionary God-idea perhaps the most profound expression with the thought that God abandons himself many times in a sequence of evolutions in which he transforms himself, accepting all the risks introduced by indeterminacy and free will in the play of evolutionary processes. God is thus not absolute, he evolves himself—he *is* evolution. Since we have called the self-organizing dynamics of a system its mind, we may now say that God is not the creator, but the mind of the universe.

This dynamic God-idea, in the deepest form of *re-ligio,* leads back to the notion of evolution—or, more precisely, an arc of evolution. If God evolved like a dissipative structure, the Buddhist *shunyata,* the origin, is itself comparable to a dissipative structure. Evolution is then the open interaction of processes in the instability phase between two structures—that phase in which novelty breaks in, the law of large numbers is rendered invalid and the fluctuations of consciousness prepare the decisions for the next autopoietic structure. As Ilya Prigogine formulated it (see Chapter 3), the evolution of a dissipative structure may itself be viewed as a giant fluctuation. Our life is such a fluctuation—and the whole universe, too.

The singularity, however, the "God-structure", is neither form nor quantity, but the non-unfolded, the totality of undifferentiated qualities. It is pure potential. Each of the great process philosophies has found a different name for it. In the oldest recorded world view, Hermetic philosophy—named after the mystic personality of Hermes Trismegistos who allegedly lived long before Moses in Egypt and made a double career as a god, Toth in Egypt and Hermes in Greece—this wholeness resting in itself is called the "all". In Buddhism, as already mentioned, its name is *shunyata* which is frequently and erroneously translated as "void". A version of our century is the "extensive continuum" in the process philosophy of Alfred North Whitehead (1969). In a comprehensive process philosophy, the God-idea is rooted even more deeply than is Spinoza's non-dualistic God in the energy stream of evolution—it is anchored in the origin from which a yearning for peace touches us like a remote memory and to which we may return within ourselves in the *re-ligio*.

Beyond this level of quietness at which evolution is poised, man has never been able to penetrate, even in moments of the highest mystic rapture. In this level are rooted art and wisdom and generally every creative process. At the horizons of our rational, emotional and spiritual knowledge we always face the decision between two metaphysical alternatives: a meaning which can basically never be grasped, corresponding to an ordering principle and often expressed in God-ideas—and a lack of meaning, which, equally, can never be grasped. Meaning becomes the moving force of self-transcendence, which reaches beyond the limited horizon and longs for the whole. "Might nature ultimately fathom itself?", asks Goethe. The assumption of random initial conditions before a background devoid of meaning, however, becomes the expression of an urge to reduce reality to models which—as Galileo's contemporary Gianbattista Vico already emphasized—we understand and find meaningful for no other reason but because we have made them ourselves. The objective urge to understand leads to the most profound subjective experience, the subjective urge for security to pseudo-objectivity in the detail which fails to grasp the whole. Here, a last dualism becomes dissolved: understanding is not static knowledge, but itself an evolutionary process in which subjectivity and objectivity are complementary to each other.

In the human sphere, however, such a new synthesis means hope instead of fear, the termination of the alienation of man from a world whose increasingly fast change is felt as a Kafkaesque threat and yet has man itself as the driving force. The new synthesis imparts profound meaning to human life. Meaning emerges from a sense of connectedness. If we ask somebody for the meaning of his ambitions, his hectic life and his grabbing, we usually hear that it is not for himself, but for his children, that he suffers all this. This is already an act of self-transcendence. A farther-reaching urge for meaning looks at sequences of generations, peoples, civilizations, the overall evolution of mankind and even the whole universe. The need for meaning proves to be a powerful, autocatalytic factor in the evolution of human consciousness—and thus indeed of the evolution of mankind and the universe. This connectedness of our own life processes with the dynamics of an all-embracing universe has so far been accessible only to mystic experience. In the synthesis, it becomes part of science which in this way comes closer to life.

This feeling of being embedded in a universal, connected dynamics may not only remove the fear of our own biological death, but also the fear which defends the "survival of the species" as a supreme value. In self-transcendence we reach not only beyond our own limits as individuals, but also beyond the limits of mankind. The fascination held by the evolution of mankind pales in comparison with the fascination held by a universal evolution whose integral aspect we are. In such an attitude, we would not only further the conditions for our own life, but also the conditions of all life which we are capable of influencing. In his visionary science fiction novel *Childhood's End*, Arthur C. Clarke (1953) has described the fusion of the mind of humanity with universal mind. In a physical sense, mankind and the planet earth became destroyed in the process. What did it matter? The mind of humanity required these physical forms only for its childhood phase. With the capability of self-reflexion we have become the mind of a universe becoming increasingly aware of itself—whether we are the only creatures with this capability, or in the company of extraterrestrial beings, is not so important.

In a process-oriented view, the evolution of specific structures is not pre-determined. But then are functions—processes which may realize themselves in a multitude of structures—predetermined? In other words, does the evolution of mind follow a predetermined pattern? Or does such an assumption again lead to a wrong conclusion already prefigured in process thinking, just as the predetermination of structures has been prefigured by mechanistic, structure-oriented thinking? Is the formula of Eastern mysticism that the universe is made to become self-reflexive, only the expression of an inherent limitation of Eastern process philosophy?

Perhaps it is not that important to find answers to these questions at all. Our search is ultimately devoted not to a precise knowledge of the universe, but to a grasp of the role which we play in it—to the meaning of our life. The self-organization paradigm which lays open the dimensions of connectedness between all forms of unfolding of a natural dynamics, is about to deepen the recognition of such a meaning. "It is conceivable", writes Freeman J. Dyson (1971), "that life may have a larger role to play than we have yet imagined. Life may succeed against all the odds in moulding the universe to its own purposes. And the design of the inanimate universe may not be as detached from the potentialities of life and intelligence as scientists of the 20th century have tended to suppose."

As we have seen, it is not only the universe, but also the process of evolution itself which is becoming increasingly self-reflexive. What, then, is the cause of this? Is self-reference the ultimate principle which becomes expressed in structures as well as in functions and fluctuations—including the fluctuation of universal evolution in its totality? Let us remember that self-organization, that is to say mind, may be described at each of these three levels. Each level may replace the other two, but none can explain the others. At the evolutionary level of the self-reflexive mind, however, self-reference becomes self-cognition.

In the immovable all, the common origin of evolution, time and space and the not-yet-unfolded—all quality—were one. With the unfolding, symmetry breaks in time and space occur step-wise. Self-transcendence becomes possible only by way of symmetry breaks. Simultaneously, however, evolution weaves a new net of time- and space-binding which makes it possible to experience increasingly in each self-organizing system the lost unity. Just as the intense processes of the hot early phase of the universe become restored in a later phase in the self-organizing systems of stars, the intensity lost in the expansion and cooling of the universe becomes experienced in the individual himself. It is no longer the *re-ligio* to the origin which mediates this highest intensity in mystical rapture. It reappears also with the progressing of evolution. The highest meaning is in the non-unfolded as well as in the fully unfolded; both reach up to the divinity.

Gracefulness, says Kleist in his essay "On the Puppet Theatre", is most perfect where reflexion is either not present at all, or where it has gone "through an infinite"—just as the image in a concave mirror, after having disappeared into infinity, suddenly reappears close to our eyes. Gracefulness appears purest in a human body which "has either no consciousness, or infinite consciousness, that is, in the puppet or in the god.

'Thus', I said somewhat distracted, 'we have to eat again from the tree of knowledge to get back into the state of innocence?'

'Indeed', he answered 'this is the last chapter in the history of the world'."

Literature References

Many of the areas which are of importance for this book are in full development and new discoveries and concepts become known all the time. Therefore, I paid particular attention to finding summaries which are up to date. In this respect, the *Neue Zürcher Zeitung* (Zurich) with its weekly "Science and Technology" supplement proved to be the most valuable source and I include the references here, although they may not be easily accessible to interested readers. Other valuable sources of summaries were the *Scientific American* and the *CoEvolution Quarterly,* published from a houseboat at Sausalito across the bay from Berkeley.

Abraham, Ralph (1976) "Vibrations and the realization of form", in: Jantsch and Waddington, eds. (1976).

Adams, Richard Newbold (1975) *Energy and Structure: A Theory of Social Power.* Austin and London: University of Texas Press.

Alfvén, Hannes (1966) *Worlds—Antiworlds: Antimatter in Cosmology.* San Francisco: Freeman.

Allen, Peter M. (1976) "Evolution, population dynamics and stability", *Proc. Nat. Acad. Sci. (USA)* **73**, 665-668.

Allen, Peter M., Deneubourg, J. M., Sanglier, M., Boon, P. and de Palma, A. (1977) *Dynamic Urban Growth Models.* Report No. TSC-1185-3, Cambridge, Mass.: Transportation Systems Center, US Dept. of Transportation.

Ashby, W. Ross (1956) *Introduction to Cybernetics.* New York: Wiley.

Ashby, W. Ross (1960) *Design for a Brain,* 2nd ed. New York: Wiley.

Babloyantz, Agnessa (1972) "Far from equilibrium synthesis of 'prebiotic' polymers", *Biopolymers,* **11,** 2349-2356.

Ballmer, Thomas T. and Weizsäcker, Ernst von (1974) "Biogenese und Selbstorganisation", in: Weizsäcker, Ernst von, ed. (1974).

Barash, David P. (1977) *Sociobiology and Behavior.* New York and Amsterdam: Elsevier.

Barghoorn, Elso S. and Knoll, Andrew S. (1977) *Science,* 28 Oct. 1977.

Bastin, Ted, and Noyes, H. Pierre (1978) "On the physical interpretation of the combinatorial hierarchy", *Int. J. of Theoretical Physics* (forthcoming).

Bateson, Gregory (1972) *Steps to an Ecology of Mind.* San Francisco: Chandler; New York: Ballantine paperback.

Bateson, Gregory (1979) *Mind and Nature: A Necessary Unity.* New York: Dutton.

Baumann, Gilbert (1975) "Das künstliche Nervensignal: Synthese einer Zellfunktion", *Neue Zürcher Zeitung,* 29 Jan. 1975.

Bell, John S. (1976) "Testing quantum mechanics", and summary of reports presented at a "Thinkshop on Physics", held at the Ettore Majorana Centre for Scientific Culture in Erice, Sicily. *Progress in Scientific Culture,* **1,** 439-460.

Benz, A. O. (1975) "Physik des Universums", *Neue Zürcher Zeitung*, 8 Jan. 1975.

Bergson, Henri (1896) *La Matière et le mémoire*. Paris. English transl.: *Matter and Memory*, London: Allen & Unwin, 1962.

Bertalanffy, Ludwig von (1968) *General System Theory: Foundations, Development, Applications*. New York: Braziller.

Bettelheim, Bruno (1976) *The Uses of Echantment: The Meaning and Importance of Fairy Tales*. New York: Knopf; New York: Random paperback, 1977.

Boiteux, A. and Hess, B. (1974) "Oscillations in glycolysis, cellular respiration and communication", in: Faraday Symposium (1974).

Bonner, John Tyler (1959) "Differentiation in social amoebae", *Scientific American*, Dec. 1959; also in: Kennedy, ed. (1974).

Boos, Winfried (1978) "Intelligente Bakterien: Chemotaxis als primitives Modell von Reizleitungssystemen", *Neue Zürcher Zeitung*, 10 Jan. 1978.

Brain/Mind Bulletin (1977a) "'New nervous system' may effect behavior illness", *Brain/Mind Bulletin*, **2**, No. 15.

Brain/Mind Bulletin (1977b) "'Mind Mirror' EEG identifies states of awareness", *Brain/Mind Bulletin*, **2**, No. 20.

Brain/Mind Bulletin (1977c) "Left, right brain differences are more fundamental than verbal, non-verbal", *Brain/Mind Bulletin*, **2**, No. 22.

Brain/Mind Bulletin (1977d) "Art reinforces cognitive learning", *Brain/Mind Bulletin*, **2**, No. 22.

Broda, Engelbert (1975) *The Evolution of the Bioenergetic Processes*. Oxford: Pergamon Press.

Bronowski, Jacob (1970) "New concepts in the evolution of complexity: stratified stability and unbounded plans", *Zygon*, **5**, 18-35.

Bünning, E. (1977) *Die physiologische Uhr*. 3rd rev. ed. Berlin, Heidelberg, New York: Springer.

Camara, Sory (1975) "The concept of heterogeneity and change among the Mandenka", *Technological Forecasting and Social Change*, **7**, 273-284.

Cameron, A. G. W. (1975) "The origin and evolution of the solar system", *Scientific American*, Sept. 1975.

Campbell, Allan M. (1976) "How viruses insert their DNA into the DNA of the host cell", *Scientific American*, Dec. 1976.

Campbell, Joseph (1956) *Hero with a Thousand Faces*. New York: Meridian.

Castaneda, Carlos (1975) *Tales of Power*. New York: Simon & Schuster.

Charbonnier, G. (1969) *Conversations with Claude Lévi-Strauss*. London: Jonathan Cape.

Chew, Geoffrey F. (1968) "Bootstrap: a scientific idea?", *Science*, **161**, 762.

Chomsky, Noam (1976) *Reflections on Language*. New York: Pantheon.

Churchman, C. West (1968) *Challenge to Reason*. New York: McGraw-Hill.

Clarke, Arthur C. (1953) *Childhood's End*. New York: Ballantine.

Cox, Allan (1973) *Plate Tectonics and Geomagnetic Reversal*. San Francisco: Freeman.

Cudmore, L. L. Larison (1977) *The Center of Life: A Natural History of the Cell*. New York: Quadrangle.

Dassman, Raymond (1976) "Biogeographical provinces", *CoEvolution Quarterly*, No. 11, 32-37.

Dawkins, Richard (1976) *The Selfish Gene*. Oxford: Oxford Univ. Press.

Deleuze, Gilles and Guattari, Félix (1976) *Rhizome: Introduction*. Paris: Les Editions de Minuit.

Douglas-Hamilton, Iain and Oria (1975) *Among the Elephants*. New York: Viking; Bantam paperback, 1976.

Dudits, D., Rasko, I., Hadlaczky, Gy. and Lima-de-Faria, A. (1976) "Fusion of human cells with carrot protoplasts induced by polyethylene glycol", *Hereditas*, **82**, 121-124.

Dübendorfer, Andreas (1977) "Die Metamorphose der Insekten", *Neue Zürcher Zeitung*, 15 Nov. 1977.

Dunn, Edgar, S., Jr. (1971) *Economic and Social Development: A Process of Social Learning*. Baltimore and London: John Hopkins Press.

Durkheim, Emile (1912) *Les Formes élémentaires de la vie religieuse*. Paris: Alcan.

Dyson, Freeman J. (1971) "Energy in the universe", *Scientific American.* Sept. 1971; also in: Scientific American (1971).

Ebert, Rolf (1974) "Entropie und Struktur kosmischer Systeme", in: Weizsäcker, Ernst von, ed. (1974).

Echlin, Patrick (1966) "The blue-green algae", *Scientific American,* June 1966; also in: Kennedy, ed. (1974).

Eddy, John A., ed. (1978) *The New Solar Physics.* Boulder, Col.: Westview.

Eigen, Manfred (1971) "Self-organization of matter and the evolution of biological macromolecules", *Naturwissenschaften,* **58,** 465-523.

Eigen, Manfred and Schuster, Peter (1977/78) "The hypercycle: a principle of natural self-organization"; in three parts: "Part A: Emergence of the hypercycle", *Naturwissenschaften,* **64** (1977), 541-565; "Part B: The abstract hypercycle", *Naturwissenschaften,* **65** (1978), 7-41; "Part C: The realistic hypercycle", *Naturwissenschaften,* **65** (1978), 347-369. Appeared under the same title as book: Berlin, Heidelberg and New York: Springer, 1979.

Eigen, Manfred and Winkler, Ruthild (1975) *Das Spiel: Naturgesetze steuern den Zufall.* Munich and Zurich: Piper.

Eliade, Mircea (1954) *The Myth of the Eternal Return: or, Cosmos and History;* Bollingen Series XLVI. Princeton, N.J.: Princeton University Press.

Epstein, Helen (1978) "The most electric pianist around", *San Francisco Chronicle,* 10 Jan. 1978.

Erneux, T. and Herschkowitz-Kaufman, M. (1975) "Dissipative structures in two dimensions", *Biophys. Chem.* **3,** 345.

d'Espagnat, Bernard (1976) *Conceptual Foundations of Quantum Mechanics,* 2n rev. ed. Reading, Mass.: Benjamin.

Falkehag, S. Ingemar (1975) "Lignin in materials", *Applied Polymer Symposium,* No. 28, 247-257. New York: Wiley.

Faraday Symposium (1974) No. 9 *Physical Chemistry of Oscillatory Phenomena.* London: Faraday Division of the Chemical Society.

Feyerabend, Paul (1975) *Against Method: Outline of an Anarchistic Theory of Knowledge.* London: New Left Books.

Fischer, Roland (1970) "Über das Rhythmisch-Ornamentale im Halluzinatorisch-Schöpferischen", *Confinia Psychiatrica,* **13,** 1-25.

Fischer, Roland (1975/76) "Transformations of consciousness. A cartography", in two parts: "I. The perception-hallucination continuum", *Confinia Psychiatrica,* **18,** (1975), 221-244; "II. The perception-meditation continuum", *Confinia Psychiatrica,* **19,** (1976), 1-23.

Foerster, Heinz von (1973) "On constructing a reality", in: W. F. E. Preiser, ed., *Design, Research,* Vol. II. Stroudsburg: Dowden, Hutchinson & Ross.

Fong, P. (1973) "Thermodynamic and statistical theory of life: An outline", in: A. Locker, ed., *Biogenesis, Evolution, Homeostasis,* Berlin, Heidelberg and New York: Springer.

Frankl, Viktor (1978) *The Unheard Cry for Meaning: Psychotherapy and Humanism.* New York: Simon and Schuster.

Franz, Marie-Louise von (1974) *Number and Time: Reflections Leading toward a Unification of Depth Psychology and Physics.* Evanston, Ill.: Northwestern University Press.

Freeman, Walter J. (1975) *Mass Action in the Nervous System: Examination of the Neurophysiological Basis of Adaptive Behavior through the EEG.* New York: Academic Press.

Friedman, Jonathan (1975) "Dynamique et transformations du système tribal: l'exemple des Katchin", *L'Homme,* **XV,** 63-98.

Frisch, Karl von (1974) *Animal Architecture.* New York: Harcourt, Brace, Janovich.

Fuchs, O. (1977) "Physikalisch-chemische Mechanismen zur Speicherung und Wiedergewinnung von Information", *Colloid and Polymer Sci.* **225,** 398-400; summary of: *Berichte der Bunsengesellschaft,* **80,** 1041-1223.

Furtwängler, Wilhelm (1955) *Der Musiker und sein Publikum.* Zurich: Atlantis.

Gabor, Dennis (1963) *Inventing the Future.* London: Secker & Warburg; New York: Knopf, 1964; Harmondsworth, Middlesex: Penguin paperback, 1964.

Galbraith, John Kenneth (1967) *The New Industrial State.* Boston: Houghton-Mifflin.

Georgescu-Roegen, Nicholas (1971) *The Entropy Law and the Economic Process.* Cambridge, Mass.: Harvard University Press.

Gierer, A. (1974) "Hydra as a model for the development of biological form", *Scientific American*, Dec. 1974.

Glansdorff, Paul and Prigogine, Ilya (1971) *Thermodynamic Theory of Structure, Stability, and Fluctuations*. New York: Wiley-Inter-science.

Goldbeter, A. and Lefever, R. (1972) *Biophys. J.* **12**, 1302.

Goodwin, B. C. (1978) "A cognitive view of biological process", *Journal of Social and Biological Structures*, **1**, 117-125.

Gorenstein, Paul and Tucker, Wallace (1978) "Rich clusters of galaxies", *Scientific American*, Nov. 1978.

Gould, Stephen Jay (1977) *Ontogeny and Phylogeny*. Cambridge, Mass.: Belknap Press of Harvard University.

Grof, Stanislav (1975) *Realms of the Human Unconscious: Observations from LSD Research*. New York: Viking.

Haken, H. (1977) *Synergetics: Nonequilibrium Phase Transitions and Self-Organization in Physics, Chemistry and Biology*. Berlin, Heidelberg and New York: Springer.

Hall, D. O. (1978) "Solar energy use through biology—past and future", paper presented to the *World Conference on Future Sources of Organic Raw Materials*, Toronto, July 1978.

Halstead, L. B. (1975) *The Evolution and Ecology of the Dinosaurs*. London: Peter Lowe, Eurobook.

Harman, Willis W. (1974) "Humanistic capitalism: another alternative", *Journal of Humanistic Psychology*, Winter 1974.

Harnoncourt, Nikolaus (1976) "Werk und Aufführung bei Monteverdi", *Neue Zürcher Zeitung*, 31 Dec. 1976.

Hawking, S. W. (1977) "The quantum mechanics of black holes", *Scientific American*, Jan. 1977; also in: Scientific American (1977).

Hedges, R. W. (1972) *Heredity*, **28**, 39. Quoted in: Broda (1975).

Herschkowitz-Kaufman, Marcelle (1973) *Quelques aspects du comportement des systèmes chimiques ouverts loin de l'équilibre thermodynamique*. Doctoral thesis. Brussels: Université Libre de Bruxelles.

Herschkowitz-Kaufman, Marcelle and Nicolis, Grégoire (1972) "Localized spatial structures and non-linear chemical waves in dissipative systems", *J. Chem. Phys.* **56**, 1890-1895.

Hilbertz, Wolf (1975) "Evolutionary environments: notes for a manifesto", in: Frei Otto, ed., *I.L.* **13**, Stuttgart: Institut für Leichtbau, University of Stuttgart.

Hiltz, Starr Roxanne and Turoff, Murray (1978) *The Network Nation: Human Communication via Computer*. Reading, Mass.: Addison-Wesley.

Holling, C. S. (1976) "Resilience and stability of ecosystems", in: Jantsch and Waddington, eds. (1976).

Holling, C. S. and Ewing, S. (1971) "Blind man's buff: exploring the response space generated by realistic ecological simulation models", *Proc. Int. Symp. Stat. Ecol.*, New Haven, Conn.: Yale University Press.

Hönl, Helmut (1978) "Kosmologische Nichtstandardmodelle und der Ursprung der Materie im Universum", *Neue Zürcher Zeitung*, 17 May 1978.

Hubbert, M. King (1971) "The energy resources of the earth", *Scientific American*, Sept. 1971; also in: Scientific American (1971).

Huber, Martin C. E. and Tammann, G. E. (1977). "Geschichte des Universums: Kosmologische Problem in neuester Sicht", *Neue Zürcher Zeitung*, 4 Jan. 1977.

Huxley, Aldous (1954) *The Doors of Perception*. New York: Harper & Row.

Hydén Holger (1976) "The brain, learning and values", *Proc. Fifth International Conference on the Unity of the Sciences*, "The search for absolute values: harmony among the sciences", Washington, D.C. Tarrytown, N.Y.: International Cultural Foundation.

Illich, Ivan (1976) *Medical Nemesis*. New York: Pantheon; Bantam paperback, 1977.

Illich, Ivan (1978) *Toward a History of Needs*. New York: Pantheon.

Jantsch, Erich (1967) *Technological Forecasting in Perspective*. Paris: OECD.

Jantsch, Erich, ed. (1969) *Perspectives of Planning*. Proc. Bellagio Symposium on Long-range Forecasting and Planning. Paris: OECD.

Jantsch, Erich (1972) *Technological Planning and Social Futures.* London: Associated Business Programmes; New York: Halsted Press; paperback 1974.

Jantsch, Erich (1975) *Design for Evolution: Self-organization and Planning in the Life of Human Systems.* New York: Braziller.

Jantsch, Erich (1976) "Evolving images of man: dynamic guidance for the mankind process", in: Jantsch and Waddington, eds. (1976).

Jantsch, Erich and Waddington, Conrad H., eds. (1976) *Evolution and Consciousness: Human Systems in Transition.* Reading, Mass., London and Amsterdam: Addison-Wesley.

Jaynes, Julian (1976) *The Origin of Consciousness in the Breakdown of the Bicameral Mind.* Boston: Houghton-Mifflin.

Jenny, Hans (1967, 1972) *Kymatik,* 2 vols. Basle: Basileus. Vol. 1: 1967; Vol. 2: 1972.

Johansen, Robert, Vallée, Jacques and Spangler, Kathleen (1978) *Electronic Meetings: Technical Alternatives and Social Choices.* Reading, Mass.: Addison-Wesley.

Johanson, D. C., and White, T. D. (1979) "A Systematic Assessment of Early African Hominids", *Science,* **203.** 321-330.

Jonas, Hans (1969) *The Phenomenon of Life.* New York: Delta Books.

Josephson, Brian (1975) "The Tonal-Nagual model of reality", paper presented to the *First Int. Conf. on Science and Consciousness,* Fairfield, Iowa, Dec. 1975.

Jouvenel, Bertrand de (1964) *L'Art de la conjecture.* Monaco: Editions du Rocher. Engl. transl. *The Art of Conjecture,* New York: Basic Books, 1967.

Jouvenel, Bertrand de (1968) *Arcadie, essais sur le mieux vivre.* Paris: SEDEIS.

Jung, Carl Gustav (1961) *Memories, Dreams, Reflections.* New York: Vintage Books.

Jung, Carl Gustav and Pauli, Wolfgang (1954) *The Interpretation of Nature and the Psyche.* New York and London.

Junge, C. (1976) "Die Entstehung der Erdatmosphäre", *Neue Zürcher Zeitung,* 9 Nov. 1976.

Katchalsky, Aharon (1971) "Biological flow structures and their relations to chemodiffusional coupling", *Neurosci. Res. Prog. Bull.* **9,** 397-413.

Kennedy, Donald, ed. (1974) *Cellular and Organismal Biology,* from Scientific American. San Francisco: Freeman.

Korzybski, Alfred (1949) *Time-binding: The General Theory, Two Papers, 1924-26.* Lakeville, Conn.: Institute of General Semantics.

Krippner, Stanley and Rubin, Daniel, eds. (1974) *The Kirlian Aura: Photographing the Galaxies of Life.* Garden City, N.Y.: Anchor/Doubleday.

Kuhn, Hans (1973) "Modellbetrachtungen zur Frage der Entstehung des Lebens", *Jahrbuch der Max-Planck-Gesellschaft, 1973,* 104-130.

Kuhn, Thomas S. (1962) *The Structure of Scientific Revolutions.* Chicago: University of Chicago Press; 2nd enlarged ed., 1970.

Kuhn, Thomas S. (1977) *The Essential Tension: Selected Studies in Scientific Tradition and Change.* Chicago: University of Chicago Press.

Kurokawa, Kisho (1977) *Metabolism in Architecture.* Boulder, Col.: Westview.

Langer, Suzanne K. (1967, 1972) *Mind, an Essay on Human Feeling,* 2 vols. Baltimore and London: Johns Hopkins Press. Vol. 1: 1967; Vol. 2: 1972.

Laszlo, Ervin (1972) *Introduction to Systems Philosophy: Toward a New Paradigm of Conmporary Thought.* New York: Gordon and Breach; also Harper Torch Books.

Laszlo, Ervin (1974) "Goals for global society—a positive approach to the predicament of mankind", *Proc. Third Intn. Conf. Unity of the Sciences,* "Science and Absolute Values", London; Tarrytown, N.Y.: Intn. Cultural Foundation.

Laszlo, Ervin (1978) *The Inner Limits of Mankind: Heretical Reflections on Today's Values, Culture and Politics.* Oxford and New York: Pergamon Press.

Leaky, Richard E. and Lewin, Roger (1978) *People of the Lake: Mankind and its Beginnings.* Garden City, N.Y.: Anchor/Doubleday.

Leboyer, Frederick (1975) *Birth Without Violence.* London: Wildwood House.

Lefever, R. (1968) "Stabilité des structures dissipatives", *Bull. Classe Sci. Acad. Roy. Belg.* **54,** 712.

Lefever, R. and Garay, R. (1977) "A model of the immune surveillance of cancer", in: G. Bell, A. Perelson and G. Pimbley, eds., *Theoretical Immunology,* New York.

Lessing, Doris (1975) *The Memoirs of a Survivor.* New York: Knopf; Bantam paperback, 1976.

Lieber, Arnold L. (1978) *The Lunar Effect: Biological Tides and Human Emotions.* Garden City, N.Y., Anchor/Doubleday.

Lima-de-Faria, Antonio (1975) "The relation between chromomers, replicons, operons, transcription units, genes, viruses and palindromes", *Hereditas,* **8,** 249-284.

Lima-de-Faria, Antonio (1976) "The chromosome field", in five parts: "I. Prediction of the location of ribosomal cistrons", *Hereditas,* **83,** 1-22; "II. The location of 'knobs' in relation to telomeres", *Hereditas,* **83,** 23-34; "III. The regularity of distribution of cold-induced regions", *Hereditas,* **83,** 139-152; "IV. The distribution of non-disjunction, chiasmata and other properties", *Hereditas,* **83,** 175-190; "V. The distribution of chromomere gradients in relation to kinetochore and telomeres", *Hereditas,* **84,** 19-34.

Longair, M. S. and Einasto, J., eds. (1978) *The Large Scale Structure of the Universe.* Int. Astron. Union Symp. No. 79. Boston: Reidel.

Longchenpa (1976) *Kindly Bent to Ease Us,* 3 vols. Transl. and annotated by Herbert V. Guenther. *Part Two: Meditation; Part Three: Wonderment.* Emeryville, Calif.: Dharma Publishing.

Lotka, Alfred J. (1956) *Elements of Mathematical Biology,* New York: Dover.

Lowrie, W. (1976) "Paläomagnetismus", *Neue Zürcher Zeitung,* 10 Aug. 1976.

McKenna, Terence K. and Dennis, J. (1975) *The Invisible Landscape: Mind, Hallucinogens and the I Ching.* New York: Seabury Press.

MacLean, Paul D. (1973) "A triune concept of the brain and behavior", in: T. Boag and D. Campbell, eds., *The Hincks Memorial Lectures,* Toronto: University of Toronto Press.

Maeder, André (1975) "Die Evolution der Sonne", *Neue Zürcher Zeitung,* 15 Jan. 1975.

Maeder, André (1977) "Das Rätsel der Quasare", *Neue Zürcher Zeitung,* 8 Feb. 1977.

Maeder, André (1978) "Die Kosmologie von Dirac", *Neue Zürcher Zeitung,* 4 July 1978.

Margalef, Ramón (1968) *Perspectives in Ecological Theory.* Chicago: University of Chicago Press. Large extracts reprinted in *CoEvolution Quarterly,* **No. 6,** Summer 1975, 49-66.

Margulis, Lynn (1970) *Origin of Eukaryotic Cells.* New Haven, Conn.: Yale University Press.

Margulis, Lynn and Lovelock, James E. (1974) "Biological modulation of the earth's atmosphere", *Icarus,* **21,** 471-489.

Markley, O. W. (1976) "Human consciousness in transformation", in: Jantsch and Waddington, eds. (1976).

Marthaler, Daniel (1976) "Geheimnisvoller Nervenfilz", *Neue Zürcher Zeitung,* 2 Nov. 1976; summary of an article by Francis O. Schmitt, Pawati Dev and Barry H. Smith in *Science,* **192** (1976), 114-120.

Maruyama, Magoroh (1976) "Toward cultural symbiosis", in: Jantsch and Waddington, eds. (1976).

Maturana, Humberto R. (1970) *Biology of Cognition.* Report BCL 9.0. Urbana, Ill.: Biological Computer Laboratory, University of Illinois.

Maturana, Humberto R. and Varela, Francisco (1975) *Autopoietic Systems.* Report BCL 9.4. Urbana, Ill.: Biological Computer Laboratory, University of Illinois.

May, Robert M. (1973) *Stability and Complexity in Model Ecosystems.* Princeton, N.J.: Princeton Univeristy Press.

May, Robert M. (1978) "The evolution of ecological systems", *Scientific American,* Sept. 1978.

Mesarović, Mihajlo D., Macko, D. and Takahara, Y. (1970) *Theory of Hierarchical, Multilevel Systems.* New York: Academic Press.

Miller, Stanley S. and Orgel, Leslie E. (1978) *The Origins of Life on Earth.* Englewood Cliffs, N.J.; Prentice-Hall.

Monod, Jacques (1971) *Chance and Necessity.* New York: Knopf; Random paperback, 1972.

Morowitz, Harold J. (1968) *Energy Flow in Biology.* New York: Academic Press.

Morowitz, Harold J. and Tourtellotte, Mark E. (1962) "The smallest living cells", *Scientific American,* Mar. 1962; also in: Kennedy, ed. (1974).

Motz, Lloyd (1975) *The Universe: Its Beginning and End.* New York: Scribner's.

Müller, A. M. Klaus (1974) "Naturgesetz, Wirklichkeit, Zeitlichkeit", in: Weizsäcker, Ernst von, ed. (1974).

Murphy, Michael (1977) *Jacob Atabet: A Speculative Fiction*. Millbrae, Calif.: Celestial Arts.

Neue Zürcher Zeitung (1978), jd.; "Methanbakterien und Archäbakterien", *Neue Zürcher Zeitung*, 22 Aug. 1978.

Nicolis, Grégoire (1974) "Dissipative structures with applications to chemical reactions", in: H. Haken, ed., *Cooperative Phenomena*, Amsterdam: North-Holland.

Nicolis, Grégoire and Prigogine, Ilya (1971) "Fluctuations in non-equilibrium systems", *Proc. Natl. Acad. Sci. (USA)*, **68**, 2102-2107.

Nicolis, Grégoire and Prigogine, Ilya (1977) *Self-organization in Nonequilibrium Systems: From Dissipative Structures to Order Through Fluctuations*. New York: Wiley-Interscience.

O'Neill, Gerard K. (1977) *The Higher Frontier: Human Colonies in Space*. New York: Morrow.

O'Neill, Gerard K. ed. (1978) *Space-based Manufacturing from Nonterrestrial Materials*. New York: American Institute of Aeronautics and Astronautics.

Oparin, Andreas I. (1938) *Origin of Life*. New York: Macmillan; reprinted New York: Dover, 1953.

Ortega y Gasset, José (1943) *Das Wesen geschichtlicher Krisen*. Stuttgart.

Ozbekhan, Hasan (1976) "The predicament of mankind", in: C. West Churchman and Richard O. Mason, eds., *World Modeling: A Dialogue*. North-Holland/TIMS Studies in the Management Sciences, Vol. 2. Amsterdam and Oxford: North-Holland; New York: American Elsevier.

Pankow, Walter (1976) "Openness as self-transcendence", in: Jantsch and Waddington, eds. (1976).

Pattee, Howard H. (1978) "The complementarity principle in biological and social structures, *Journal of Social and Biological Structures*, **1**, 191-200.

Pauling, Linus (1977) "Vitamin C und Krebs", lecture given at the 27th Meeting of Nobel Laureates in Lindau, *Neue Zürcher Zeitung*, 19 July 1977.

Pearson, Keir (1976) "The control of walking", *Scientific American*, Dec. 1976.

Petit, Charles (1977) "The Sun, and Earth's weather", *San Francisco Chronicle*, 7 Dec. 1977.

Popper, Karl R. and Eccles, John C. (1977) *The Self and Its Brain: An Argument for Interactionism*. Berlin, Heidelberg and New York: Springer.

Pribram, Karl (1971) *Languages of the Brain*. Engleworth Cliffs, N.J.: Prentice-Hall.

Prigogine, Ilya (1973) "Irreversibility as a symmetry breaking factor", *Nature*, **248**, 67-71.

Prigogine, Ilya (1976) "Order through fluctuation: self-organization and social system", in: Jantsch and Waddington, eds. (1976).

Prigogine, Ilya (1977) "The metamorphosis of science: culture and science", paper presented to the *Conference on Science in Society*, European Community, Brussels, 1977.

Prigogine, Ilya and Stengers, Isabelle (1975) "Nature et Créativité", *Revue de l'AUPELF,* **XIII**, No. 2.

Prigogine, Ilya, Nicolis, Grégoire and Babloyantz, Agnès (1972) "Thermodynamics of evolution", *Physics Today,* **25**, 23-28 and 38-44.

Pugh, George Edgin (1977) *The Biological Origin of Human Values*. New York: Basic Books.

Riedl, Rupert (1976) *Die Strategie der Genesis: Naturgeschichte der realen Welt*. Munich and Zurich: Piper.

Rodman, John (1977) "Theory and practice in the environmental movement: notes towards an ecology of experience", *Proc. Sixth Int. Conf. Unity of the Sciences*, "The search for absolute values in a changing world", San Francisco; Tarrytown, N.Y.: Int. Cultural Foundation.

Rosenfeld, Leon (1967) in: S. Rosenthal, ed., *Niels Bohr: His Life and Work as Seen by His Friends and Colleagues*, New York; quoted in: Feyerabend (1975), p. 24.

Sagan, Carl (1977) *The Dragons of Eden: Speculation on the Evolution of Human Intelligence,* New York: Random House.

Sauter, Karl and Bertschy, Hanspeter (1977) "Die Wollbilder der Huichol-Indianer"; *Neue Zürcher Zeitung*, 22 Oct. 1977.

Scheving, Lawrence E. (1977) "Chronobiologie: Die Dimension der Zeit in Biologie und Medizin", *Neue Zürcher Zeitung*, 1 Mar. 1977.

Schoch, Max (1977) "Das Unvollkommene und das Leben: Eranos-Tagung 1977", *Neue Zürcher Zeitung*, 13 Oct. 1977.

Schopf, J. William (1978) "The evolution of the earliest cells", *Scientific American*, Sept. 1978.

Schramm, David N. and Clayton, Robert N. (1978) "Did a supernova trigger the formation of the solar system?", *Scientific American*, Oct. 1978.

Schumacher, E. F. (1973) *Small is Beautiful: Economics as if People Mattered.* New York: Harper & Row.

Schurian, Walter (1978) "Bilder als Systeme der Entwicklung", in: Ernst Fuchs, *Im Zeichen der Sphinx: Schriften und Bilder*, Munich: Deutscher Taschenbuchverlag.

Schurig, Volker (1976) *Die Entstehung des Bewusstseins.* Frankfurt and New York: Campus.

Scientific American (1971) *Energy and Power.* San Francisco: Freeman

Scientific American (1977) *Cosmology + 1.* San Francisco: Freeman.

Scrimshaw, Nevin S. and Young, Vernon R. (1976) "The requirements of human nutrition", *Scientific American*, Sept. 1976.

Senghaas, Dieter (1977) *Weltwirtschaftsordnung und Entwicklungspolitik: Plädoyer für Dissoziation.* Frankfurt am Main: Suhrkamp.

Shainberg, David (1973) *The Transforming Self: New Dimensions in Psychoanalytic Process.* New York: Intercontinental Medical Book Corp.

Shannon, Claude E. and Weaver, Warren (1949) *The Mathematical Theory of Communications.* Urbana, Ill.: University of Illinois Press.

Smuts, Jan Christiaan (1926) *Holism and Evolution.* Republished New York: Viking (1967).

Soleri, Paolo (1973) *Matter Becoming Spirit: The Arcology of Paolo Soleri.* Garden City, N.Y.: Anchor/Doubleday.

Solzhenitsyn, Aleksandr (1971) *One Day in the Life of Ivan Denissovich.* New York: Farrar, Straus & Giroux.

Solzhenitsyn, Aleksandr (1976) *Lenin in Zurich.* New York: Farrar, Straus & Giroux.

Spencer Brown, G. (1969) *Laws of Form.* London: Allen & Unwin; New York: Julian Press, 1972; New York: Bantam paperback, 1974.

Spurgeon, Bud (1977) "Tesla", *CoEvolution Quarterly*, **No. 16,** Winter 1977/78.

Staehelin, Theophil (1976) "Hoffnung an der Grippefront", *Neue Zürcher Zeitung*, 20 Jan. 1976.

Starr, Chauncey (1971) "Energy and power", *Scientific American*, Sept. 1971; also in: Scientific American (1971).

Stebbins, G. L. (1973) "Evolution of morphogenetic patterns", *Brookhaven Symp. Biol.* **25**, 227-243; quoted in: Gould (1977), p. 407.

Stegmüller, Wolfgang (1975) *Hauptströmungen der Gegenwartsphilosophie*, Band II, Stuttgart: Kröner.

Steinlin, Uli W. (1977) "Kugelsternhaufen", *Neue Zürcher Zeitung*, 26 Apr. 1977.

Stent, Gunther S. (1972) "Cellular communication", *Scientific American*, Sept. 1972; also in: Kennedy, ed. (1974).

Stent, Gunther S. (1975) "Explicit and implicit semantic content of the genetic information", *Proc. Fourth Int. Conf. Unity of the Sciences*, "The centrality of science and absolute values", New York; Tarrytown, N.Y.: Int. Cultural Foundation.

Strom, Richard G., Miley, George K. and Oort, Jan (1975) "Giant radio galaxies", *Scientific American*, Aug. 1975.

Stumm, Werner, ed. (1977) *Global Chemical Cycles and their Alterations by Man.* Berlin: Dahlem Konferenzen/Abakon.

Taylor, John (1975) *Superminds: A Scientist Looks at the Paranormal.* New York: Viking.

Thom, René (1972) *Stabilité Structurelle et Morphogenèse.* Reading, Mass.: Bejamin. Engl. transl.: *Structural Stability and Morphogenesis*, Reading, Mass.: Benjamin, 1975.

Thomas, Lewis (1974) *The Lives of a Cell: Notes of a Biology Watcher.* New York: Viking; Bantam paperback, 1975.

Thorn, Fritz (1976) "Die Triebfedern der Apathie: Zur Rolle des Publikums im Gegenwartstheater", *Neue Zürcher Zeitung*, 26 Mar. 1976.

Thorpe, Willard H. (1976) "Science and man's need for meaning", *Proc. Fifth Int. Conf. Unity of the Sciences*, "The search for absolute values: harmony among the sciences", Washington,

D.C.; Tarrytown, N.Y.: Int. Cultural Foundation.

Toulmin, Stephen (1977) "From form to function: philosophy and history of science in the 1950s and now", *Daedalus*, **106,** 143-162.

Tromp, Solco W. (1972) "Possible effects of extra-terrestrial stimuli on colloidal systems and living organisms", *Proc. 5th Int. Biometeor. Congr.*, Noordwijk; Amsterdam: Swets and Zeitlinger.

Trueb, Lucien (1974) "Energiesystem und Energieprobleme", *Neue Zürcher Zeitung*, 16 July 1974.

Tryon, Edward P. (1973) "Is the universe a vacuum fluctuation?", *Nature*, **246,** 396-397.

Turnbull, Colin (1972) *The Mountain People.* New York: Simon & Schuster.

Valentine, James W. (1978) "The evolution of multicellular plants and animals", *Scientific American*, Sept. 1978.

Vallée, Jacques (1977) "The priest, the well and the pendulum", *CoEvolution Quarterly*, **No. 16,** Winter 1977/78.

Varela, Francisco J. (1975) "A calculus of self-reference", *Int. Journal of General Systems*, **2,** 5.

Varela, Francisco, Maturana, Humberto R. and Uribe, Ricardo (1974) "Autopoiesis: the organization of living systems, its characterization and a model", *Biosystems*, **5,** 187-196.

Vickers, Geoffrey (1968) *Value Systems and Social Process.* London: Tavistock; New York: Basic Books; Pelican paperback, 1970.

Vickers, Geoffrey (1973) *Making Institutions Work.* London: Associated Business Programmes.

Volterra, Vito (1926) "Variazioni e fluttuazioni del numero d'individui in specie animali conviventi", *Mem. Accad. Lincei*, **2,** 31-113.

Waddington, Conrad H. (1975) *The Evolution of an Evolutionist.* Edinburgh: Edinburgh University Press; Ithaca, N.Y.: Cornell University Press.

Watson, J. A., Nel, J. J. C. and Hewitt, P. H. (1972) "Behavioral changes in founding pairs of the termite *Hodotermes mossambicus*", *Journal of Insect Physiology*, **18,** 373-387; quoted in: Thomas (1974).

Watson, Lyall (1973) *Supernature.* London: Hodder & Stoughton; Garden City, New York: Doubleday Coronet and Bantam paperbacks, 1974.

Weinberg, Steven (1977) *The First Three Minutes: A Modern View of the Origin of the Universe.* New York: Basic Books.

Weizsäcker, Carl Friedrich von (1974) "Evolution und Entropiewachstum", in: Weizsäcker, Ernst von, ed. (1974).

Weizsäcker, Christine U. von (1975) "Die umweltfreundliche Emanzipation", in: *Humanökologie* (Int. Tagung für Humanökologie), Vienna: Georgi.

Weizsäcker, Ernst von, ed. (1974) *Offene Systeme I: Beiträge zur Zeitstruktur von Information, Entropie und Evolution.* Stuttgart: Klett.

Weizsäcker, Ernst von (1974) "Erstmaligkeit und Bestätigung als Komponenten der pragmatischen Information", in: Weizsäcker, Ernst von, ed. (1974).

Whitehead, Alfred North (1933) *Adventures of Ideas.* New York: Macmillan; reprinted New York: Free Press, 1967

Whitehead, Alfred North (1969) *Process and Reality.* New York: Free Press.

Wilson, Edward O. (1975) *Sociobiology: The New Synthesis.* Cambridge, Mass.: Belknap Press of Harvard University.

Wilson, Edward O. (1978) *On Human Nature.* Cambridge, Mass.: Harvard University Press.

Winfree, Arthur (1978) in: Henry Eyring, ed., *Theoretical Chemistry*, **4,** New York: Academic Press.

Wolpert, Lewis (1978) "Pattern formation in biological development", *Scientific American*, Oct. 1978.

Zeeman, E. Christopher (1977) *Catastrophe Theory: Selected Papers 1972-1977.* Reading, Mass., London and Amsterdam: Addison-Wesley.

Zeleny, Milan (1977) "Self-organisation of living systems", *Int. Journal of General Systems*, **4,** 13-28.

Zeleny, Milan and Pierre, Norbert A. (1976) "Simulation of self-renewing systems", in: Jantsch and Waddington, eds. (1976).

Zhabotinsky, A. M. (1974) *Self-oscillating Concentrations.* Moscow: Nauka.

Zwahlen, R. (1978) "Kagera—eine ökologische und ökonomische Herausforderung", *Neue Zürcher Zeitung,* 7 Mar. 1978.

Zwicky, Fritz (1966) *Entdecken, Erfinden, Forschen im morphologischen Weltbild.* Munich: Droemer-Knaur.

Name Index

Subject Index